T0329575

Dynamic System Modelling and Analysis with MATLAB and Python

IEEE Press
445 Hoes Lane
Piscataway, NJ 08854

Dynamic System Modelling and Analysis with MATLAB and Python

For Control Engineers

Jongrae Kim
University of Leeds
Leeds, UK

IEEE Press Series on Control Systems Theory and Applications
Maria Domenica Di Benedetto, Series Editor

Published by John Wiley & Sons, Inc., Hoboken, New Jersey.
Published simultaneously in Canada.

Library of Congress Cataloging-in-Publication Data Applied for:

Hardback: 9781119801627

Cover Design: Wiley
Cover Images: © Bocskai Istvan/Shutterstock

Set in 9.5/12.5pt STIXTwoText by Straive, Chennai, India

To Miyoung

Contents

Preface

This book is for control engineers to learn dynamic system modelling and simulation and control design and analysis using MATLAB or Python. The readers are assumed to have the undergraduate final-year level of knowledge on ordinary differential equations, vector calculus, probability, and basic programming.

We have verified all the MATLAB and Python codes in the book using MATLAB R2021a and Python 3.8 in Spyder, the scientific Python development environment. To reduce the confusion in running a particular program, most of the programs are independent on their own. Organizing programming with multiple files is left as an advanced skill for readers to learn after reading this book.

Leeds, West Yorkshire, England, UK *Jongrae Kim*
30 November 2021

Acknowledgements

I have learned dynamic modelling and simulation through my undergraduate and post-graduate education and research projects in the past 30 years. Hence, this book will not be possible without having my teachers, supervisors, and collaborators. I thank Dr Jinho Kim, Professor John L. Crassidis, Professor João P. Hespanhna, Professor Declan G. Bates, Dr Daizhan Cheng, Professor Kwang-Hyun Cho, Professor Frank Pollick, and Dr Rajeev Krishnadas.

Jongrae Kim

Acronyms

DCM	direction cosine matrix
DNA	deoxyribonucleic acid
EKF	extended Kalman filter
KF	Kalman filter
LHS	left-hand side
LTI	linear time-invariant
mRNA	messenger RNA
mRNAP	messenger RNA polymerase
N2L	Netwton's second law of motion
ODE	ordinary differential equation
pdf	probability density function
PI	proportional integral
QUEST	quaternion estimation algorithm
RHS	right-hand side
RNA	ribonucleic acid

About the Companion Website

This book is accompanied by a companion website.

www.wiley.com/go/kim/dynamicmodeling

This website includes:

- The solutions for the problems listed in the chapters and the program codes used in Python and MATLAB softwares.

1 Introduction

1.1 Scope of the Book

This book is for advanced undergraduate students, post-graduate students, or engineers to acquire programming skills for dynamic system modelling and analysis using control theory. The readers are assumed to have a basic understanding of computer programming, ordinary differential equations (ODE), vector calculus, and probability.

Most engineering curricula at the undergraduate level include only an elementary-level programming course in the early of the undergraduate years. Only a handful of self-motivated engineering students acquire advanced level programming skills mainly from self-study through tedious time-consuming practices and trivial mistakes. As modern engineering systems such as aircraft, satellite, automobile, or autonomous robots are implemented through inseparable tight integration of hardware systems and software algorithms, the demand for engineers having fluent skills in dynamic system modelling and algorithm design is increasing. In addition, the emergence of interdisciplinary areas merging the experimental domain with mathematical and computational approaches such as systems biology, synthetic biology, or computational neuroscience further increases the necessity of the engineers who understand dynamics and are capable of computational implementations of dynamic models.

This book aims to fill the gap in learning practical dynamic modelling, simulation, and analysis skills in aerospace engineering, robotics, and biology. Learning programming in the engineering or biology domain requires not only domain knowledge but also a robust conceptual understanding of algorithm design and implementation. It is not, of course, the skills to learn in 14 days or less as many online courses claim. To be confident in dynamic system modelling and analysis takes more than several years of practice and dedication. This book provides the starting point of the long journey for the readers to equip and prepare better for real engineering and scientific problems.

Dynamic System Modelling and Analysis with MATLAB and Python: For Control Engineers,
First Edition. Jongrae Kim.

Companion Website: www.wiley.com/go/kim/dynamicmodeling

1.2 Motivation Examples

1.2.1 Free-Falling Object

Newton's second law of motion is given by

$$\sum_i F_i = \frac{d}{dt}(mv) \tag{1.1}$$

where F_i is the i-th external force in Newtons (N) acting on the object characterized by the mass, m, in kg, d/dt is the time derivative, t is the time in seconds, v is the velocity in m/s, and mv is the momentum of the object. Newton's second law states that *the sum of all external forces is equal to the momentum change per unit of time.*

Consider a free-falling object shown in Figure 1.1. There exists only one external force, i.e. the gravitational force acting downwards in the figure. Hence, the left-hand side of (1.1) is simply given by $\sum_i F_i = F_g$, where F_g is the gravitational force. Introduce the additional assumption that the object is within the reasonable range from the sea level. With the assumption, the gravitational force, F_g, is known to be proportional to the mass, and the proportional constant is the gravitational acceleration constant, g, which is equal to 9.81 m/s^2 in the sea level. Therefore, $F_g = mg$. Replace the left-hand side of (1.1), i.e. $\sum_i F_i$, by $F_g = mg$ provides

$$mg = F_g = \sum_i F_i = \frac{d}{dt}(mv) \tag{1.2}$$

where the downward direction is set to the positive direction, which is the opposite of the usual convention. *It highlights that establishing a consistent coordinate system at the beginning of modelling is vital in dynamic system simulation.*

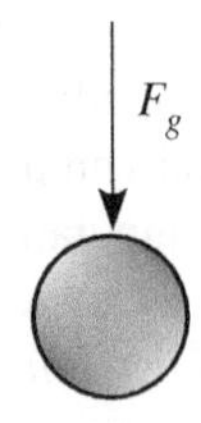

Figure 1.1 Free-falling object.

From the kinematic relationship between the velocity, v, and the displacement, x, we have

$$\frac{dx}{dt} = v$$

where the origin of x is at the initial position of the object, m, and the positive direction of x is downwards in the figure. The right-hand side of (1.2) becomes

$$mg = F_g = \sum_i F_i = \frac{d}{dt}(mv) = \frac{d}{dt}\left(m\frac{dx}{dt}\right)$$

Finally, the leftmost and the rightmost terms are equal to each other as follows:

$$mg = \frac{d}{dt}\left(m\frac{dx}{dt}\right)$$

and it is expanded as follows:

$$mg = \frac{dm}{dt}\frac{dx}{dt} + m\frac{d^2x}{dt^2}$$

Using the short notations, $\dot{m} = dm/dt$, $\dot{x} = dx/dt$, and $\ddot{x} = d^2x/dt^2$, and after rearrangements, the governing equation is given by

$$\ddot{x} = g - \frac{\dot{m}}{m}\dot{x} \tag{1.3}$$

For purely educational purposes, assume that the mass change rate is given by

$$\dot{m} = -m + 2 \tag{1.4}$$

We can identify now that there are three independent time-varying states, which are the position, x, the velocity, $\dot{x}$, and the mass, m. All the other time-varying states, for example, $\ddot{x}$ and $\dot{m}$, can be expressed using the independent state variables. Define the state variables as follows:

$$x_1 = x$$

$$x_2 = \dot{x}$$

$$x_3 = m$$

Obtain the time derivative of each state expressed in the state variable as follows:

$$\dot{x}_1 = \dot{x} = x_2 \tag{1.5a}$$

$$\dot{x}_2 = \ddot{x} = g - \frac{-m+2}{m}\dot{x} = g - \frac{-x_3+2}{x_3}x_2 \tag{1.5b}$$

$$\dot{x}_3 = \dot{m} = -m + 2 = -x_3 + 2 \tag{1.5c}$$

and this is called *the state-space form*.

Let the initial conditions be equal to $x_1(0) = x(0) = 0.0$ m, $x_2(0) = \dot{x}(0) = 0.5$ m/s, and $x_3(0) = m(0) = 5$ kg. Equation (1.5) can be written in a compact form using the matrix–vector notations. Define the state vector, $\mathbf{x}$, as follows:

$$\mathbf{x} = \begin{bmatrix} x_1 \\ x_2 \\ x_3 \end{bmatrix}$$

and the corresponding state-space form is written as

$$\dot{\mathbf{x}} = \mathbf{f}(\mathbf{x}) = \begin{bmatrix} x_2 \\ g + (x_3 - 2)(x_2/x_3) \\ -x_3 + 2 \end{bmatrix} \tag{1.6}$$

The second-order differential equation, (1.3), and the first-order differential equation, (1.4), are combined into the first-order three-dimensional vector differential equation, (1.6). Any higher order differential equations can be transformed into the first-order multi-dimensional vector differential equation, $\dot{\mathbf{x}} = \mathbf{f}(\mathbf{x})$. Numerical integration methods such as Runge–Kutta integration (Press et al., 2007) solves the first-order ODE. They can solve any high-order differential equations by transforming them into the corresponding first-order multi-dimensional differential equation.

1.2.1.1 First Program in Matlab

We are ready to solve (1.6) with the initial condition equal to $\mathbf{x}(0) = [0.0\ 0.5\ 5.0]^T$, where the superscript T is the transpose of the vector. We solve the differential equation from $t = 0$ to $t = 5$ seconds using Matlab. Matlab includes many numerical functions and libraries to be used for dynamic simulation and analysis. A numerical integrator is one of the functions already implemented in Matlab. Hence, the only task we have to do for solving the differential equation is to learn how to use the existing functions and libraries in Matlab. The complete programme to solve the free-falling object problem is given in Program 1.1. Producing Figure 1.2 is left as an exercise in Exercise 1.1.

```
1  clear;
2
3  grv_const = 9.81; % [m/s^2]
4  init_pos = 0.0; %[m]
5  init_vel = 0.5; % [m/s]
6  init_mass = 5.0; %[kg]
7
8  init_time = 0; % [s]
9  final_time = 5.0; % [s]
10 time_interval = [init_time final_time];
11
12 x0 = [init_pos init_vel init_mass];
13 [tout,xout] = ode45(@(time,state) free_falling_obj(time,state,
       grv_const), time_interval, x0);
14
15 figure(1);
16 plot(tout,xout(:,1))
17 ylabel('position [m]');
```

```
18 xlabel('time [s]');
19
20 figure(2);
21 plot(tout,xout(:,2))
22 ylabel('velocity [m/s]');
23 xlabel('time [s]');
24
25 figure(3);
26 plot(tout,xout(:,3))
27 ylabel('m(t) [kg]');
28 xlabel('time [s]');
29
30 function dxdt = free_falling_obj(time,state,grv_const)
31     x1 = state(1);
32     x2 = state(2);
33     x3 = state(3);
34
35     dxdt = zeros(3,1);
36     dxdt(1) = x2;
37     dxdt(2) = grv_const + (x3-2)*(x2/x3);
38     dxdt(3) = -x3 + 2;
39 end
```

Program 1.1 (Matlab) Free-falling object

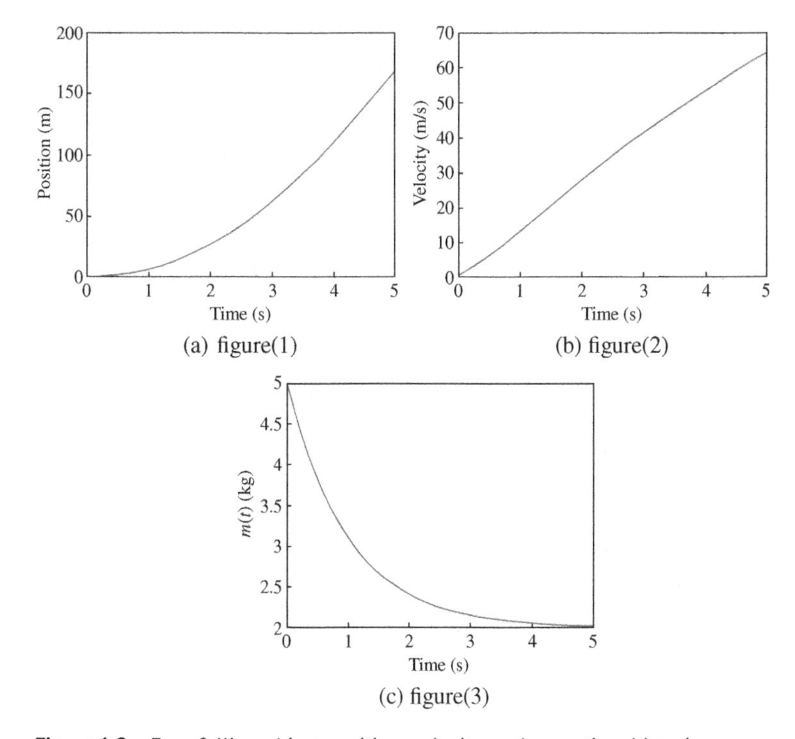

(a) figure(1)

(b) figure(2)

(c) figure(3)

Figure 1.2 Free-falling object position, velocity, and mass time histories.

Now, we study the first program line by line. The m-script starts with the command 'clear'. The clear command removes all variables in the workspace. In the workspace, there would be some variables defined and used in previous activities. They may have the same names but different meanings and values in the current calculation. For example, the gravitational acceleration 'grv_const' in the third line is undefined in the current program and uses a variable of the same name used to analyse objects falling on the moon. A falling object program in the Moon was executed earlier, and 'grv_const' is still in the workspace. Without the clear command, the incorrect constant is used in the program producing wrong results. Hence, it is recommended to clear the workspace before starting new calculations. We must be careful, however, that the clear command erases all variables in the workspace. Before the clear command, we check if all values, which might be generated from a long computer simulation, were saved.

From line 3 to line 12, several constants are defined. Based on the equations we have seen earlier, it is tempting to write a code as follows:

```
g = 9.81
x = 0.0
v = 0.5
t = [0 5]
x0 = [x v m]
```

Program 1.2 (Matlab) Poor style constant definitions

These seem to look compact and closer to the equations we derived. It is a bad habit to write a program in this way. The list of problems in the above programming style is as follows:

- It defines a variable with a single character, 'g', 'x', 'v', etc. Using a single character variable might cause confusion on the meaning of the variable and lead to using them in wrong places with incorrect interpretations.
- Numerical numbers are written without units. There is no indication of units of the numerical values, e.g. 9.81, is it m/s^2 or ft/s^2?
- It uses magic numbers. What do the numbers, 0 and 5, mean in defining 't'?

Program 1.1 uses a better style. The initial position is defined using the variable name, 'init_pos', whose value is 0.0 and the unit is in metres. Appropriately named variables reduce mistakes and confusion in the program. Program 1.1 indicates the corresponding unit for each numerical value, e.g. the 'init mass' value 5.0 is in kg. We understand the meaning of each variable by its name. The texts after '%' are the comments, where we could add various information such as the unit of each numerical value.

In line 13, the built-in Runge–Kutta integrator, *ode45()*, is used to integrate the differential equation provided by the function, 'free_falling_obj', at the end of the m-script. Frequently, each function is saved as a separate m-script. It could also be included in the m-script for the cases that the functions might be used in the specific m-script only. To include functions in the m-script, they must be placed at the end of the m-script as in this example.

Functions in Matlab begin with the keyword *function* and close with the keyword *end*. In line 30, 'dxdt' is the return variable of the function and 'free_falling_obj' is the function name. The function has three input arguments. A function can have any input argument used by the function. This particular function, 'free_falling_obj', is not an ordinary function, however. This is the function to describe the ODE. The function is to be passed into the built-in integrator, *ode45*. The first two arguments of the function for *ode45* must be time and states, i.e. t and $\mathbf{x}$ in (1.6).

In lines 31–33, the variable 'state' is assumed to be a three-dimensional vector, and each element of the vector corresponds to the states, x_1, x_2, and x_3. In line 35, the return variable 'dxdt' is initialized as [0 0 0] by the built-in function *zeros(3,1)*. *zeros(m,n)* creates the m × n matrix filled in zeros. Lines 36, 37, and 38 define the state-space form ODE, (1.6).

The function works perfectly well without the initialization line for 'dxdt', line 35. However, it is not good programming if line 35 is removed. Without the initialization, 'dxdt' in line 36 is a one-dimensional scalar value. In the next lines, it becomes a two-dimensional value and a three-dimensional value. Each line, the size of 'dxdt' changes, and this requires the computer to find additional memory to store the additional value. This could increase the total computation time longer and could be noticeably longer if this function is called a million times or more. Hence, it is better to acquire all the required memory ahead as in line 35.

Efficiency vs. development cycle: We strive to create efficient programs, but the prototyping phase requires a fast development cycle.

It is vital to have the habit of being conscious of the efficiency of algorithm implementation. On the other hand, try not to overthink the efficiency of the program. Script languages such as Matlab and Python are for rapid implementation and testing. Hence, it needs a proper balance between optimizing codes and saving the development time.

Now, we are ready to solve the differential equation using the built-in numerical integrator, *ode45*. *ode45* stands for ODE with Runge–Kutta fourth- and fifth-order methods. Details of the Runge–Kutta integration methods can be found in Press et al. (2007).

Recall, the following line from Program 1.1:

```
13 [tout,xout] = ode45(@(time,state) free_falling_obj(time,state,
       grv_const), time_interval, x0);
```

When we use ode45, the input argument starts with @ symbol, which is the function handle. The function handle, @, is used when we pass function A, e.g. 'free_falling_obj', to function B, e.g. *ode45*, where function B would call function A multiple times. With the function handle, we can control or construct the function to be passed with some flexibility. '@(time,state)' explicitly indicates that the function to be passed has two arguments, 'time' and 'state', and they will be passed between *ode45* and 'free_falling_obj' function in the specific order, i.e. 'time' be the first and 'state' be the second argument. This order is required by the integrator, ode45.

With the function handle, we can take some freedom to order the function arguments differently in the function definition of 'free_falling_obj'. For example, we could write the function as follows:

```
function dxdt = free_falling_obj(time,grv_const,state)
    x1 = state(1);
    x2 = state(2);
    x3 = state(3);

    dxdt = zeros(3,1);
    dxdt(1) = x2;
    dxdt(2) = grv_const + (x3-2)*(x2/x3);
    dxdt(3) = -x3 + 2;
end
```

and the integration part is updated to follow the updated function definition as follows:

```
[tout,xout] = ode45(@(time,state) free_falling_obj(time,grv_const,
    state), time_interval, x0);
```

The program works the same as the ones before the modifications. Also, we notice that we have an additional input argument, 'grv_const'. Similarly, we could add more input parameters if they are necessary. As long as the first argument, 'time', and the second argument, 'state', are indicated in the function handle, the function can have any number of input arguments in any order to pass to the integrator, *ode45*.

Once the integration is completed, the results return to two output variables, 'tout' and 'xout'. Execute the command, *whos*, in the Matlab command prompt, the following information is displayed:

```
>> whos
  Name              Size              Bytes  Class     Attributes

  final_time        1x1                   8  double
  grv_const         1x1                   8  double
  init_mass         1x1                   8  double
  init_pos          1x1                   8  double
  init_time         1x1                   8  double
  init_vel          1x1                   8  double
  time_interval     1x2                  16  double
  tout             61x1                 488  double
  x0                1x3                  24  double
  xout             61x3                1464  double
```

The first column shows all variables created including the two output results from the integrator. The second column shows the size of each variable: 'tout' is 61 rows and 1 column and 'xout' is 61 rows and 3 columns. Hence, each row of 'xout' corresponds to the time instance of the corresponding row values of 'tout'. Why is the number of row 61? This is determined by the integrator automatically to adjust the integration accuracy and computation time. We can assign the number of rows or the number of time steps explicitly, and this is covered in the later chapters. The three columns of 'xout' correspond to the state, x, $\dot{x}$, and m. The first column of 'xout' is for x, the second column of 'xout' is for $\dot{x}$, and the last column of 'xout' is for m.

By executing the following line in the Matlab command prompt, we can print out all values of $x(t)$ in the command window:

```
>> xout(:,1)
```

where ':' indicates all rows. If we want to see the values of x from the 11th row to the 15th row, then

```
>> xout(11:15,1)
```

Similarly, the time history of $\dot{x}$ is xout(:,2) and the time history of m is xout(:,3).

The plot command in Matlab plots the results as follows:

```
plot(tout, xout(:,1))
```

Before plotting each figure, open a new figure window using *figure(1)*, *figure(2)*, and *figure(3)*, respectively. The label for each axis is created using the commands *xlabel* and *ylabel* for the horizontal and the vertical axes, respectively, where each axis must indicate what quantity and what units are used.

1.2.1.2 First Program in Python

Program 1.3 solves the free-falling object differential equation. The program is remarkably similar to the Matlab script in Program 1.1. There are, however, many differences between the two languages.

```
from numpy import linspace
from scipy.integrate import solve_ivp

grv_const = 9.81 # [m/s^2]
init_pos = 0.0 # [m]
init_vel = 0.5 # [m/s]
init_mass = 5.0 #[kg]

init_cond = [init_pos, init_vel, init_mass]

init_time = 0 # [s]
final_time = 5.0 # [s]
num_data = 100
tout = linspace(init_time, final_time, num_data)


def free_falling_obj(time, state, grv_const):
    x1, x2, x3 = state
    dxdt = [x2,
            grv_const + (x3-2)*(x2/x3),
            -x3 + 2]
    return dxd


sol = solve_ivp(free_falling_obj, (init_time, final_time),
    init_cond, t_eval=tout, args=(grv_const,))
xout = sol.y

import matplotlib.pyplot as plt
plt.figure(1)
plt.plot(tout,xout[0,:])
plt.ylabel('position [m]');
plt.xlabel('time [s]');

plt.figure(2);
plt.plot(tout,xout[1,:])
plt.ylabel('velocity [m/s]');
plt.xlabel('time [s]');

plt.figure(3);
plt.plot(tout,xout[2,:])
plt.ylabel('m(t) [kg]');
plt.xlabel('time [s]');
```

Program 1.3 (Python) Free-falling object

On lines 4 through 14, the constants are defined with the proper naming and the units indicated in the comments. In Python, comments are placed after #.

The first two lines shown are not trivial to understand for the beginners of the Python language. Python has many packages, and each package is a collection of functions. There are several different ways to load these functions and the first line in the program,

```
1 from numpy import linspace
```

shows one of the methods. *from* and *import* are the keywords in Python. It loads the function *linspace* from the library called *numpy*. *numpy* is one of the scientific and engineering libraries and includes many useful functions such as matrix manipulations, and maths functions.

Numpy vs. scipy: The two packages are very similar and have many common functions. The execution speed of numpy is faster than scipy; in general, as numpy is written in C-language while scipy is written in Python. Scipy, however, has more specialized functions, which are not implemented in numpy.

We might wonder why each function is manually loaded before it is used, unlike in Matlab. This is one of the design principles of the Python language. If all functions are pre-loaded or they are automatically searched and loaded when they are used, then the search time or the size of the memory storing the function lists is long or larger. Hence, it is more efficient to load the functions manually when they are used.

The function linspace has three input arguments, for example, line 14 generates an array of numerical values starting from the initial time, 0.0, to the final time, 5.0, whose number of elements is equal to 'num data', 100. Unlike the integrator in Matlab, the Python integrator, discussed shortly later, needs the explicit time lists as one of the input arguments.

In the second line, the numerical integrator, *solve_ivp*, is loaded

```
2 from scipy.integrate import solve_ivp
```

This is slightly different from the way to load a function shown in the first line. *scipy* is another science and engineering function library. Some library divides the functions in the library into several categories. *integrate* is one of the categories in the scipy library. To access the functions under the category, *integrate*, the period is used after the library name, i.e. *scipy.integrate*. The numerical integrator, *solve_ivp*, is defined in the integrate category of the scipy library. If we try to load the function using *from scipy import solve_ivp*, it cannot find the integrator and generates an import error.

The ODE are defined between lines 17 and 22. The first line of the function definition begins with the keyword, '**def**', the function name, 'free_falling_obj', the three input arguments, and the colon, ':' as follows:

```
def free_falling_obj(time, state, grv_const):
```

In general, the function to be defined could have any input arguments. The function to be passed to *solve_ivp*, however, must have the first two input arguments, time and state, in this order. *solve_ivp* assumes that the first arguments and the second argument of the function passed are t and $\mathbf{x}$ in $\dot{\mathbf{x}} = d\mathbf{x}/dt$ in (1.6). The main body of the function is between the line below the function heading and the *return* line. Those lines that belong to the main part of the function are indented. The indentation in Python is not a decoration to simply improve the readability as in many other programming languages. The indentation in Python is the way to indicate which lines belong to the function body. The following is the first line of the function body:

```
    x1, x2, x3 = state
```

where 'state' is presumed to have three elements, and they are assigned to the three new variables on the left-hand side of the equal sign, 'x1', 'x2', and 'x3'. Instead of unpacking the three elements one by one, it unpacks all the three elements in one line.

'dxdt' is the list element in Python. In the list, each element is separated by the comma, ','. Finally, 'dxdt' becomes the return value of the function by the keyword, *return*, and the function is passed to the integrator, *solve_ivp*.

The first input argument of the integrator is the function name describing the ODE. The second one is the integration time interval. The third one is the initial condition. 't eval' is the list of time points, where the solution, $x(t)$, is stored to the output of the integrator. The last one is the arguments, whose name is reserved by *args*. As the function 'free_falling_obj' has the additional input variable apart from the time and the state, i.e. 'grv_const', this value must be sent to 'solve_ivp'. *args* is the input variable of 'solve_ivp' to pass additional input variables. 'grv const' is passed to the integrator by 'arg=(grv const,)'. The data type of *args* is a tuple. (1.3, 4.2, 4.3) or (1.3, 2.3) is a tuple. When there is only one element in a tuple, for example, (1.2,), the comma at the end must not be omitted. (1.2) is interpreted as floating-point 1.2, not a tuple. To make it a tuple, it must be (1.2,). Hence, there is the comma after 'grv const' in 'args=(grv const,)'.

Similar to Matlab, typing ‘whos’ at the command prompt in Python prints out the following list to the screen:

```
Variable            Type          Data/Info
-----------------------------------------------
final_time          float         5.0
free_falling_obj    function      <function free_falling_obj
grv_const           float         9.81
init_cond           list          n=3
init_mass           float         5.0
init_pos            float         0.0
init_time           int           0
init_vel            float         0.5
linspace            function      <function linspace at 0x7f
num_data            int           100
plt                 module        <module 'matplotlib.pyplo<...
sol                 OdeResult     message: 'The solver su<.
solve_ivp           function      <function solve_ivp at 0x7f
tout                ndarray       100: 100 elems, type `float64`,
xout                ndarray       3x100: 300 elems, type `float64
```

The solution of the ODE is stored in ‘sol’, whose type is OdeResult, and it includes various information about the integration results. Typing ‘sol’ in the command prompt and hitting enter shows what variables are in ‘sol’. We can access $\mathbf{x}(t)$ through ‘sol.y’. To avoid keep adding the dot to access $\mathbf{x}(t)$ inside ‘sol’, create a new variable, ‘xout’, and store ‘sol.y’ into ‘xout’. We can also see from the variable list that the size of ‘xout’ is 3×100. Each of the rows corresponds to $x(t)$, $\dot{x}(t)$, and $m(t)$, respectively.

To plot the results, a plotting library must be loaded. *matplotlib* is the most widely used plotting library in Python. More specifically, plot functions under *matplotlib.pyplot* category are the most frequently used. Load the functions as follows:

```
import matplotlib.pyplot
```

The way to access the functions under a specific category is using the dot next to the package name. *matplotlib.pyplot* means that we want to access the functions under the sub-category called *pyplot* in *matplotlib* instead of loading all functions in *matplotlib*. Now, we can use the *plot* command in *pyplot* as follows:

```
matplotlib.pyplot.plot(tout, xout[0,:])
```

This is inconvenient as the name becomes very long. To reduce the length of the name, *pyplot* is loaded as follows:

```
import matplotlib.pyplot as plt
```

After the keyword *as*, any convenient name we would call it could be used. By convention or almost standard, *matplotlib.pyplot* is called 'plt'. Hence, the long name to call 'plot' is shortened to

```
plt.plot(tout, xout[0,:])
```

This plots $x(t)$ vs. time t. Unlike Matlab, array indices in Python start at 0, not 1. The first row of 'xout' is 'xout[0,:]', the second row of 'xout' is 'xout[1,:]', and so forth. *xlabel* and *ylabel* commend work the same way as the ones in Matlab.

1.2.2 Ligand–Receptor Interactions

Ligand–receptor interactions are one of the most common interactions in biomolecular systems. As shown in Figure 1.3, the ligands, L, bind to the receptors, R, which spread on the cell boundary, form the ligand–receptor complex, C, and the complex evokes further reactions through various cascade signalling pathways inside the cell. L is produced with the rate given by a function of time, $f(t)$. From the control point of view, $f(t)$ is considered as the input, R is the internal state, and the concentration of C is the output of the ligand–receptor interactions.

The following molecular interactions describe the interactions between L, R, C, and $f(t)$:

$$\mathrm{R} + \mathrm{L} \xrightarrow{k_{\mathrm{on}}} \mathrm{C} \tag{1.7a}$$

$$\mathrm{C} \xrightarrow{k_{\mathrm{off}}} \mathrm{R} + \mathrm{L} \tag{1.7b}$$

$$\mathrm{R} \xrightarrow{k_{\mathrm{t}}} \emptyset \tag{1.7c}$$

$$\mathrm{C} \xrightarrow{k_{\mathrm{e}}} \emptyset \tag{1.7d}$$

$$\mathrm{f(t)} \xrightarrow{1} \mathrm{L} \tag{1.7e}$$

$$Q_{\mathrm{R}} \xrightarrow{1} \mathrm{R} \tag{1.7f}$$

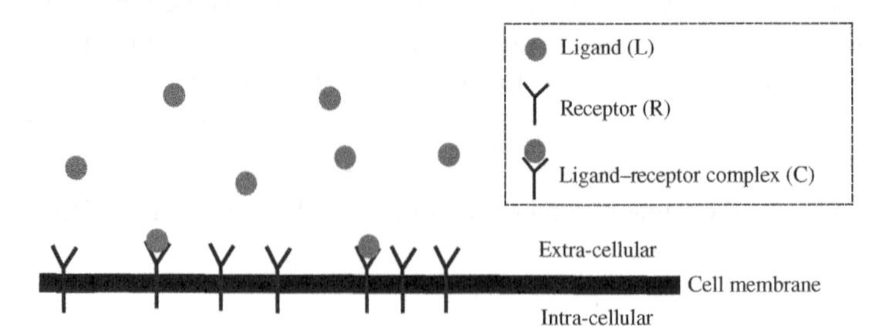

Figure 1.3 Ligand–receptor interactions form ligand–receptor complex.

where k_{on} and k_{off} are the reaction rates of binding or unbinding the receptor and the ligand, R and L, respectively, to form or destroy the complex, C, the receptor is destroyed with the rate of k_t, the complex is also destroyed with the rate of k_e, $f(t)$ is the stimulus that produces the ligand at the unit rate, and Q_R is the internal receptor generation at the unit rate.

We derive a set of ODE using the molecular interactions. To this end, we introduce the following two assumptions:

- All the molecules and the sources are uniformly distributed in the reaction space
- There are a sufficient number of molecules for every molecular species to consider concentration alone.

The first assumption makes the modelling being ODE. Otherwise, partial differential equations with the spatial coordinates are solved. Solving partial differential equations is computationally a lot more challenging than solving ODE. The second assumption indicates that the population of each molecular species is far away from 0. The randomness of molecular interactions and the integer nature of the number of molecules are ignored in the modelling.

Molecular interactions are stochastic. The probability of the occurrence of each reaction is calculated in stochastic simulations. We will discuss the details of stochastic modelling and simulation in the later chapter. On the other hand, deterministic simulations are performed by assuming a large number of molecules. The average molecular numbers show deterministic trajectories, where the random fluctuations are negligible.

Consider the receptor, R, which is directly involved in the three reactions. L binds to R and becomes C in (1.7a). The concentration of R is decreased by this reaction. The change rate is proportional to the concentrations of R and L as follows:

$$\frac{d[\text{R}]}{dt} \propto -[\text{R}] \times [\text{L}] \tag{1.8}$$

where $[\cdot]$ is the concentration of the molecules. The proportional constant is given by k_{on} in the reaction. The concentration unit is nanomolar (nM). Molar is equal to $N/(N_A V)$, where N is the number of molecules, N_A is Avogadro's number equal to 6.022×10^{23}, and V is the reaction space volume in litres.

In (1.7b), C is decomposed into R and L. The concentration of R is increased by this reaction. The decreasing rate is proportional to the concentration of C as follows:

$$\frac{d[\text{R}]}{dt} \propto [\text{C}] \tag{1.9}$$

where the proportional constant is k_{off}. The receptor is destroyed by itself at the rate of k_t as follows:

$$\frac{d[\text{R}]}{dt} \propto -[\text{R}] \tag{1.10}$$

Finally, in (1.7f), R is created at the rate of Q_R:

$$\frac{d[\mathrm{R}]}{dt} \propto [Q_R] \tag{1.11}$$

where the proportional constant is 1.

Combining (1.8)–(1.11) as follows: Shankaran et al. (2007)

$$\frac{d[\mathrm{R}]}{dt} = -k_{\mathrm{on}}[\mathrm{R}][\mathrm{L}] + k_{\mathrm{off}}[\mathrm{C}] - k_t[\mathrm{R}] + [Q_R] \tag{1.12}$$

Similarly, the following differential equations are established for L and C:

$$\frac{d[\mathrm{L}]}{dt} = -k_{\mathrm{on}}[\mathrm{R}][\mathrm{L}] + k_{\mathrm{off}}[\mathrm{C}] + [f(t)] \tag{1.13a}$$

$$\frac{d[\mathrm{C}]}{dt} = k_{\mathrm{on}}[\mathrm{R}][\mathrm{L}] - k_{\mathrm{off}}[\mathrm{C}] - k_e[\mathrm{C}] \tag{1.13b}$$

where $k_{\mathrm{off}} = 0.24$ [1/min], $k_{\mathrm{on}} = 0.0972$ [1/(min nM)], $k_t = 0.02$ [1/min], $k_e = 0.15$ [1/min], and $[f(t)] = 0.0$ [nM/min], i.e. no external stimulation. The values are the ones for the epidermal growth factor receptor (EGFR), which plays an important role in understanding tumour formation and growth.

Because of Q_R in $d[\mathrm{R}]/dt$, R would increase to infinity, which does not coincide with the reality as there would be the possible maximum number of receptors to be present in the cell. It is known that the maximum number of receptors for the EGFR is around 100,000 (Wee and Wang, 2017, Carpenter and Cohen, 1979). As the volume of the reaction space is given by $4 \times 10^{-10}\ell$ in Shankaran et al. (2007), the maximum concentration of R is $10,000/(N_A V) \approx 0.415$ nM. We model Q_R as follows:

$$[Q_R] = \begin{cases} 0.0166\ [\mathrm{nM/min}], & \text{for } [\mathrm{R}] \leq [\mathrm{R}]_{\mathrm{max}} \\ 0, & \text{otherwise} \end{cases} \tag{1.14}$$

where $[\mathrm{R}]_{\mathrm{max}}$ is equal to 0.415 nM.

The initial conditions for the following simulation are set as follows: $[\mathrm{R}(0)] = 0.1$ nM, $[\mathrm{L}(0)] = 0.0415$ nM, and $[\mathrm{C}(0)] = 0$ nM. In biomolecular network simulations, we must confirm that the molecular quantities such as the number of molecules or the concentrations must be non-negative. [C] at the beginning of the simulation could become negative if the time rate is negative. In the above initial conditions, [C] is strictly increasing because $d[\mathrm{C}(0)]/dt = k_{\mathrm{on}}[\mathrm{R}(0)][\mathrm{L}(0)]$ is positive at the beginning. As we can see from (1.13b), $d[\mathrm{C}]/dt$ is only negative when [C] is high enough, i.e. $[C] > k_{\mathrm{on}}[R][L]/(k_{\mathrm{off}} + k_e)$.

The Matlab script to simulate the EGFR concentration kinetics is given in Program 1.4.

```
1  clear;
2
3  init_receptor = 0.1; % [nM]
4  init_ligand = 0.0415; %[nM]
5  init_complex = 0.0; %[kg]
6
7  init_time = 0; % [min]
8  final_time = 180.0; % [min]
9  time_interval = [init_time final_time];
10
11 kon = 0.0972; % [1/(min nM)]
12 koff = 0.24; % [1/min]
13 kt = 0.02; %[1/min]
14 ke = 0.15; % [1/min]
15
16 ft = 0.0; % [nM/min]
17 QR = 0.0166; % [nM/min]
18 R_max = 0.415; %[nM]
19
20 sim_para = [kon koff kt ke ft QR R_max];
21
22 x0 = [init_receptor init_ligand init_complex];
23 [tout,xout] = ode45(@(time,state) RLC_kinetics(time,state,sim_para)
       , time_interval, x0);
24
25 figure(1); clf;
26 subplot(311);
27 plot(tout,xout(:,1))
28 ylabel('Receptor [nM]');
29 xlabel('time [min]');
30 axis([time_interval 0 0.5]);
31 subplot(312);
32 plot(tout,xout(:,2))
33 ylabel('Ligand [nM]');
34 xlabel('time [min]');
35 axis([time_interval 0 0.05]);
36 subplot(313);
37 plot(tout,xout(:,3))
38 ylabel('Complex [nM]');
39 xlabel('time [min]');
40 axis([time_interval 0 0.004]);
41
42 function dxdt = RLC_kinetics(time,state, sim_para)
43     R = state(1);
44     L = state(2);
45     C = state(3);
46
47     kon = sim_para(1);
48     koff = sim_para(2);
49     kt = sim_para(3);
50     ke = sim_para(4);
51     ft = sim_para(5);
52     QR = sim_para(6);
```

```
53        R_max = sim_para(7);
54
55        if R > R_max
56            QR = 0;
57        end
58
59        dxdt = zeros(3,1);
60        dxdt(1) = -kon*R*L + koff*C - kt*R + QR;
61        dxdt(2) = -kon*R*L + koff*C + ft;
62        dxdt(3) = kon*R*L - koff*C - ke*C;
63    end
```

Program 1.4 (Matlab) EGFR receptor, ligand, and complex kinetics

Figure 1.4 shows the simulation results. The receptor concentration increases almost linearly at the beginning and fluctuates later around the maximum concentration limit. The ligand–receptor reaction steadily consumes the ligand when they bind together and become the ligand–receptor complex. The complex has a peak concentration that occurred around 20 minutes and then slowly decayed.

Figure 1.5 shows the simulation results of the Python program, Program 1.5. Unlike the figure commands in Matlab for Figure 1.4, plotting subfigures in *matplotlib* is not as simple as in Matlab. We need advanced features in *matplotlib*. The advanced features of *subplots* in *matplotlib* are introduced in detail later in Program 2.2. As we notice in the figure, the figure fonts are too small to read. How to adjust the figure font sizes is also discussed in Program 2.2.

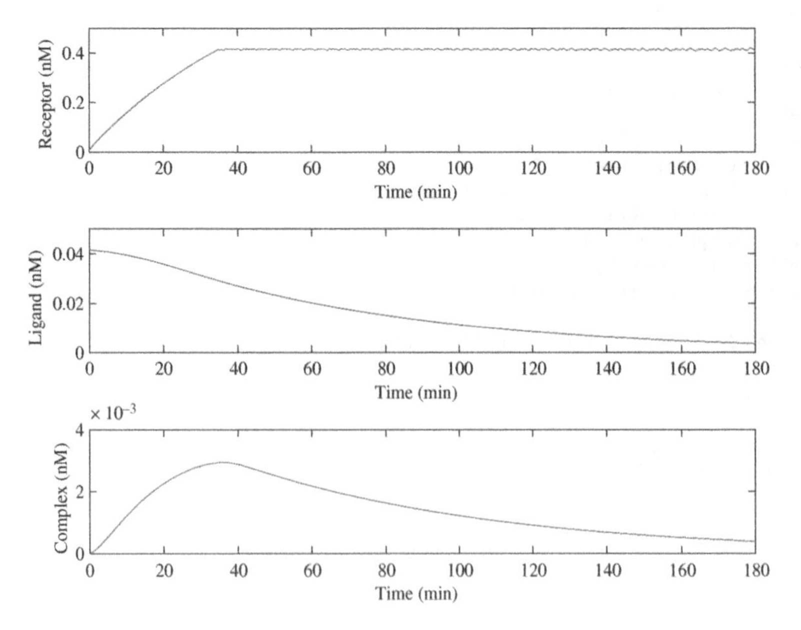

Figure 1.4 (Matlab) EGFR receptor, ligand, and complex time histories.

Program 1.5 uses two different integrators, i.e. *solve_ivp* and *odeint*. The ODE includes the discontinuous part, Q_R, given in (1.14). *odeint* cannot handle the differential equations with the discontinuity, and the solutions diverge. *solve_ivp* returns the correct numerical results. We recommend using *solve_ivp* instead of *odeint*.

```
1  from numpy import linspace
2  from scipy.integrate import solve_ivp
3
4
5  init_receptor = 0.01 #[nM]
6  init_ligand = 0.0415 #[nM]
7  init_complex = 0.0 #[kg]
8
9  init_time = 0 #[min]
10 final_time = 180.0 #[min]
11 time_interval = [init_time, final_time]
12
13 kon = 0.0972 #[1/(min nM)]
14 koff = 0.24 #[1/min]
15 kt = 0.02 #[1/min]
16 ke = 0.15 #[1/min]
17
18 ft = 0.0 #[nM/min]
19 QR = 0.0166 #[nM/min]
20 R_max = 0.415 #[nM]
21
22 sim_para = [kon, koff, kt, ke, ft, QR, R_max]
23
24 init_cond = [init_receptor, init_ligand, init_complex]
25
26
27 num_data = int(final_time*10)
28 tout = linspace(init_time, final_time, num_data)
29
30
31 def RLC_kinetics(time,state,sim_para):
32     R, L, C = state
33
34     kon, koff, kt, ke, ft, QR, R_max = sim_para
35
36     if R > R_max:
37         QR = 0
38
39     dxdt = [-kon*R*L + koff*C - kt*R + QR,
40             -kon*R*L + koff*C + ft,
41             kon*R*L - koff*C - ke*C]
42     return dxdt
43
44 sol_out = solve_ivp(RLC_kinetics, (init_time, final_time),
       init_cond, args=(sim_para,))
45
46 tout = sol_out.t
47 xout = sol_out.y
```

```
48
49 from scipy.integrate import odeint
50 xout_odeint = odeint(RLC_kinetics, init_cond, linspace(init_time,
       final_time, num_data), args=(sim_para,), tfirst=True)
51
52 import matplotlib.pyplot as plt
53 plt.figure(1)
54 plt.plot(tout,xout[0,:])
55 plt.ylabel('Receptor [nM]')
56 plt.xlabel('time [min]')
57 plt.axis([0, final_time, 0, 0.5])
58
59 plt.figure(2)
60 plt.plot(tout,xout[1,:])
61 plt.ylabel('Ligand [nM]')
62 plt.xlabel('time [min]')
63 plt.axis([0, final_time, 0, 0.05])
64
65 plt.figure(3)
66 plt.plot(tout,xout[2,:])
67 plt.ylabel('Complex [nM]')
68 plt.xlabel('time [min]')
69 plt.axis([0, final_time, 0, 0.004])
```

Program 1.5 (Python) EGFR receptor, ligand, and complex kinetics

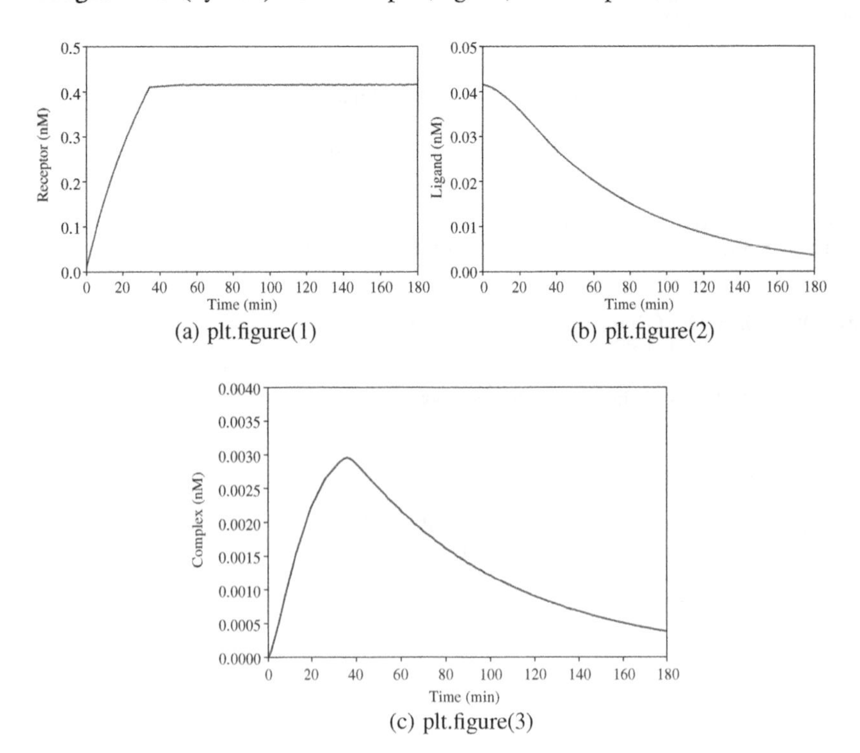

Figure 1.5 (Python) EGFR receptor, ligand, and complex time histories.

1.3 Organization of the Book

Chapters 2 and 3 cover the dynamics, control, and estimation algorithms of autonomous vehicles. Chapters 4 and 5 cover modelling and analysis of biological systems. Each of the chapters provides examples and exercises. We discuss additional readings and topics in the last chapter, Chapter 6.

Exercises

Exercise 1.1 (Matlab) Run Matlab, open the editor, type Program 1.1, save it as an m-script, execute the m-script in the Matlab command prompt, and obtain Figure 1.2.

Exercise 1.2 (Matlab) Using the *ode45* results from Program 1.1, plot Figure 1.6 using the *subplot* command in Matlab. Hint: Check the help for *subplot* in Matlab.

Exercise 1.3 (Python) Plot Figure 1.6 using the functions under *matplotlib.pyplot* in Python.

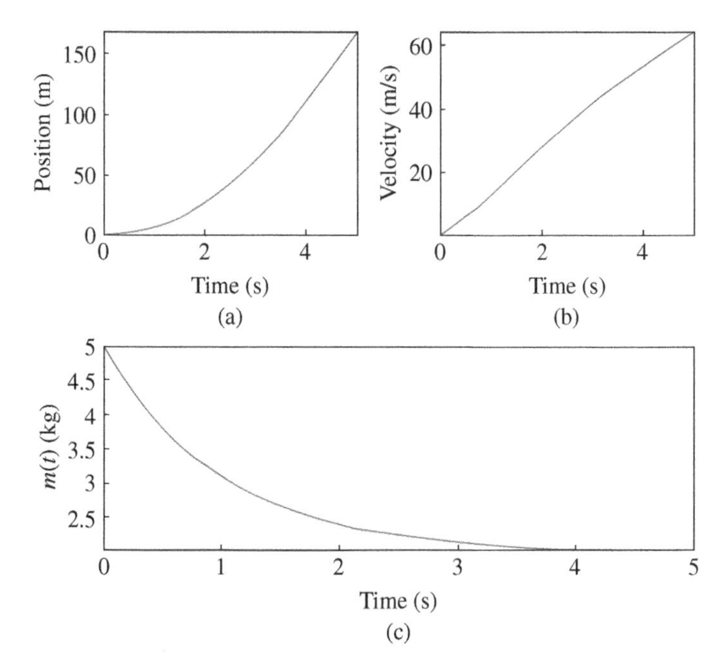

Figure 1.6 The time histories of (a) position (x), (b) velocity ($\dot{x}$), and (c) mass (m).

Exercise 1.4 Derive (1.13) from the molecular interactions in (1.7).

Exercise 1.5 (Python) What is the purpose of 'tfirst=True' in the arguments of *odeint* in Program 1.5?

Exercise 1.6 (Matlab/Python) Run the EGFR kinetic simulation 1000 times using the Matlab or the Python script, randomly selecting the initial concentration values in the following range: $[R(0)] \in [0, 0.2]$ nM, $[L(0)] \in [0, 0.05]$ nM, and $[C(0)] \in [0, 0.01]$ nM. Check if the concentrations are always positive.

Bibliography

G. Carpenter and S. Cohen. Epidermal growth factor. *Annual Review of Biochemistry*, 48(1):193–216, 1979. https://doi.org/10.1146/annurev.bi.48.070179.001205. PMID: 382984.

W.H. Press, S.A. Teukolsky, W.T. Vetterling, and B.P. Flannery. *Numerical Recipes 3rd Edition: The Art of Scientific Computing*. Cambridge University Press, 2007. ISBN 9780521880688.

Harish Shankaran, Haluk Resat, and H. Steven Wiley. Cell surface receptors for signal transduction and ligand transport: a design principles study. *PLOS Computational Biology*, 3(6):1–14, 2007. https://doi.org/10.1371/journal.pcbi.0030101.

Ping Wee and Zhixiang Wang. Epidermal growth factor receptor cell proliferation signaling pathways. *Cancers*, 9(5), 2017. ISSN 2072-6694. https://doi.org/10.3390/cancers9050052. https://www.mdpi.com/2072-6694/9/5/52.

2

Attitude Estimation and Control

Attitude is one of the fundamental properties of objects moving in a three-dimensional space. It is vital information for the satellite to point its camera in the desired direction, autonomous humanoid robot to balance its body, aerial vehicle to stabilize its attitude, and so forth.

2.1 Attitude Kinematics and Sensors

As shown in Figure 2.1, the rotation about the single axis, i.e. the z-axis perpendicular to the plane defined by x and y axes, can be *interpreted as a particle moving on the unit circle in the two-dimensional space.* The coordinates of the particle are equal to $(\cos\theta, \sin\theta)$, where θ is the angle measured from the positive x-axis in the anti-clockwise direction. As the movement of the particle is constrained on the perimeter of the unit circle, the coordinates of the particle are satisfied with the algebraic equation, $x^2 + y^2 = 1$, which is the equation for the unit circle centred at the origin.

The single-axis rotation about the z-axis is summarized as follows: the axis of rotation is equal to $\mathbf{k} = [0, 0, 1]^T$, which is the unit vector towards the positive z-axis, the coordinates of the particle are $(\cos\theta, \sin\theta)$, and the constraint is $(\sin\theta)^2 + (\cos\theta)^2 = 1$.

Consider pointing a telescope to observe a star in the sky as shown in Figure 2.2. The telescope is at the centre of the unit sphere, and the star is at the surface of the sphere. The initial pointing direction of the telescope is the positive x-axis, i.e. $\mathbf{i} = [1, 0, 0]^T$. We want to direct the telescope to the star indicated by the vector, $\mathbf{r}_2$.

The required rotation for pointing the telescope to the star is a two-axis rotation. The rotation angles are the azimuth angle, α, and the elevation angle, β, in the figure. Rotate the telescope α about $\mathbf{k}$, and it points $\mathbf{r}_1$ after the rotation. In addition, rotate it from $\mathbf{r}_1$ pointing direction with β about the axis obtained by $\mathbf{r}_1 \times \mathbf{k}$,

Dynamic System Modelling and Analysis with MATLAB and Python: For Control Engineers,
First Edition. Jongrae Kim.

Companion Website: www.wiley.com/go/kim/dynamicmodeling

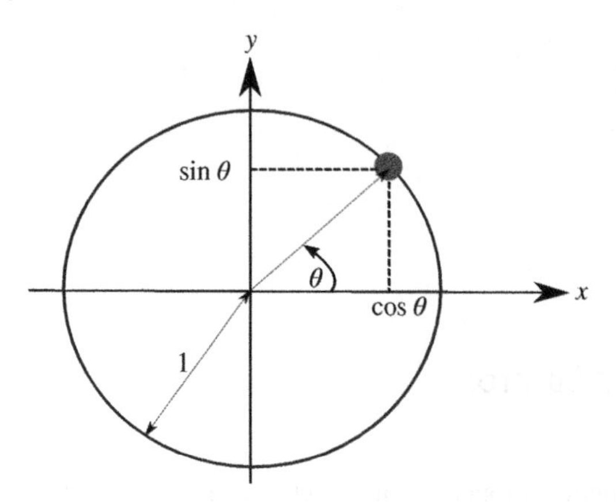

Figure 2.1 Single-axis rotation about the axis perpendicular to the surface defined by the $x-y$ axes.

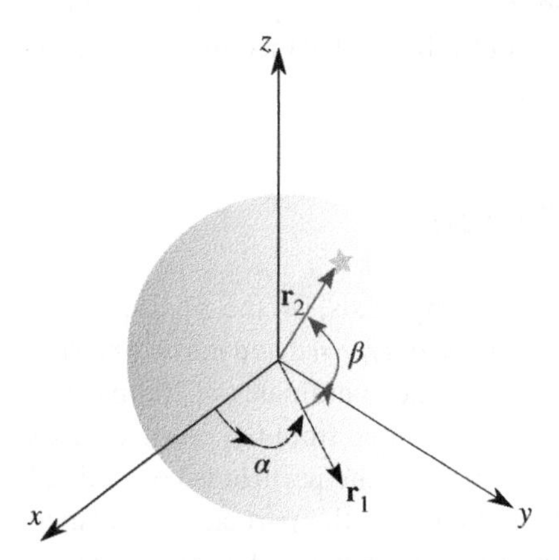

Figure 2.2 The two-axis rotation is equivalent to the single-axis rotation about the axis perpendicular to the surface defined by $\mathbf{r}_1$ and $\mathbf{r}_2$.

where × is the vector cross product. *The rotations about the two axes are equivalent to a particle moving on the surface of the unit sphere in the three-dimensional space.* The particle position given by the vector, $\mathbf{r}_2 = [r_x, r_y, r_z]^T$, must satisfy $r_x^2 + r_y^2 + r_z^2 = 1$ as it is on the surface of the unit sphere. The two-step rotation from the initial pointing, **i**, to the final pointing, $\mathbf{r}_2$, can be achieved by the single rotation about the axis defined by $\mathbf{e} = \mathbf{i} \times \mathbf{r}_2$ with the rotation angle, θ, equal to

$$\theta = \cos^{-1}\left(\mathbf{i} \cdot \mathbf{r}_2\right) \tag{2.1}$$

where $(\cdot)$ is the vector dot product. The single-step rotation exists for any star on the surface of the unit sphere. However, it does not mean that the two-axis rotation is the same as the fixed single-axis rotation discussed earlier. Unlike the fixed single-axis rotation, where the rotation axis, **e**, is fixed to **k**, the axis to achieve a single-axis rotation to point a star on the unit sphere surface changes depending on the position of the star.

We extend the same logic for the general object rotations in a three-dimensional space. *The rotation about three axes is equivalent to a particle moving on the surface of the unit sphere in a four-dimensional space.* The constraint to be satisfied is the squared sum of the four coordinates of the particle given by $\mathbf{q} = [q_1, q_2, q_3, q_4]^T$ equal to 1, i.e.

$$\mathbf{q}^T\mathbf{q} = q_1^2 + q_2^2 + q_3^2 + q_4^2 = 1; \tag{2.2}$$

We can achieve any three-axis rotation with the corresponding single-axis rotation where the rotation axis and the angle are given by $\mathbf{e}$ and θ, respectively.

$\mathbf{q}$ is defined using $\mathbf{e}$ and θ as follows:

$$\mathbf{q} = \begin{bmatrix} q_1 \\ q_2 \\ q_3 \\ q_4 \end{bmatrix} = \begin{bmatrix} \mathbf{q}_{13} \\ q_4 \end{bmatrix} = \begin{bmatrix} \mathbf{e}\sin\frac{\theta}{2} \\ \cos\frac{\theta}{2} \end{bmatrix} = \begin{bmatrix} e_1\sin\frac{\theta}{2} \\ e_2\sin\frac{\theta}{2} \\ e_3\sin\frac{\theta}{2} \\ \cos\frac{\theta}{2} \end{bmatrix} \tag{2.3}$$

where the rotation axis, $\mathbf{e}$, is the unit vector and is equal to $[e_1, e_2, e_3]^T$. Equation (2.3) defines the quaternion, $\mathbf{q}$. The quaternion is one of the most frequently used attitude parameterization methods. The rotation angle, θ, is divided by 2 in the definition. The half-angle leads to a simple algebraic relation when the governing equation for time-varying $\mathbf{q}$, i.e. attitude kinematics, is derived.

Consider a tumbling three-dimensional object with the angular velocity, $\boldsymbol{\omega}$.

$$\boldsymbol{\omega} = \begin{bmatrix} \omega_x & \omega_y & \omega_z \end{bmatrix}^T \tag{2.4}$$

The triple arrowhead in Figure 2.3 indicates the $\boldsymbol{\omega}$ vector, where ω_x, ω_y, and ω_z are the instantaneous angular velocity of the object at the current time towards each

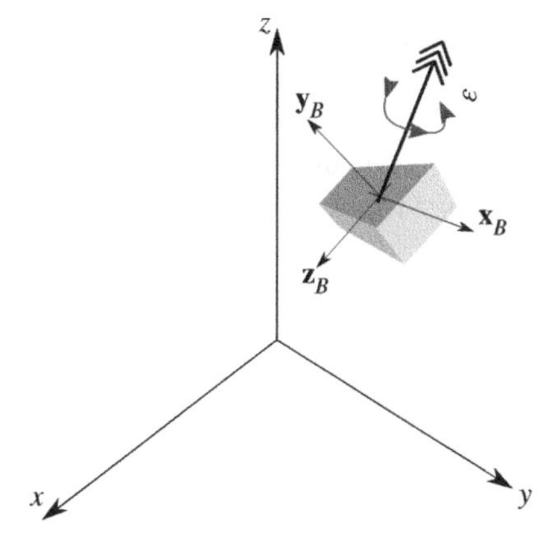

Figure 2.3 The object is tumbling in a three-dimensional space, where the body coordinates $\mathbf{x}_B$, $\mathbf{y}_B$, and $\mathbf{z}_B$ are attached to the object.

body axis, $\mathbf{x}_B$, $\mathbf{y}_B$, and $\mathbf{z}_B$, respectively, and their unit is [rad/s]. The gyroscopes attached to the object measures the three components of the angular velocity vector, where the gyroscope sensing directions are aligned to the body axes.

The quaternion kinematics is given by Crassidis and Junkins (2011)

$$\frac{d\mathbf{q}}{dt} = \frac{1}{2}\Omega(\boldsymbol{\omega})\mathbf{q} \tag{2.5}$$

where

$$\Omega(\boldsymbol{\omega}) = \begin{bmatrix} 0 & \omega_z & -\omega_y & \omega_x \\ -\omega_z & 0 & \omega_x & \omega_y \\ \omega_y & -\omega_x & 0 & \omega_z \\ -\omega_x & -\omega_y & -\omega_z & 0 \end{bmatrix} \tag{2.6}$$

and this is written in a compact form as follows:

$$\Omega(\boldsymbol{\omega}) = \begin{bmatrix} -[\boldsymbol{\omega}\times] & \boldsymbol{\omega} \\ -\boldsymbol{\omega}^T & 0 \end{bmatrix} \tag{2.7}$$

where

$$[\boldsymbol{\omega}\times] = \begin{bmatrix} 0 & -\omega_z & \omega_y \\ \omega_z & 0 & -\omega_x \\ -\omega_y & \omega_x & 0 \end{bmatrix} \tag{2.8}$$

2.1.1 Solve Quaternion Kinematics

Consider the angular velocity given by

$$\boldsymbol{\omega} = \begin{bmatrix} 0.1\sin(2\pi \times 0.005t) \\ 0.05\cos(2\pi \times 0.1t + 0.2) \\ 0.02 \end{bmatrix} \text{[rad/s]} \tag{2.9}$$

where t is the time in seconds, ω_x and ω_y oscillate with the frequency equal to 0.005 and 0.1 Hz, respectively, and the $\mathbf{z}_B$ axis rotates with the constant angular velocity, 0.02 rad/s.

2.1.1.1 MATLAB

Modify Program 1.1 and solve the quaternion kinematic equation, (2.5), as follows:

```
1 clear;
2
3 init_time = 0; % [s]
4 final_time = 60.0; % [s]
5 time_interval = [init_time final_time];
6
7 q0 = [0 0 0 1]';
```

```
8  [tout,qout] = ode45(@(time,state) dqdt_attitude_kinematics(time,
       state), time_interval, q0);
9
10 figure;
11 plot(tout,qout(:,1),'b-',tout,qout(:,2),'r--',tout,qout(:,3),'g-.',
       tout,qout(:,4),'m:')
12 ylabel('quaternion');
13 xlabel('time [s]');
14 legend('q1','q2','q3','q4');
15 set(gca,'FontSize',14);
16
17 function dqdt = dqdt_attitude_kinematics(time,state)
18     q_true = state(:);
19
20     w_true(1) = 0.1*sin(2*pi*0.005*time); % [rad/s]
21     w_true(2) = 0.05*cos(2*pi*0.01*time + 0.2); %[rad/s]
22     w_true(3) = 0.02; %[rad/s]
23     w_true = w_true(:);
24
25     wx = [ 0             -w_true(3)   w_true(2);
26            w_true(3)     0            -w_true(1);
27            -w_true(2)    w_true(1)    0];
28
29     Omega = [   -wx           w_true;
30                 -w_true'      0];
31
32     dqdt = 0.5*Omega*q_true;
33 end
```

Program 2.1 (MATLAB) Solve $d\mathbf{q}/dt$ for $\boldsymbol{\omega}$ given by (2.9)

The quaternion time history is shown in Figure 2.4. Whenever a figure is created in MATLAB, all properties of the figure are stored in the automatically generated variable called *gca*. One of the properties is the font size of the characters in the figure, and it can be changed using the command, *set*, as follows:

```
set(gca,'FontSize',14);
```

where the default font size, 12 pt (point font), is changed to 14 pt.

Recall that the quaternion must satisfy the unit norm condition, (2.2). The *ode45* function does not care constraints. It solves the differential equation given by (2.5) as if there is no constraint. It is not trivial to integrate differential equations with constraints. There is a way, however, to control the speed of the error growth. Define the unit norm error as follows:

$$(\mathbf{q}\text{ unit norm error}) = \log\left|\mathbf{q}^T\mathbf{q} - 1\right| \tag{2.10}$$

where $\log(\cdot)$ is the natural logarithm.

The two options in *ode45* to adjust the numerical errors are the relative tolerance, *RelTol*, and the absolute tolerance, *AbsTol*. *ode45* adjusts the integration

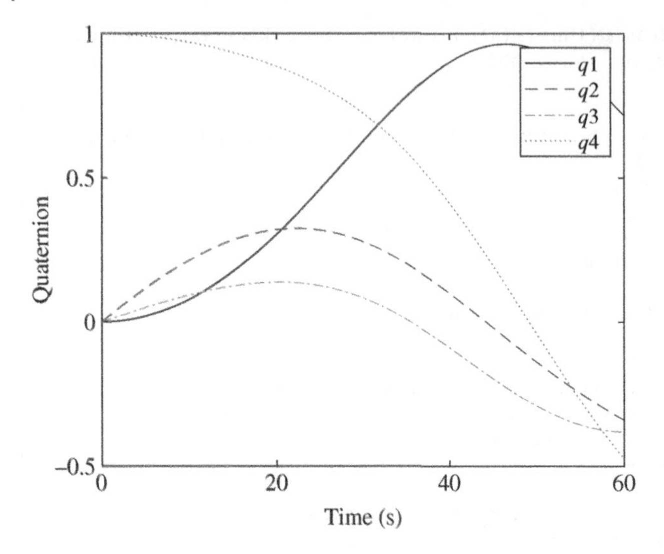

Figure 2.4 The quaternion time history for ω given by (2.9).

interval according to these two values. For the differential equation given by $\dot{x} = f(x)$, *ode45* compares the value of $f(x)$ at the previous integration at t with the current at $t + \Delta t$. If the difference, $|f[x(t + \Delta t)] - f[x(t)]|$, is larger than *RelTol*, it reduces Δt so that the difference is smaller than *RelTol*. Similarly, *ode45* compares $|f[x(t + \Delta t)]|$ with zero. If it is greater than *AbsTol*, it reduces Δt so that $|f[x(t + \Delta t)]|$ becomes smaller than *AbsTol*. To adjust the tolerances, the *odeset* function is used before *ode45* is called as follows:

```
ode_options = odeset('RelTol',1e-3,'AbsTol',1e-6);
```

and the option is passed to *ode45* as follows:

```
[tout,qout] = ode45(@(time,state) dqdt_attitude_kinematics(time,
    state), time_interval, q0, ode_options);
```

We cannot, however, reduce these two tolerances arbitrarily small. Unreasonably, small Δt slows down the integration speed or causes the round-off error. The small numbers could be too small, and the computer cannot distinguish them from zero. Then, the numerical error would increase, and this is called the round-off error in the computer. As long as the tolerance remains within a reasonable range, the smaller the tolerance, the smaller the numerical integration error.

Figure 2.5 compares three different cases, where the relative tolerances are as shown in the labels and the absolute tolerances are 1000 times smaller than the relative tolerances. As time increases, the error gradually increases. At the end of the simulation time, 6000 seconds, the error for the relative tolerance equal to 0.001

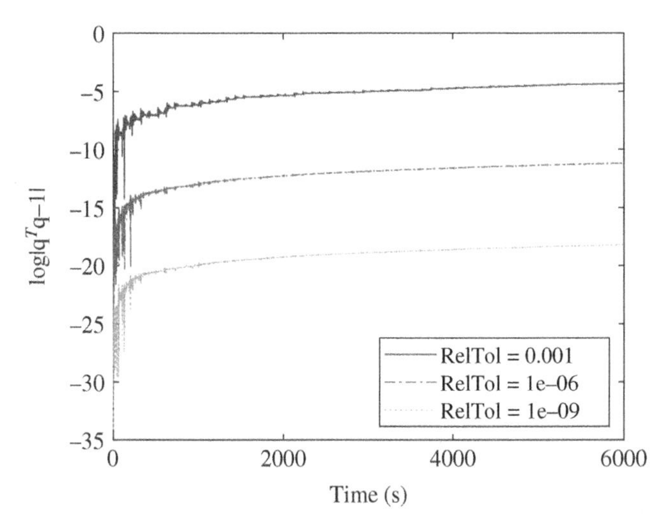

Figure 2.5 The time history of the quaternion unit norm error, (2.10), for three different tolerance settings for ode45.

reaches around $e^{-4.33} \approx 0.132$. Hence, all interpretations of the rotations should not be based on values less than this. For example, if we compare two quaternions, only differences much larger than 0.132 have a meaningful interpretation based on the numerical solution. Differences less than 0.132 would be numerical artefacts.

2.1.1.2 Python

Recall the first python program, Program 1.3, and modify it to solve the quaternion kinematic equation given by (2.5), where the angular velocity is given by (2.9).

```
1  import numpy as np
2  from numpy import linspace
3  from scipy.integrate import solve_ivp
4
5  init_time = 0 # [s]
6  final_time = 60.0 # [s]
7  num_data = 1000
8  tout = linspace(init_time, final_time, num_data)
9
10 q0 = np.array([0,0,0,1])
11
12 def dqdt_attitude_kinematics(time, state):
13     quat = state
14     w_true = np.array([0.1*np.sin(2*np.pi*0.005*time), #[rad/s]
15                        0.05*np.cos(2*np.pi*0.01*time + 0.2), #[rad
                              /s]
16                        0.02]) #[rad/s]
17
```

```
18        wx=np.array([[0,                -w_true[2],        w_true[1]],
19                     [w_true[2],        0,                 -w_true[0]],
20                     [-w_true[1],       w_true[0],         0]])
21
22        Omega_13 = np.hstack((-wx,np.resize(w_true,(3,1))))
23        Omega_4  = np.hstack((-w_true,0))
24        Omega = np.vstack((Omega_13, Omega_4))
25
26        dqdt = 0.5*(Omega@quat)
27
28        return dqdt
29
30
31  sol = solve_ivp(dqdt_attitude_kinematics, (init_time, final_time),
        q0, t_eval=tout)
32  qout = sol.y
33
34  import matplotlib.pyplot as plt
35
36  fig, ax = plt.subplots()
37  ax.plot(tout,qout[0,:],'b-',tout,qout[1,:],'r--',tout,qout[2,:],'g
        -.',tout,qout[3,:],'m:')
38
39  fig.set_figheight(6) # size in inches
40  fig.set_figwidth(8)  # size in inches
41
42  xtick_list = np.array([0,10,20,30,40,50,60])
43  ax.set_xticks(xtick_list)
44  ax.set_xticklabels(xtick_list,fontsize=14)
45
46  ytick_list = np.array([-0.5,0.0,0.5,1.0])
47  ax.set_yticks(ytick_list)
48  ax.set_yticklabels(ytick_list,fontsize=14)
49
50  ax.legend(('q1','q2','q3','q4'),fontsize=14, loc='upper right')
51  ax.axis((0,60,-0.5,1.0))
52  ax.set_xlabel('time [s]',fontsize=14)
53  ax.set_ylabel('quaternion',fontsize=14)
```

Program 2.2 (Python) Solve $d\mathbf{q}/dt$ for $\boldsymbol{\omega}$ given by (2.9)

Be careful for the index numbering, the array index in Python starts from 0. $[\boldsymbol{\omega}\times]$, (2.8), is defined as follows:

```
        wx=np.array([[0,                -w_true[2],        w_true[1]],
                     [w_true[2],        0,                 -w_true[0]],
                     [-w_true[1],       w_true[0],         0]])
```

where w_true[0] $= \omega_x$, w_true[1] $= \omega_y$, and w_true[2] $= \omega_z$.

Each row of a two-dimensional matrix is defined using box brackets, '[ ]'. Commas separate the elements in a row and also different rows. Another two box brackets construct a two-dimensional matrix. For example, [[2.0, −3.0, 1.5], [0.0, 5.2, 9.8]] defines the 2×3 matrix.

The plot parts in Program 2.2 look very different from the plot commands used in Program 1.3. The commands used in Program 1.3, i.e. 'plt.plot()', 'plt.xlabel()', and 'plt.ylabel()', provide convenient ways to plot simple figures. To have the capability for fine-tuning figures such as adjusting font size, changing the tick intervals for each axis and so forth, we must use these plot command styles shown in Program 2.2. Run the following lines directly in the iPython command prompt:

```
In [21]: import numpy as np
In [22]: import matplotlib.pyplot as plt
In [23]: x=np.linspace(1,10,100)
In [24]: y0=2*x
In [25]: y1=10+10*(x**2)
In [26]: fig ,(ax0,ax1)=plt.subplots(nrows=2,ncols=1)
In [27]: ax0.plot(x,y0)
Out[27]: [<matplotlib.lines.Line2D at 0x7f9cf864ed90>]
In [28]: ax1.plot(x,y1,'r--')
Out[28]: [<matplotlib.lines.Line2D at 0x7f9cf9c45250>]
```

'fig, (ax1, ax2) = plt.subplots(nrows=2,ncols=1)' creates two sub-figures in the figure placed in the two rows and one column format. The return variable, fig, indicates the whole figure, and ax0 and ax1 indicate the first and the second sub-figures, respectively. In Program 2.2, the figure size is set to 6″ high and 8″ wide using 'fig.set_figheight(6)' and 'fig.set_figwidth(8)', where the lengths are in inches.

The plot command for drawing on the first sub-figure is 'ax0.plot()', and the plot command for the second figure is 'ax1.plot()'. Similarly, commands for the tick intervals and labels, the font sizes, and the legend are indicated by 'ax0' or 'ax1' for each sub-figure in the example. In Program 2.2, the ticks for each axis can be manually set using the 'ax.set_xticks' command, and the labels for the ticks can also be manually set using 'ax.set_xticklables'. Similarly, the ticks and the labels for the y-axis are assigned using 'set_yticks' and 'set_yticklabels'. In addition, the font size and the location of the legend can be controlled using the additional arguments in the 'ax.legend' command, and the font size for x-axis or y-axis labels can be changed using the fontsize value in 'ax.set_xlabel' or 'ax.set_ylabel'.

Figure 2.6 shows the quaternion time history calculated by the python program.

Figure 2.7 shows that the error for the relative tolerance equal to 0.001 increases around $e^{-7.24} \approx 0.0007$ at the time equal to 6000 seconds. Compared to the error

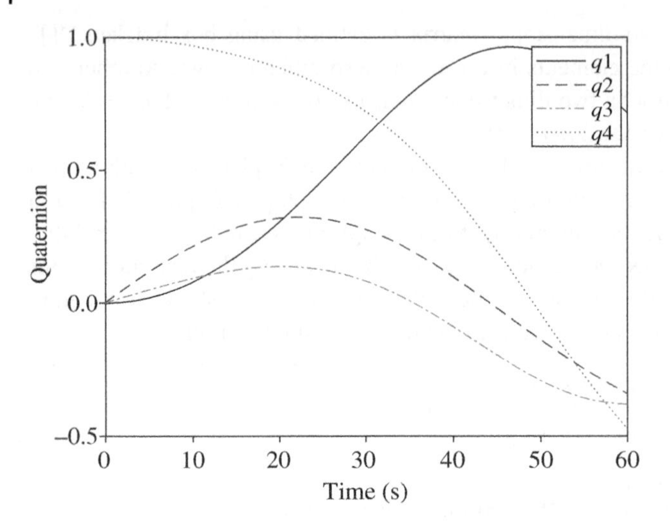

Figure 2.6 The quaternion time history for $\boldsymbol{\omega}$ given by (2.9).

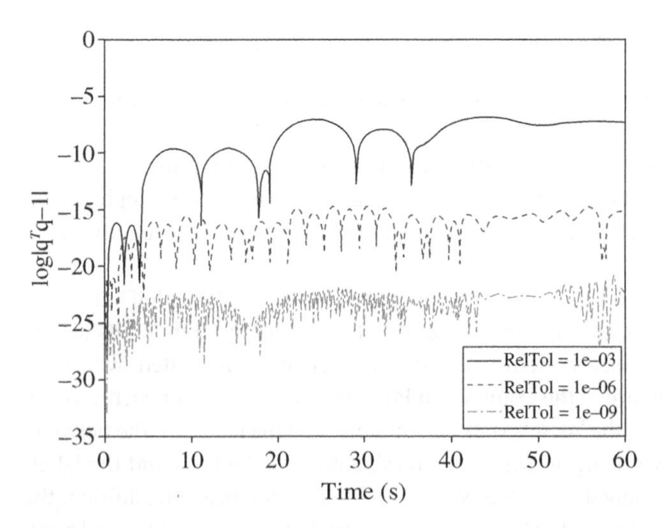

Figure 2.7 The time history of the quaternion unit norm error, (2.10), for three different tolerance settings for ode45.

in MATLAB, it is less than 1000 times smaller. It would be, however, not correct to conclude that *solve_ivp* in the Scipy is superior to *ode45* in MATLAB. The algorithms have different ways of controlling the numerical error, and the numerical error is not necessarily always the same as the example. The important message is that we must be aware of how the errors would be propagated as the time increases for the given tolerance levels.

2.1.2 Gyroscope Sensor Model

A rate gyro measures the angular velocity, and two different types of stochastic noises corrupt its measurement as follows:

$$\tilde{\boldsymbol{\omega}} = \boldsymbol{\omega} + \boldsymbol{\beta} + \boldsymbol{\eta}_v \tag{2.11}$$

where $\tilde{\boldsymbol{\omega}}$ is the gyro measurement output, and the bias drift, $\boldsymbol{\beta}$, and the white noise, $\boldsymbol{\eta}_v$, corrupt the true angular velocity, $\boldsymbol{\omega}$. The measurement gives the sum of three values, i.e. the truth and two random noises, and they cannot be distinguished. We do not exactly know the truth, $\boldsymbol{\omega}$, but the corrupted information from the sensor.

2.1.2.1 Zero-Mean Gaussian White Noise

The zero-mean Gaussian white noise, $\boldsymbol{\eta}_v$, is one of the typical types of sensor noise. The zero-mean indicates that the noise has a mean value of zero. Its distribution is Gaussian or normal. The white noise signifies that the same strength of signals for all frequencies as the term is derived from white light, which includes all visible frequency lights with equal strength. The following two equations express these properties of the noise:

$$\begin{aligned} E\left\{\boldsymbol{\eta}_v(t)\right\} &= 0 \\ E\left\{\boldsymbol{\eta}_v(t_1)\boldsymbol{\eta}_v^T(t_2)\right\} &= \sigma_v^2\delta(t_1 - t_2)I_3 \end{aligned}$$

for all time, t, and any t_1 and t_2 in $[0, \infty)$, where $E(\cdot)$ is the expectation, σ_v^2 is the variance of the noise, i.e. the strength of the noise, $\delta(t_1 - t_2)$ is the Dirac delta function equal to 1 only if $t_1 = t_2$ and zero otherwise, and I_3 is the 3×3 identify matrix. As the off-diagonal terms of I_3 are all zero, the white noise for each axis is independent or not correlated with each other.

For brevity, consider a one-dimensional random number, $x(t)$, with the following properties:

$$\begin{aligned} E\left\{x(t)\right\} &= 0 \\ E\left\{x(t_1)x(t_2)\right\} &= \sigma^2\delta(t_1 - t_2) \end{aligned}$$

for all time, t, and any t_1 and t_2 in $[0, \infty)$. The extension of the discussion below to the three-dimensional random vector $\boldsymbol{\eta}_v(t)$ is trivial. Let the probability density function (pdf) of x at t be equal to $p(x)$, and the expectation of $x(t)$, i.e. the mean value of x at the time t, is given by

$$E\left\{x(t)\right\} = \int_{\Omega} x(t)p(x)dx \tag{2.12}$$

The expectation is the weighted integration of the variable by the probability density function, where Ω is the sampling space of the random variable, $x(t)$.

2.1.2.2 Generate Random Numbers

The function, *randn*, in MATLAB generates random numbers with a mean and variance of 0 and 1, respectively. Run the following lines in the MATLAB command prompt:

```
>> x = randn(1,100);
>> mean(x)
ans =
  -0.2711
>> var(x)
ans =
   1.1052
```

Program 2.3 (MATLAB) Generate 100 random numbers, x, whose mean and variance are equal to 0 and 1, respectively

The mean value and the variance of x printed on the screen are different whenever the commands are executed. When *randn* is called, it generates a different set of 100 random numbers drawn from Gaussian distribution, whose mean value and variance are equal to 0 and 1, respectively. The mean and the variance calculated using the samples, 'x', are only approximately close to 0 and 1, respectively. As the number of the samples increases, they converge to the true values.

Gaussian distribution is also called the normal distribution and the 'n' at the end in the function name, *randn*, stands for the normal distribution. Be careful to use the correct random number generator; *rand* function generates the uniformly distributed random numbers between 0 and 1. The sensor noise is typically modelled as the normal distribution rather than the uniform distribution.

Similarly, in Python, *randn* under the numpy.random package is used to generate the random numbers as follows:

```
In [54]: import numpy as np

In [55]: x=np.random.randn(100)

In [56]: x.mean()
Out[56]: -0.05332928410865288

In [57]: x.var()
Out[57]: 0.8078225617520309
```

Program 2.4 (Python) Generate 100 random numbers, x, whose mean and variance are equal to 0 and 1, respectively

In Python, every variable created is an object in object-oriented programming. When an object is created, various methods attach to the object. 'x' is the object, and *mean()* and *var()* are the methods to calculate the mean value and the variance of 'x'. To call each method, put '.' and the method name after 'x', e.g. x.mean() for calculating the mean of x. There is another function to generate random numbers with the normal distribution. The function is under *numpy.random* package, called *numpy.random.normal*. This function is equivalent to *numpy.random.randn* apart from some slight differences in the format of input arguments.

How to generate the random number, z, with the mean equal to 0.5 and the variance equal to 0.2 using *randn* in MATLAB is shown in Program 2.5.

```
>> mean_z = 0.5;
>> var_z = 0.2;
>> z = mean_x + sqrt(var_x)*randn(1,100);
>> mean(z)
ans =
   0.6137
>> var(z)
ans =
   0.1917
```

Program 2.5 (MATLAB) Generate 100 random numbers, z, whose mean and variance are equal to 0.5 and 0.2, respectively

Similarly, using Python, z is generated in Program 2.6.

```
In [58]: import numpy as np

In [59]: mean_z=0.5

In [60]: var_z=0.2

In [61]: z=mean_z+np.sqrt(var_z)*np.random.randn(100)

In [62]: z.mean()
Out[62]: 0.4834311699410189

In [63]: z.var()
Out[63]: 0.24051712417906854
```

Program 2.6 (Python) Generate 100 random numbers, z, whose mean and variance are equal to 0.5 and 0.2, respectively

As the number of samples increases, the mean and variance approach the given true value.

We validate how to generate the random number, z, using the random number, x, as follows:

$$z = \mu + \sqrt{\sigma^2} x \tag{2.13}$$

where x is the random variable whose mean and variance are equal to 0 and 1, respectively. The mean value of z is given by

$$E(z) = E(\mu + \sqrt{\sigma^2}x) = \mu + \sqrt{\sigma^2}E(x) = \mu + \sqrt{\sigma^2} \times 0 = \mu \tag{2.14}$$

where the expectations of the deterministic values, μ and σ, are equal to the values themselves. From the definition of the variance, the variance of z becomes

$$\begin{aligned}\sigma^2 &= E(z^2) - [E(z)]^2 = E\left[(\mu + \sqrt{\sigma^2}x)^2\right] - \mu^2 \\ &= \mu^2 + 2\mu\sqrt{\sigma^2}E(x) + \sigma^2 E(x^2) - \mu^2 = \sigma^2\end{aligned} \tag{2.15}$$

where E(x) and E(x^2) are equal to 0 and 1, respectively, by the definitions.

In the above examples, the 100 random numbers, x, generated by *randn* is drawn from the following probability density function, $p(x)$:

$$p(x) = \frac{1}{\sqrt{2\pi\sigma^2}} e^{-\frac{(x-\mu)^2}{2\sigma^2}} \tag{2.16}$$

where μ and σ are the mean and the variance of x equal to 0 and 1, respectively. The probability if the random number x belongs to the interval $[x_k, x_{k+1}]$ is given by

$$\Pr[x_k \le x \le x_{k+1}] = \int_{x=x_k}^{x=x_{k+1}} p(x)dx \tag{2.17}$$

```
1  clear;
2
3  % true probability density function (pdf)
4  var_x = 1;
5  mean_x = 0;
6  Omega_x = linspace(-5,5,1000);
7  px = (1/(sqrt(2*pi*var_x)))*exp(-(Omega_x-mean_x).^2/(2*var_x));
8
9  figure(1); clf;
10 plot(Omega_x,px,'LineWidth',2);
11 hold on;
12
13 % generate N random numbers with the mean zero and the variance 1
       using
14 % randn
15 N_all = [100 10000];
16 x_bin = linspace(-5,5,30);
17 dx=mean(diff(x_bin));
18 line_style = {'rs-' 'go-'};
19 for idx=1:length(N_all)
20      N_trial = N_all(idx);
21      x_rand = randn(1,N_trial);
22
23      % number of occurance of x_rand in x_bin
```

```
24     N_occur = histcounts(x_rand,x_bin);
25
26     figure(1);
27     plot(x_bin(1:end-1)+dx/2, N_occur/(dx*N_trial),line_style{idx})
          ;
28 end
29
30 figure(1);
31 set(gca,'FontSize',14);
32 xlabel('Random Variable x Sampling Space: $\Omega_x,'Interpreter','
      latex');
33 ylabel('probability density function');
34 legend('True $p(x),'N=100','N=10,000','Location','northeast','
      Interpreter','latex');
```

Program 2.7 (MATLAB) Compare the true pdf with the approximated pdf generated by *randn*

To check if the random numbers generated by *randn* are indeed from the normal distribution, we estimated the pdf by counting the number of random numbers fallen into each of the intervals, and the estimated pdf is compared with the true pdf. Let the number of the random numbers in $[x_k, x_{k+1}]$ be N_k. The estimated pdf value at x in the interval, $\hat{p}(x)$, is derived from (2.17) as follows:

$$[\text{LHS of (2.17)}] \approx \frac{N_k}{N_{\text{total}}} \tag{2.18a}$$

$$[\text{RHS of (2.17)}] \approx \hat{p}(x)\Delta x_k \tag{2.18b}$$

where $\Delta x_k = x_{k+1} - x_k$, N_{total} is the total number of samples equal to 100 in this example, and $\hat{p}(x)$ is assumed to be constant for $x \in [x_k, x_{k+1}]$. Hence,

$$\hat{p}(x) = \frac{N_k}{\Delta x_k N_{\text{total}}} \quad \text{for } x \in [x_k, x_{k+1}] \tag{2.19}$$

Program 2.7 draws Figure 2.8, which shows the true pdf, $p(x)$, and two approximated pdf for N_{total} equal to 100 and 10,000. As the total number of random numbers generated increases, the estimated pdf, $\hat{p}(x)$, converges to the true pdf. 30 bins are generated below between $x = -5$ and $x = 5$, counting how many x generated by *randn* belong to each bin.

```
x_bin = linspace(-5,5,30);
```

For the generated random number, 'x_rand', and the bin list, 'x_bin', the number of occurrences for each bin is counted using *histcounts* command:

```
N_occur = histcounts(x_rand,x_bin);
```

The dimension of N_occur is one less than the dimension of x_bin as the i-th element of N_occur corresponds to the interval given by the i-th and the $(i+1)$-th elements of x_bin. Hence, when plotting N_occur with respect to x, the middle point of each bin is used as follows:

```
plot(x_bin(1:end-1)+dx/2, N_occur/(dx*N),line_style{idx});
```

For each N_total-cases inside the loop, two different line styles are defined in the strings, 'rs-' and 'go-', which produce the square marked line and the circle marked line, respectively, in the figure. The lines are displayed in red or green if printed in colour. To make a list to include these strings, the cell data format surrounded by the curly bracket, { }, is used.

```
line_style = {'rs-' 'go-'};
```

The following line in Program 2.9 makes the x-axis label in Figure 2.8 is in the mathematical fonts instead of the normal fonts. In the xlabel command, the interpreter is indicated as latex.[1]

```
xlabel('Random Variable x Sampling Space: $\Omega_x$','Interpreter'
    ,'latex');
```

LaTeX is a typesetting system widely used in writing mathematical papers and books. A draft version of this book is also written using LaTeX. In MATLAB,

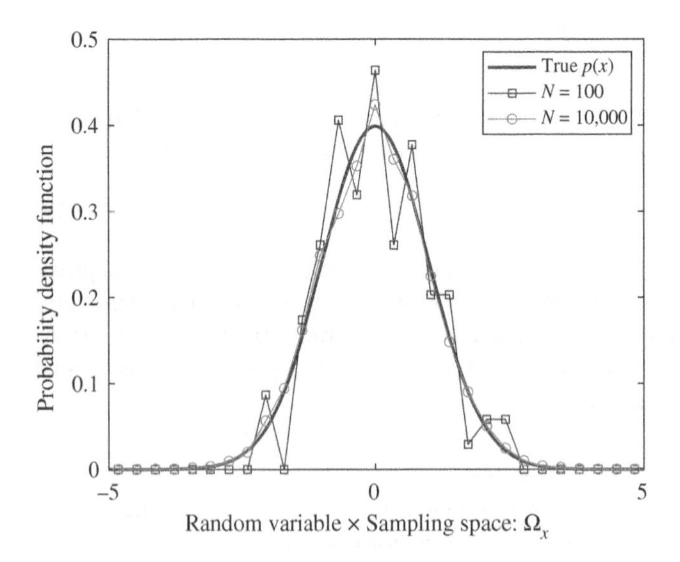

Figure 2.8 The probability density function of the random number generated by *randn* in MATLAB and the comparison with the true probability density function, $p(x)$.

1 It is pronounced as Lay-Tech.

mathematical symbols of LaTeX can be used in the labels by indicating the interpreter, 'latex'. In LaTeX, the characters surrounded by '$' are interpreted as mathematical expressions and '\Omega_x' is appeared as 'Ω_x' in the axis label. More information on LaTeX can be found in The LaTeX Project Team (2020).

The corresponding Python program to Program 2.7 is given in Program 2.8. The labels interpreted as raw LaTeX expressions start with 'r' and the single quotation mark. In addition, maths symbols are surrounded by $ signs. Random numbers from the normal distribution with mean 0 and variance 1 are generated using *numpy.random.randn*. As numpy is imported in line 1 of the program, randn can be called *np.random.randn*. It is, however, convenient to import *numpy.random* as 'rp' so that randn can be called in a compact way, i.e. *rp.randn*.

One of the main differences in syntax between Python and MATLAB is the presence of the comma in Python to distinguish the elements in an array or a list. Two numbers, 100 and 1000 in 'N_all' array, are separated by a comma, Two line styles in 'line_style' are also separated by a comma.

Parts included in *for-loop* are distinguished by indents the same way as indents define the body of functions in Python. The lines between line 29 to line 22 belongs to the for-loop. The for-loop in line 22 is a frequently used programming pattern used in Python.

To print each element of 'x = [1 2 3 4]' in MATLAB,

```
x = [1 2 3 4];
for idx = 1:length(x)
  disp(x)
end
```

In Python, a for-loop is implemented with the keyword, *in*, and a colon ':' at the end of the line as follows:

```
x = [1, 2, 3, 4]
for x_val in x:
  print(x_val)
End
```

No index number has to be explicitly generated as 'idx' in the MATLAB program. In the for-loop, each array value in 'x' is sequentially assigned to 'x_val'. Similarly, it can be done for two lists using the *zip* command as shown in line 22, where each value of 'N_all' and 'line_style' is assigned to 'N_trial' and 'lnsty', respectively.

In line 26, *np.histogram* calculates the number of occurrences for the given bin, 'x_bin', and the return values are stored in 'N_occur' as a tuple data format. 'N_occur[0]' is the array including the occurrence for each bin, and 'N_occur[1]' is the array including the bin list. The *fig.savefig* command in the last line saves the figure in a specified format by the file name extension, e.g. pdf (Portable Document Format).

```python
import numpy as np
from numpy import linspace
import numpy.random as rp

import matplotlib.pyplot as plt

# true probability density function (pdf)
var_x = 1;
mean_x = 0;
Omega_x = linspace(-5,5,1000);
px = (1/(np.sqrt(2*np.pi*var_x)))*np.exp(-(Omega_x-mean_x)**2/(2*
    var_x));

fig, ax = plt.subplots(nrows=1,ncols=1)
ax.plot(Omega_x,px,linewidth=3)

# generate N random numbers with the mean zero and the variance
# 1 using numpy.random.randn
N_all = np.array([100,10000])
x_bin = linspace(-5,5,30)
dx=np.mean(np.diff(x_bin))
line_style = ['rs-','go-']
for N_trial, lnsty in zip(N_all,line_style):
    x_rand = rp.randn(1,N_trial)

    # number of occurrence of x_rand in x_bin
    N_occur = np.histogram(x_rand,bins=x_bin)
    N_occur = N_occur[0]

    ax.plot(x_bin[0:-1]+dx/2, N_occur/(dx*N_trial),lnsty);

ax.set_xlabel(r'Random Variable x Sampling Space: $\Omega_x$',
    fontsize=14)
ax.set_ylabel('probability density function',fontsize=14)
ax.legend((r'True $p(x)$','N=100','N=10,000'),loc='upper right',
    fontsize=14)

fig.savefig('compare_mu_sgm2_true_estimated_python.pdf')
```

Program 2.8 (Python) Compare the true pdf with the ones approximated by the random numbers generated by *numpy.random.randn*

2.1.2.3 Stochastic Process

The zero-mean white noise is a stochastic process. A stochastic process is a time-dependent process. The random number generating procedure shown previously is not a stochastic process as time is not introduced. *Distinguishing between a stochastic process and a random number based on whether it is a process with time or not* is an important concept in implementing stochastic process simulations.

Consider the following stochastic process:

$$E[x(t)] = \mu(t) \tag{2.20a}$$

$$E\left[x(t_1)x(t_2)\right] = [\sigma(t)]^2\delta(t_1 - t_2) \tag{2.20b}$$

where the mean and the variance are time-varying, the pdf of $x(t)$ is given by (2.16), and μ and σ are time-varying. Whenever the time t is fixed, e.g. $t = 2.5$ seconds, it is one of the cases of generating the random numbers examples shown in Programs 2.3–2.8.

In computer simulations, the continuous time is approximated in a discrete sampling sequence as follows:

$$\left\{t_0, t_1, t_2, \dots, t_{n-1}, t_n\right\} \tag{2.21}$$

where n is a positive integer, t_0 is the initial time, t_n is the final time of the simulation, and we assume that

$$t_k = t_{k-1} + \Delta t \tag{2.22}$$

for $k = 1, 2, \dots, n-1, n$, i.e. the time interval between two sampling times is constant Δt. $\mu(t_k)$ and $\left[\sigma(t_k)\right]^2$ are the corresponding mean and variance for each instance, respectively.

Let $\Delta t = 0.1$ seconds, $n = 100$, $\mu(t_k)$, and $\sigma(t_k)$ given by

$$\mu(t_k) = -2 + \frac{4k}{n} \tag{2.23a}$$

$$\sigma(t_k) = 0.1 + \frac{1.4k}{n} \tag{2.23b}$$

One specific time history of $x(t)$ generated is called a *realization of the stochastic process*. The stochastic process is implemented in MATLAB and Python in the following two paragraphs.

2.1.2.4 MATLAB

Figure 2.9 shows the five realizations of $x(t)$. All five realizations start around -2 at $t = 0$, when the mean and the variance are equal to -2 and 0.1, respectively. As the time increases, the mean value increases linearly to $+2$ at $t = 10$ seconds.

The variance increases with time, and the five realizations of $x(t)$ spread wider as the time increases. For each fixed time, t_k, the realizations are random numbers as in the previous section with the mean and the variance constants. We can calculate the mean and the variance at the fixed time using the mean and the variance functions in MATLAB or Python shown in Programs 2.3–2.8.

Five realizations are too small to have good estimates of the mean and the variance. The number of realization increases to 1000. Figure 2.10 compares the estimated values with the true mean and variance. The MATLAB program is given in Program 2.9.

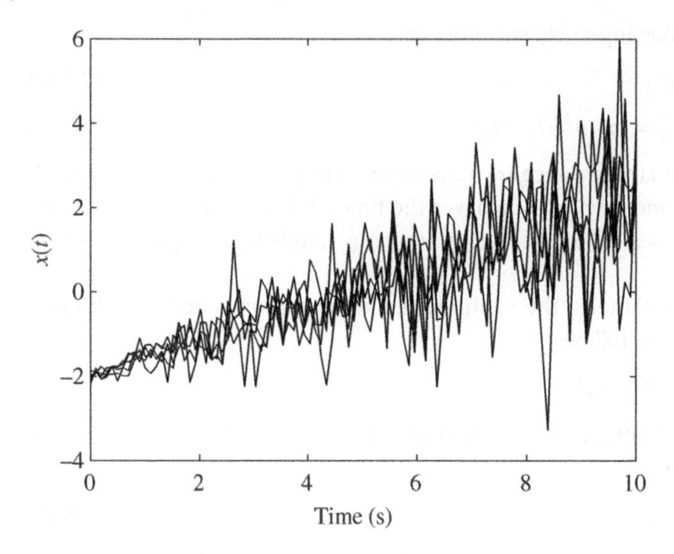

Figure 2.9 Five realizations of $x(t)$ whose mean and variance are given by (2.23).

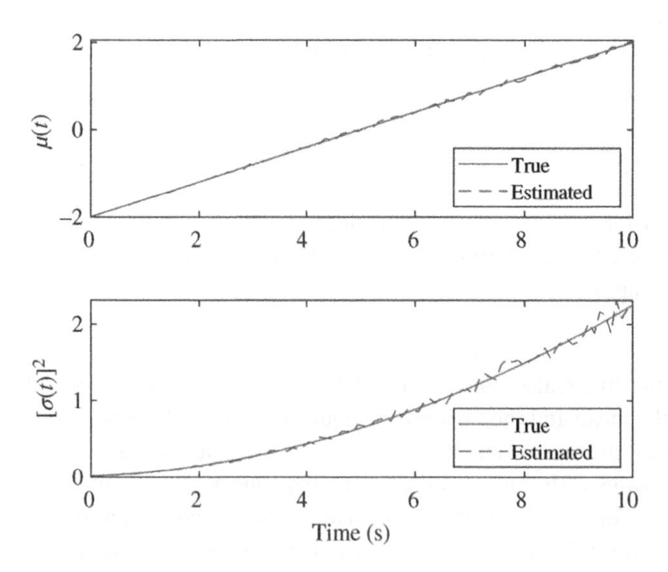

Figure 2.10 Compare the true $\mu(t)$ and $[\sigma(t)]^2$ with the estimated values using 1000 realizations.

Be careful to execute the line to plot all realizations of $x(t)$ in the following line:

```
% plot(time,x_rand_all,'k-');
```

It is commented out to prevent accidental execution of the line with the large values of 'N_sample' and/or 'N_realize'. It would consume the whole memory of the computer to complete the plot. It is also difficult to kill the plotting procedures in the middle of the execution.

```
1  clear;
2
3  % numer of time samplng & number of stochastic process trial
4  N_sample = 100;
5  N_realize = 1000;
6
7  % time
8  dt = 0.1; % [seconds]
9  time_init = 0;
10 time_final = dt*N_sample;
11 time = linspace(time_init,time_final,N_sample);
12
13 % declare memory space for x_rand_all to include all trials
14 x_rand_all = zeros(N_realize,N_sample);
15
16 % time varying mean and sqrt(variance) at the time instance
17 mu_all = linspace(-2,2,N_sample);
18 sigma_all = linspace(0.1,1.5,N_sample);
19
20 % for a fixed time instance, generate the random numbers
21 % with the mean and the variance at the fixed time
22 for idx=1:N_sample
23     mu_t = mu_all(idx);
24     sigma_t = sigma_all(idx);
25
26     x_rand = mu_t+sigma_t*randn(N_realize,1);
27     x_rand_all(:,idx) = x_rand;
28 end
29
30 % plot all trials with respect to the time
31
32 % Warning: this part is only executed with the small N_trial,
33 % e.g., 5,
34 % the plot takes really long and causing the computer crashed
35 % with the large N_trial, e.g., 1000
36 % figure; clf;
37 % plot(time,x_rand_all,'k-');
38 % set(gca,'FontSize',14);
39 % xlabel('time [s]');
40 % ylabel('x(t)');
41
42 % approximate mean and variance from the realisation
43 % and compare with the true
```

```
44 mu_approx = mean(x_rand_all);
45 sigma2_approx = var(x_rand_all);
46 figure;
47 subplot(211);
48 plot(time,mu_all);
49 hold on;
50 plot(time,mu_approx,'r--');
51 set(gca,'FontSize',14);
52 ylabel('$\mu(t)$','Interpreter','latex');
53 legend('True','Estimated','Location','southeast');
54 subplot(212);
55 plot(time,sigma_all.^2);
56 hold on;
57 plot(time,sigma2_approx,'r--');
58 set(gca,'FontSize',14);
59 ylabel('$[\sigma(t)]^2$','Interpreter','latex');
60 xlabel('time [s]');
61 legend('True','Estimated','Location','southeast');
```

Program 2.9 (MATLAB) Realizations of the stochastic process $x(t)$ given by (2.23) and estimation of the mean and the variance

In Program 2.9, the i-th row of 'x_rand_all' is the i-th realization of $x(t)$ and the j-th column of 'x_rand_all' corresponds to $x(t)$ for t is fixed to t_j for $i = 1, 2, \ldots$, N_realize and $j = 1, 2, \ldots$, N_sample. Continuing Program 2.10 from Program 2.9 calculates the pdf for each time instance using the histcounts function and stores the pdf in each column of 'px_all', two-dimensional matrix. How the pdf changes over time is shown using surf command, which draws the two-dimensional surface indicated by 'px_all' and the coordinates are indicated by the sampling space, x, and the time series as shown in Figure 2.11.

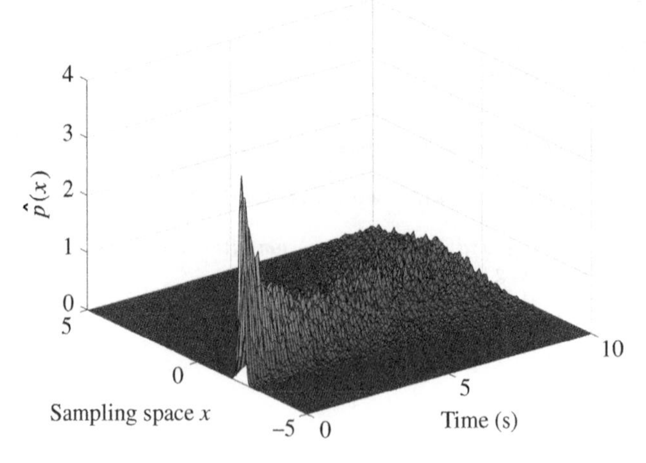

Figure 2.11 The estimated pdf, $\hat{p}(x)$, shows the complete picture of the Gaussian distributions over time.

As shown in line 37 in Program 2.9, the plot command finds the correct dimension to plot figures. Hence, the following two lines in the MATLAB command prompt generates the same plot: 'plot(time, x_rand_all)' or 'plot(time,x_rand_all′)', where 'x_rand_all' in the second plot command is transposed by '( )′'.

This automatic manipulation in MATLAB would cause some confusion, e.g. the size of 'x_rand_all' is 100×100. Each row is the realizations of $x(t)$ at a fixed time and each column is one realization of $x(t)$ with respect to time, respectively. We might not ensure which direction of the matrix is drawn with respect to the time vector. Hence, *it is a good practice to ensure that the size of the matrix is not square when the row and the column have different physical interpretations.*

```
1  % (continue from Program 2.9)
2  % esimate the pdf for each instance using N-trials at each instance
3  N_bin = 100;
4  x_bin = linspace(-5,5,N_bin);
5  dx=mean(diff(x_bin));
6  px_all = zeros(N_bin-1,N_sample);
7  for jdx=1:N_sample
8        x_rand = x_rand_all(:,jdx);
9        N_occur = histcounts(x_rand,x_bin);
10       px_at_t = N_occur/(dx*N_realize);
11       px_all(:,jdx) = px_at_t(:);
12 end
13
14 % plot the estimated pdf
15 figure;
16 surf(time,x_bin(1:end-1)+0.5*dx,px_all);
17 set(gca,'FontSize',14);
18 xlabel('time [s]');
19 ylabel('x');
20 zlabel('$\hat{p}(x)$','Interpreter','latex');
```

Program 2.10 (MATLAB) Plot the pdf with respect to time

2.1.2.5 Python

Five realizations of $x(t)$ given by (2.23) are drawn in Program 2.11. The for-loop in the program needs some attention to understand.

```
for idx, (mu_t, sigma_t) in enumerate(zip(mu_all,sigma_all)):
    x_rand = mu_t+sigma_t*rp.randn(N_realize)
    x_rand_all[:,idx] = x_rand
```

The for-loop not only substitutes the values of 'mu_all' and 'sigma_all' one by one into 'mu_t' and 'sigma_t' but also assigns the numerical index number to 'idx'. *enumerate* generates the index number and passes it into the variable, idx. For example, two arrays, a and b, have four elements. Each element in a or b is substituted into a_now or b_now, and idx stores the index number. In the following commands in the Python command prompt, print 'idx', 'a_now', and 'b_now' in the screen.

```
In [1]: import numpy as np

In [2]: a=np.array([1,2,3,4])

In [3]: b=['x1','x2','x3','x4']

In [4]: for idx, (a_now, b_now) in enumerate(zip(a,b)):
   ...:     print(idx, a_now, b_now)
   ...:
0 1 x1
1 2 x2
2 3 x3
3 4 x4
```

The plot command from *matplotlib* in Python might generate the plot with excessive empty space. The manual adjustment of the axis limitations using *ax.set* provides a tight fit of the plots in the figure window. *xlim* and *ylim* in *ax.set* specify the axis range. The values for *xlim* and *ylim* must be in the tuple format. It is common for function arguments in Python to be in the tuple format, where the values are in the bracket, '()', and separated by the comma. To prevent accidental attempts of plotting for the large N_realize, we add the if-condition so that the plot parts are only executed if N_realise is less than 10.

```
1  import numpy as np
2  from numpy import linspace
3  import numpy.random as rp
4
5  import matplotlib.pyplot as plt
6
7  # numer of time samplng & number of stochastic process trial
8  N_sample = 100
9  N_realize = 5
10
11 # time
12 dt = 0.1 # [seconds]
13 time_init = 0
14 time_final = dt*N_sample
15 time = linspace(time_init,time_final,N_sample)
16
17 # declare memory space for x_rand_all to include all trials
18 x_rand_all = np.zeros((N_realize,N_sample))
19
20 # time varying mean and sqrt(variance) at the time instance
21 mu_all = linspace(-2,2,N_sample)
22 sigma_all = linspace(0.1,1.5,N_sample)
23
24 # for a fixed time instance, generate the random numbers
```

```
25 # with the mean and the variance at the fixed time
26 for idx, (mu_t, sigma_t) in enumerate(zip(mu_all,sigma_all)):
27     x_rand = mu_t+sigma_t*rp.randn(N_realize)
28     x_rand_all[:,idx] = x_rand
29
30 # plot all trials with respect to the time
31
32 # Warning: this part is only executed with the small N_trial,
33 # e.g., 5
34 # the plot takes really long and causing the computer crashed
35 # with the large N_trial, e.g., 1000
36 if N_realize < 10:
37
38     fig, ax = plt.subplots(nrows=1,ncols=1)
39     ax.plot(time,x_rand_all.transpose(),'k-')
40     ax.set_xlabel('time [s]',fontsize=14)
41     ax.set_ylabel(r'$x(t)$',fontsize=14)
42     ax.set(xlim=(0, time_final),ylim=(-4,6))
```

Program 2.11 (Python) Realizations of the stochastic process $x(t)$ given by (2.23) and estimation of the mean and variance

The same array including the integer from 1 to 5 is generated in numpy and MATLAB as follows:

```
# numpy
a = np.array([1,2,3,4,5])
```

```
% matlab
a = [1 2 3 4 5]
```

One-dimensional array in numpy has only one index and starts from 0, i.e. 'a[0]' equal to 1, 'a[1]' equal to 2, and so forth. One-dimensional array in MATLAB has both the one-dimensional index starting from 1 and the two-dimensional index indicating the row and the column numbers, i.e. 'a(2)' can be accessed by the first row and the second-column element of 'a', 'a(1,2)'. The *a.shape* command in the Python command prompt prints (5,), which indicates that the array has five elements and a one-dimensional index. *rp.rand* in line 27 in Program 2.11 has only one argument, N_realise, and it generates a one-dimensional array including the 'N_realise' random numbers. One-dimensional array in numpy does not have the information whether it is a row vector array or a column vector array as in MATLAB. In the next line of the program, the one-dimensional array, 'x_rand', is stored in the idx-column of 'x_rand_all', which is a two-dimensional array, without checking whether 'x_rand' is a column vector or a row vector. This is automatically completed as long as two sizes are matched to each other, i.e. the number of elements of 'x_rand' is equal to the number of rows of 'x_rand_all'.

There is no automatic data manipulation for the plotting commands from matplotlib. In line 39 in Program 2.11, 'x_rand_all' is transposed as the plot command requires the dimension of time and the first dimension of 'x_rand_all' to be matched. For example, 'N_sample = 1000' and 'N_realize = 5', the 'x_rand_all.shape()' commands return the shape of the matrix equal to (5,100), while 'time.shape()' prints out (100,). To make the first size element of 'x_rand_all' equal to 100, it needs to be transposed using 'x_rand_all.transpose()'. *It is always recommended to generate a none-square matrix to prevent to interpreting or plotting the data for the wrong axis.* If 'x_rand_all' is a square matrix, then the plot would succeed to produce a plot but it draw for the wrong axis.

Unlike the *surf* command in MATLAB, the *plot_surface* command in matplolib needs the full list of the coordinates for the two-dimensional matrix data, 'x_bin_matrix', to be drawn in a three-dimensional space. The coordinate for each element of the two-dimensional matrix is generated using *meshgrid* in numpy. For example,

$$\text{a} = \begin{bmatrix} 0 & 1 & 2 \end{bmatrix}, \ \text{b} = \begin{bmatrix} 0 & 1 & 2 & 3 & 4 \end{bmatrix}, \ \text{C\_mat} = \begin{bmatrix} 0 & 1 & 2 \\ 3 & 4 & 5 \\ 6 & 7 & 8 \\ 9 & 10 & 11 \\ 12 & 13 & 14 \end{bmatrix} \tag{2.24}$$

are generated by

```
In [1]: a=np.arange(3)
In [2]: b=np.arange(5)
In [3]: C_mat=np.reshape(np.arange(15) ,(5,3))
```

where the 'C_mat[i,j]' element corresponds to 'a[i]' and 'b[j]', and the following line

```
In [4]: A_mat, B_mat = meshgrid(a,b)
```

generates 'A_mat' and 'B_mat' equal to

$$\text{A\_mat} = \begin{bmatrix} 0 & 1 & 2 \\ 0 & 1 & 2 \\ 0 & 1 & 2 \\ 0 & 1 & 2 \\ 0 & 1 & 2 \end{bmatrix}, \ \text{B\_mat} = \begin{bmatrix} 0 & 0 & 0 \\ 1 & 1 & 1 \\ 2 & 2 & 2 \\ 3 & 3 & 3 \\ 4 & 4 & 4 \end{bmatrix} \tag{2.25}$$

The coordinates '(a[i],b[j])' for 'C_mat[i,j]' are given by 'A_mat[i,j]' and 'B_mat[i,j]'. Then, the surface plot for 'C_mat' is drawn as follows:

```
fig=plt.figure()
ax=plt.axes(projection='3d')
ax.plot_surface(A_mat,B_mat,C_mat)
```

Finding the purpose of the other options in *plot_surface*, *rstride*, *cstride*, and *cmap* is left as an exercise. Plotting the pdf shown in Figure 2.12 is left as an exercise.

2.1.2.6 Gyroscope White Noise

The zero-mean white noise of the gyroscope, $\boldsymbol{\eta}_v$, in (2.11) is implemented by extending (2.20) into a vector form as follows:

$$E\left[\boldsymbol{\eta}_v(t)\right] = 0_{3\times 1} \tag{2.26a}$$

$$E\left[\boldsymbol{\eta}_v(t_1)\boldsymbol{\eta}_v^T(t_2)\right] = \begin{bmatrix} \sigma_{vx}^2 & 0 & 0 \\ 0 & \sigma_{vy}^2 & 0 \\ 0 & 0 & \sigma_{vz}^2 \end{bmatrix} \delta(t_1 - t_2) \tag{2.26b}$$

where $0_{3\times1}$ is the 3×1 vector whose elements are all zeros, and σ_{vx}, σ_{vy}, and σ_{vz} are the standard deviations of the white noise for x, y, and z directions of the gyroscope, respectively. As the noises for the three directions are not correlated, the off-diagonal terms are all zeros. Generate three random variables whose mean zero and variance are equal to σ_x, σ_y, and σ_z, respectively, for a fixed time and repeat it for every instance implements $\boldsymbol{\eta}_v(t)$ time series.

Unlike the time-varying mean and variance case, where multiple realizations are required to estimate the mean and the variance for a fixed time, the mean and the variance of the white noise can be calculated using the sampled data from a single realization over a period of time, which is long enough. This is an intuitive concept of the *ergodicity* of white noise. A more precise statistical definition of ergodicity requires the deeper understanding of statistics (Shanmugan and Breipohl, 1988).

```
1  # (continue from Program 2.11)
2  # estimate the mean, the variance and the  pdf for each instance
3  # using N-trials at each instance
4
5  # approximate mean and variance from the realisation
6  # and compare with the true
7  mu_approx = np.mean(x_rand_all, axis=0);
8  sigma2_approx = np.var(x_rand_all, axis=0)
9  fig_ms, (ax_ms_0, ax_ms_1) = plt.subplots(nrows=2,ncols=1)
10 ax_ms_0.plot(time, mu_all)
11 ax_ms_0.plot(time, mu_approx, 'r--')
12 ax_ms_0.set_ylabel(r'$\mu(t)$', fontsize=14)
13 ax_ms_0.legend(('True', 'Estimated'),loc='upper left', fontsize=14)
14
15 ax_ms_1.plot(time, sigma_all**2);
16 ax_ms_1.plot(time, sigma2_approx, 'r--');
17 ax_ms_1.set_ylabel(r'$[\sigma(t)]^2$', fontsize=14);
18 ax_ms_1.set_xlabel('time [s]', fontsize=14);
19 ax_ms_1.legend(('True', 'Estimated'),loc='upper left', fontsize=14);
20
21 # estimate the pdf for each instance using N-trials at each
       instance
```

```
22 N_bin = 100
23 x_bin = np.linspace(-5,5,N_bin)
24 dx=np.mean(np.diff(x_bin))
25 px_all = np.zeros((N_bin-1,N_sample))
26 for jdx in range(N_sample):
27     x_rand = x_rand_all[:,jdx]
28     N_occur = np.histogram(x_rand,bins=x_bin)
29     N_occur = N_occur[0]
30     px_at_t = N_occur/(dx*N_realize)
31     px_all[:,jdx] = px_at_t
32 
33 # plot the estimated pdf
34 time_matrix, x_bin_matrix = np.meshgrid(time,x_bin[0:-1])
35 
36 fig_3d = plt.figure()
37 ax_3d = plt.axes(projection='3d')
38 ax_3d.plot_surface(time_matrix, x_bin_matrix, px_all, rstride=1,
       cstride=1, cmap='viridis')
39 ax_3d.set_xlabel('time [s]',fontsize=14)
40 ax_3d.set_ylabel(r'sampling space $x$',fontsize=14)
41 ax_3d.set_zlabel(r'$\hat{p}(x)$',fontsize=14)
```

Program 2.12 (Python) Plot the mean, the variance, and the pdf with respect to time

2.1.2.7 Gyroscope Random Walk Noise

Another type of random noise corrupting the gyro measurements in (2.11) is the bias, $\boldsymbol{\beta}$. The bias is modelled as the random walk: the difference between $\boldsymbol{\beta}(t_k)$ and $\boldsymbol{\beta}(t_{k-1})$ for $0 \leq t_{k-1} < t_k$ is the independent random increment, and it follows the normal distribution whose mean and variance are given by

$$E\left[\boldsymbol{\beta}(t_k) - \boldsymbol{\beta}(t_{k-1})\right] = 0_{3\times1} \tag{2.27a}$$

$$E\left\{\left[\boldsymbol{\beta}(t_k) - \boldsymbol{\beta}(t_{k-1})\right]\left[\boldsymbol{\beta}(t_k) - \boldsymbol{\beta}(t_{k-1})\right]^T\right\} = \text{diag}\begin{bmatrix}\sigma_{\beta x}^2\\ \sigma_{\beta y}^2\\ \sigma_{\beta z}^2\end{bmatrix}\Delta t_k \tag{2.27b}$$

where $\sigma_{\beta x}$, $\sigma_{\beta y}$, and $\sigma_{\beta z}$ are positive constants and Δt_k is equal to $t_k - t_{k-1}$. Rigorous mathematical definitions and discussions about the random walk can be found in Van Kampen (2007).

The following equation simulates the random walk (Figure 2.12):

$$\boldsymbol{\beta}(t_k) = \boldsymbol{\beta}(t_{k-1}) + \Delta\boldsymbol{\beta}(t_k) \tag{2.28}$$

where t_k is the sampling instance that the gyro measurement is obtained, and $\Delta\boldsymbol{\beta}(t_k)$ is the random increment. To simulate the random increment, firstly, the random increment is implemented by

$$\Delta\boldsymbol{\beta}(t_k) = \boldsymbol{\eta}_u(t_k)\Delta t_k \tag{2.29}$$

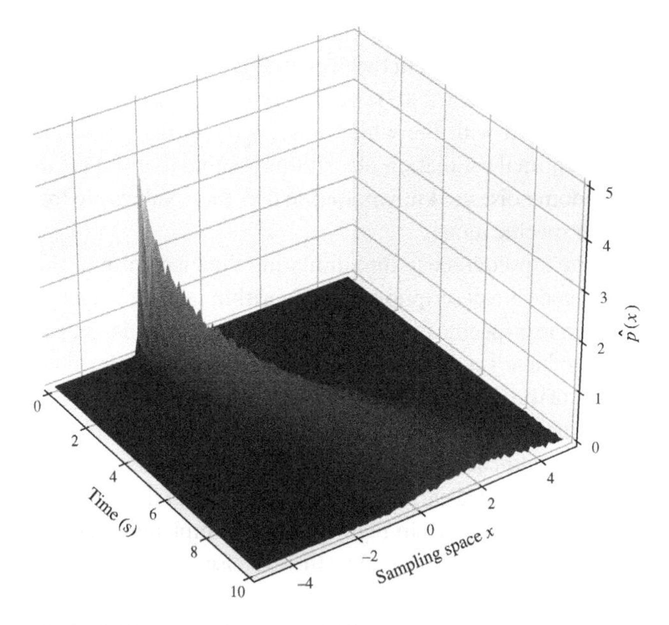

Figure 2.12 The estimated pdf, $\hat{p}(x)$, plot using Python.

where $\boldsymbol{\eta}_u(t_k)$ is a 3×1 random vector at t_k, whose each element is a random number generated from the normal distribution. Secondly, to match the mean of the random increment given in (2.27a) of $\boldsymbol{\eta}_u(t_k)$ must satisfy

$$E\left[\Delta\boldsymbol{\beta}(t_k)\right] = E\left[\boldsymbol{\eta}_u(t_k)\Delta t_k\right] = E\left[\boldsymbol{\eta}_u(t_k)\right]\Delta t_k = 0_{3\times 1} \tag{2.30}$$

The mean value of each element of $\boldsymbol{\eta}_u(t_k)$ for a fixed time t_k must be equal to zero, i.e. $E\left[\boldsymbol{\eta}_u(t_k)\right] = 0_{3\times 1}$. Finally, to match the covariance of the random increment given in (2.27b), the variance of $\boldsymbol{\eta}_u(t)$ must satisfy

$$E\left[\Delta\boldsymbol{\beta}(t_k)\Delta\boldsymbol{\beta}^T(t_k)\right] = E\left[\boldsymbol{\eta}_u(t_k)\Delta t_k\, \boldsymbol{\eta}_u^T(t_k)\Delta t_k\right]$$
$$\Downarrow$$
$$\text{diag}\begin{bmatrix}\sigma_{\beta x}^2\\ \sigma_{\beta y}^2\\ \sigma_{\beta z}^2\end{bmatrix}\Delta t_k = E\left[\boldsymbol{\eta}_u(t_k)\boldsymbol{\eta}_u^T(t_k)\right]\left(\Delta t_k\right)^2 \tag{2.31}$$

Hence, the covariance of the random number, $\boldsymbol{\eta}_u$, is given by

$$E\left[\boldsymbol{\eta}_u(t_k)\boldsymbol{\eta}_u^T(t_k)\right] = \text{diag}\begin{bmatrix}\sigma_{ux}^2\\ \sigma_{uy}^2\\ \sigma_{ux}^2\end{bmatrix} = \text{diag}\begin{bmatrix}\sigma_{\beta x}^2\\ \sigma_{\beta y}^2\\ \sigma_{\beta z}^2\end{bmatrix}\frac{1}{\Delta t_k} \tag{2.32}$$

For example, the standard deviation of the first element in $\eta(t_k)$, σ_{ux}, is equal to $\sigma_{\beta x}/\sqrt{\Delta t_k}$. A gyro noise characteristic given in the unit of $(°/s)/\sqrt{s}$=$°/s^{3/2}$ is originated from this relationship (Woodman, 2007).

Note that $\eta_u(t)$ is not correlated with the white noise, $\eta_v(t)$, in the gyroscope sensor model, (2.11). Most optimal estimation algorithms assume that the mean and the variance of the random noise are known. In practice, these values are frequently found in the sensor specifications.

We are now ready to write a pseudo-code for simulating the bias noise in the gyro measurement. Pseudo-code is a description of an algorithm in plain language without any tight connection to a specific programming language. Pseudo-code is for the simulator designers to have a clear picture of the algorithm, and it is useful to design an initial structure of the simulation program. Translating a pseudo-code to a specific programming language, e.g. MATLAB or Python, is rather straightforward. Algorithm 2.1 is the pseudo-code for generating the gyro bias noise whose mathematical descriptions are provided earlier. The 10-realization of the bias time history using MATLAB or Python is shown in Figure 2.13. The implementation of the algorithm for each MATLAB or Python is left as the exercise.

Algorithm 2.1 Gyro bias noise, $\boldsymbol{\beta}(t_k)$, simulation

1: Set $\sigma_{\beta x}$, $\sigma_{\beta y}$, $\sigma_{\beta z}$, and Δt_k, n.b.: Δt_k is usually set to a constant
2: Initialize $\boldsymbol{\beta}(t_0)$, e.g. using a random number generator
3: **for** $k = 1, 2, \dots$ **do**
4:     **for** $\ell = x, y, z$ **do**
5:         Generate $\eta_{u\ell} \sim N(0, \sigma^2_{\beta\ell}/\Delta t_k)$, See (2.30) and (2.32)
6:     **end for**
7:     $\boldsymbol{\eta}_u(t_k) \leftarrow \begin{bmatrix} \eta_{ux} & \eta_{uy} & \eta_{uz} \end{bmatrix}^T$
8:     $\Delta\boldsymbol{\beta}(t_k) \leftarrow \boldsymbol{\eta}_u(t_k)\Delta t_k$, See (2.29)
9:     $\boldsymbol{\beta}(t_k) \leftarrow \boldsymbol{\beta}(t_{k-1}) + \Delta\boldsymbol{\beta}(t_k)$, See (2.28)
10:     $t_k \leftarrow t_{k-1} + \Delta t_k$
11: **end for**

It is important to use only the SI units in the main parts of all simulation implementations. All non-SI units given must be changed to the corresponding SI units at the beginning of the program. Using the SI units only for the rest of the program implementation can significantly reduce unit related mistakes. Keep in mind that all dynamic equations are derived based on appropriate unit assumptions. The assumption we use here for most cases is that variables are in the SI units. *The computer does not have the unit information but has the numerical values only.* It does not recognize that 0.1 is in degrees or radians. After the simulation completes, some quantities could be converted into non-SI units for some purposes.

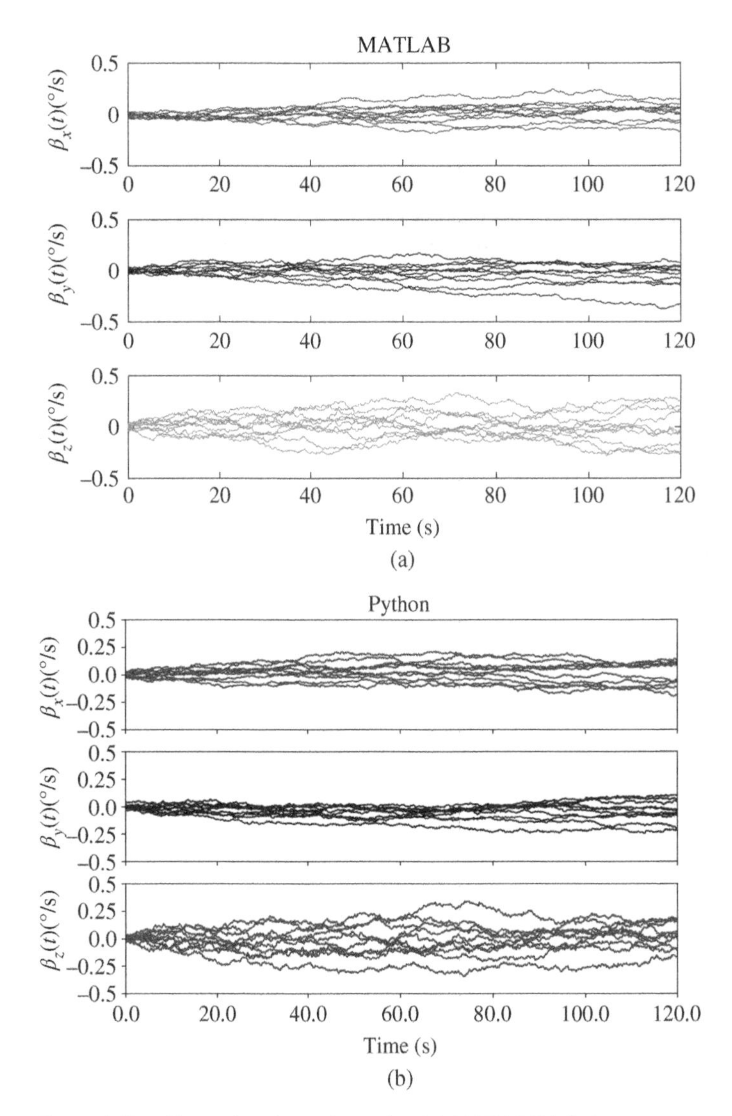

Figure 2.13 Bias noise simulation using (a) MATLAB/(b) Python.

For example, all angles must be in radians during the simulation, but they could be converted into degrees for the visualization purpose.

2.1.2.8 Gyroscope Simulation

Given the angular velocity, $\boldsymbol{\omega}(t)$, as the function of time in (2.9), the gyroscope sensor in (2.11) has the following noise characteristics indicated in the sensor specifications: $\sigma_{\beta x} = 0.05\ (°/\text{s})/\sqrt{\text{s}}$, $\sigma_{\beta y} = 0.04\ (°/\text{s})/\sqrt{\text{s}}$, $\sigma_{\beta z} = 0.06\ (°/\text{s})/\sqrt{\text{s}}$, $\sigma_v = 0.01\ °/\text{s}$, and $\Delta t_k = 0.05$ seconds. In addition, the initial bias, $\boldsymbol{\beta}(t_0 = 0)$, is taken from the uniform distribution between $-0.05\ °/\text{s}$ and $+0.05\ °/\text{s}$.

The simulation result of the gyroscope measurements is shown in Figures 2.14 and 2.15. See Program 2.13 for the MATLAB gyroscope simulation. The Python program for simulating the gyroscope measurement is given in Program 2.14. The measurements indicated in the dashed lines are drifting away from the true

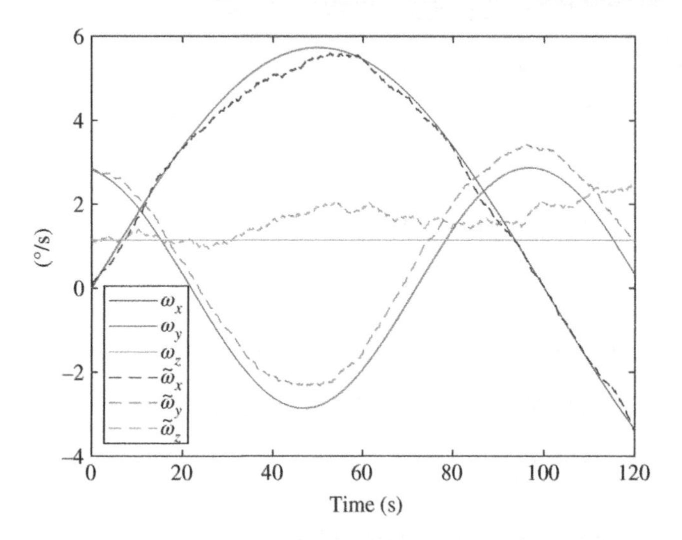

Figure 2.14 (MATLAB) Gyroscope measurement simulation.

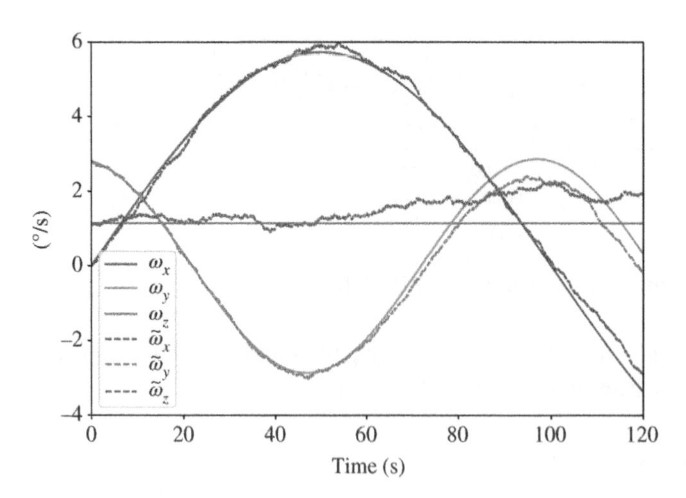

Figure 2.15 Gyroscope measurement simulation using Python Program 2.14.

angular velocity. If the measurements are directly used to obtain the quaternion by numerically integrating (2.5), where $\boldsymbol{\omega}$ is replaced by $\tilde{\boldsymbol{\omega}}$, the calculated quaternion quickly diverges from the true quaternion. No matter how expensive and accurate the gyroscope is, the gyroscope sensor measurement alone is not enough to prevent the divergence of the calculated quaternion from the true quaternion. An additional sensor directly providing the attitude measurement is needed.

```
1  clear;
2
3  %% Set initial values & change non-SI units into the SI Units
4  dt = 0.05; % [seconds]
5  time_init = 0;
6  time_final = 120;
7  time = time_init:dt:time_final;
8  N_sample = length(time);
9
10 % standard deviation of the bias, sigma_beta_xyz
11 sigma_beta_xyz = [0.05 0.04 0.06]; % [degrees/sqrt(s)]
12 sigma_beta_xyz = sigma_beta_xyz*(pi/180); % [rad/sqrt(s)]
13 sigma_eta_xyz = sigma_beta_xyz/sqrt(dt);
14
15 % standard devitation of the white noise, sigma_v
16 sigma_v = 0.01; %[degrees/s]
17 sigma_v = sigma_v*(pi/180); %[rad/s]
18
19 % initial beta(t)
20 beta = (2*rand(3,1)-1)*0.05; % +/- 0.03[degrees/s]
21 beta = beta*(pi/180); % [radians/s]
22
23 % prepare the data store
24 w_all = zeros(N_sample,3);
25 w_measure_all = zeros(N_sample,3);
26
27 %% main simulation loop
28 for idx=1:N_sample
29
30     time_c = time(idx);
31     w_true(1,1) = 0.1*sin(2*pi*0.005*time_c); % [rad/s]
32     w_true(2,1) = 0.05*cos(2*pi*0.01*time_c + 0.2); %[rad/s]
33     w_true(3,1) = 0.02; %[rad/s]
34
35     % beta(t)
36     eta_u = sigma_eta_xyz(:).*randn(3,1);
37     dbeta = eta_u*dt;
38     beta = beta + dbeta;
39
40     % eta_v(t)
41     eta_v = sigma_v*randn(3,1);
42
43     % w_tilde
44     w_measurement = w_true + beta + eta_v;
45
```

```
46     % store history
47     w_all(idx,:) = w_true(:)';
48     w_measure_all(idx,:) = w_measurement(:)';
49
50 end
51
52 % plot in degrees/s
53 figure;
54 plot(time, w_all*(180/pi));
55 hold on;
56 plot(time, w_measure_all*(180/pi),'--');
57 set(gca,'FontSize',14);
58 ylabel('$[^\circ/s]$','Interpreter','latex');
59 xlabel('time [s]','Interpreter','latex');
60 legend('$\omega_x$','$\omega_y$,'$\omega_z$, ...
61     '$\tilde{\omega}_x$,'$\tilde{\omega}_y$,'$\tilde{\omega}_z$,
       ...
62     'Interpreter','latex','Location','SouthWest');
```

Program 2.13 (MATLAB) Gyroscope simulation with white noise and bias noise

```
1 import numpy as np
2 import matplotlib.pyplot as plt
3
4 # Set initial values & change non-SI units into the SI Units
5 dt = 0.05 # [seconds]
6 time_init = 0
7 time_final = 120 # [seconds]
8 N_sample = int(time_final/dt) + 1
9 time = np.linspace(time_init,time_final, N_sample)
10
11 # standard deviation of the bias, sigma_beta_xyz
12 sigma_beta_xyz = np.array([0.05, 0.04, 0.06]) # [degrees/sqrt(s)]
13 sigma_beta_xyz = sigma_beta_xyz*(np.pi/180) # [rad/sqrt(s)]
14 sigma_eta_xyz = sigma_beta_xyz/np.sqrt(dt)
15
16 # standard devitation of the white noise, sigma_v
17 sigma_v = 0.01 #[degrees/s]
18 sigma_v = sigma_v*(np.pi/180) #[rad/s]
19
20 # initial beta(t)
21 beta = (2*np.random.rand(3)-1)*0.03 # +/- 0.03[degrees/s]
22 beta = beta*(np.pi/180) # [radians/s]
23
24 # prepare the data store
25 w_all = np.zeros((N_sample,3))
26 w_measure_all = np.zeros((N_sample,3))
27
28 # main simulation loops
29 for idx in range(N_sample):
30
31     time_c = time[idx]
32     w_true = np.array([0.1*np.sin(2*np.pi*0.005*time_c),#[rad/s]
```

```
33                    0.05*np.cos(2*np.pi*0.01*time_c + 0.2), #[rad/s]
34                    0.02 #[rad/s]
35                        ])
36        # beta(t)
37        eta_u = sigma_eta_xyz*np.random.randn(3)
38        dbeta = eta_u*dt
39        beta = beta + dbeta
40        # eta_v(t)
41        eta_v = sigma_v*np.random.randn(3)
42        # w_tilde
43        w_measurement = w_true + beta + eta_v
44        # store history
45        w_all[idx,:] = w_true
46        w_measure_all[idx,:] = w_measurement
47
48  # plot all realization of beta in degrees/s
49  fig, ax = plt.subplots(nrows=1,ncols=1)
50  ax.plot(time,w_all*180/np.pi)
51  ax.plot(time,w_measure_all*180/np.pi,'--')
52  ax.set_ylabel(r'$[^\circ/s]$',fontsize=14);
53  ax.set_xlabel(r'time [s]',fontsize=14);
54  ax.legend((r'$\omega_x$',r'$\omega_y$',r'$\omega_z$',
55        r'$\tilde{\omega}_x$',r'$\tilde{\omega}_y$',r'$\tilde{\omega}
            _z$'),
56              fontsize=14, loc='lower left')
57  ax.set(xlim=(0, time_final),ylim=(-4,6))
58  fig.set_size_inches(9,6)
59  fig.savefig('gyro_measurement_python.pdf',dpi=250)
```

Program 2.14 (Python) Gyroscope simulation with white noise and bias noise

2.1.3 Optical Sensor Model

One of the common sensors used to provide attitude measurement directly is an optical sensor, e.g. camera or star sensor. These sensors identify a priori known objects and compare the direction of the objects in the sensor measurement with the known directions of the objects in the reference coordinates.

To model a star sensor, it needs to understand the principle of vector observation illustrated in Figure 2.16. The attitude of the body coordinates, B, indicated by $\mathbf{x}_B, \mathbf{y}_B$, and $\mathbf{z}_B$ with respect to the reference coordinates indicated by $\mathbf{x}, \mathbf{y}$, and $\mathbf{z}$ is expressed in the quaternion, $\mathbf{q}$. The vector pointing towards #1 star, $\mathbf{r}^1$, can be expressed by the coordinates in the body coordinates or the reference coordinates as follows:

$$\mathbf{r}^1 \rightarrow \mathbf{r}^1_R = \begin{bmatrix} x & y & z \end{bmatrix}^T \tag{2.33}$$

$$\mathbf{r}^1 \rightarrow \mathbf{r}^1_B = \begin{bmatrix} x_B & y_B & z_B \end{bmatrix}^T \tag{2.34}$$

where $\mathbf{r}^1_R$ and $\mathbf{r}^1_B$ are the same vector but written in the two different coordinates, $\{R\}$ and $\{B\}$. $\mathbf{r}^1_R$ is usually stored in the on-board computer of satellite as a part of the star catalogue database. The star sensor with an identification algorithm

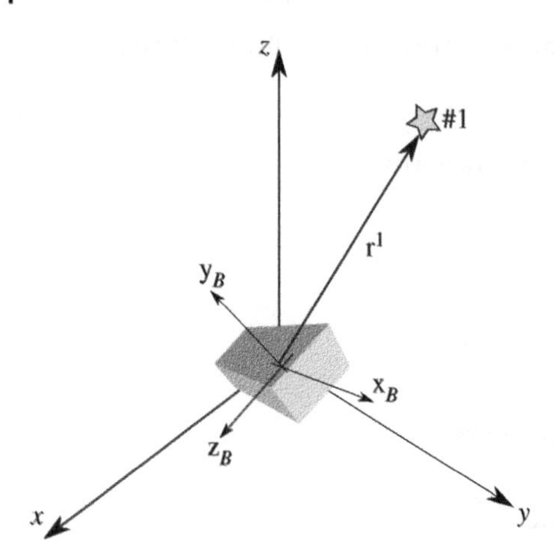

Figure 2.16 Identified star in the reference and the bode coordinates.

detects star #1, and its direction in the sensor coordinates is given by $\mathbf{r}_B^1$, where the sensor coordinates are assumed to be the same as the body coordinates. As $\mathbf{r}_R^1$ and $\mathbf{r}_B^1$ are vectors pointing the direction of the star, their magnitudes are assumed to be normalized, i.e. $\|\mathbf{r}\|_R^1 = 1$ and $\|\mathbf{r}\|_B^1 = 1$.

Their coordinates are equated using the direction cosine matrix as follows:

$$\begin{bmatrix} x \\ y \\ z \end{bmatrix}_R = \begin{bmatrix} c_{11} & c_{12} & c_{13} \\ c_{21} & c_{22} & c_{23} \\ c_{31} & c_{32} & c_{33} \end{bmatrix} \begin{bmatrix} x_B \\ y_B \\ z_B \end{bmatrix}_B \tag{2.35}$$

where $c_{ij} = \cos\theta_{ij}$ and θ_{ij} are the angles between $\mathbf{x}$, $\mathbf{y}$, or $\mathbf{z}$ and $\mathbf{x}_B$, $\mathbf{y}_B$, or $\mathbf{z}_B$. In a compact form, it is written as

$$\mathbf{r}_B^1 = C_{BR}\mathbf{r}_R^1 \tag{2.36}$$

where C_{BR} is the direction cosine matrix converting a vector in $\{R\}$ to $\{B\}$. The quaternion and the direction cosine matrix are two different ways to express attitude information. They are equivalent to each other and one-to-one conversion exists as follows:

$$C_{BR}(\mathbf{q}) = \left(q_4^2 - \mathbf{q}_{13}^T\mathbf{q}_{13}\right) I_3 + 2\mathbf{q}_{13}\mathbf{q}_{13}^T - 2q_4\left[\mathbf{q}_{13}\times\right] \tag{2.37}$$

where $\left[\mathbf{q}_{13}\times\right]$ is defined by (2.8) (Wie, 2008).

The conversion from the direction cosine matrix to the quaternion is performed using Algorithm 2.2 (Schaub and Junkins, 2003). Finding the maximum value in line 2 in Algorithm 2.2 is to prevent the divisions in lines 3–6 with a small value in the numerator. *All divisions in computer programs must be done with extreme care. If the denominator in a division is equal to or close to zero, the result becomes too large for the finite floating points in the computer to be contained.* MATLAB or Python, for example, returns *inf* for $1/10^{-309}$, where *inf* stands for the infinity, while $1/10^{-308}$ returns 10^{308}. The boundary value for separating the finite and the infinity values in the computer varies depending on the computer and/or software. To check if the value of a variable is infinity or not, *isinf* in MATLAB or *numpy.isinf* in numpy Python is used. It returns the logical value type 1 in MATLAB or True in Python if the number is infinity and the logical type 0 in MATLAB or False in Python if it is considered as a finite number.

Line 8 in Algorithm 2.2 is to provide the shortest rotational manoeuvre. q_4 is equal to $\cos(\theta/2)$ and $\cos(\theta/2)$ is greater than or equal to 0 for $|\theta| \leq 180^\circ$. Negative q_4 implies that $|\theta|$ is greater than 180°. Then, the same attitude can be achieved by the opposite direction rotation axis, i.e. $-\mathbf{e}$, with the rotational angle equal to $\pi - \theta$ radians. Also, notice that $\cos[(\pi - \theta)/2] = -\cos(\theta/2)$. Figure 2.17 shows that the same attitude for the 275° rotation about the rotation axis $\mathbf{e}$ can be achieved by the 85° rotation about the opposite axis, $-\mathbf{e}$.

Given the star direction, $\mathbf{r}_R^1$, and the star observation, $\mathbf{r}_B^1$, three algebraic equations are established using (2.35). As there are nine unknowns, c_{ij} for $i, j = 1, 2, 3$, in (2.35), six more equations are required to determine the nine unknowns. Consider that another star, #2 star, is identified, and it provides

$$\mathbf{r}_B^2 = C_{BR}(\mathbf{q})\mathbf{r}_R^2 \tag{2.38}$$

where it is assumed that $\mathbf{r}_R^1$ and $\mathbf{r}_R^2$ are not parallel to each other. Hence, additional three independent equations are provided by (2.38). Once two non-parallel direction stars are identified, the third vector can be established using the vector cross product as follows:

$$\left(\mathbf{r}_B^1 \times \mathbf{r}_B^2\right) = C_{BR}(\mathbf{q})\left(\mathbf{r}_R^1 \times \mathbf{r}_R^2\right) \tag{2.39}$$

The simplest star tracker model is given by Crassidis (2002)

$$\tilde{\mathbf{r}}_B^i = \mathbf{r}_B^i + \mathbf{v}^i \tag{2.40}$$

where $\mathbf{v}^i$ is a 3×1 noise vector, which follows the zero-mean Gaussian distribution, i.e.

$$E(\mathbf{v}^i) = \mathbf{0}_3 \tag{2.41a}$$

$$E\left[\mathbf{v}^i\left(\mathbf{v}^i\right)^T\right] = \sigma_s^2 I_{3\times3} \tag{2.41b}$$

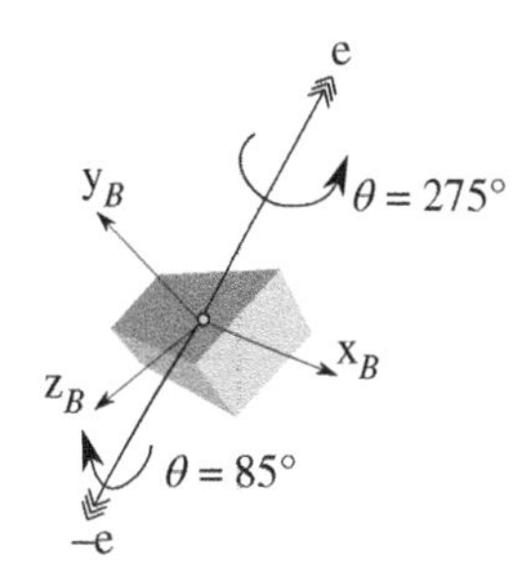

Figure 2.17 The shortest path rotation.

Algorithm 2.2 The quaternion from the direction cosine matrix

1: Calculate a_i using the given c_{ij} for $i, j = 1, 2, 3$ in (2.35)

$$a_1 = (1 + c_{11} - c_{22} - c_{33})/4$$
$$a_2 = (1 + c_{22} - c_{11} - c_{33})/4$$
$$a_3 = (1 + c_{33} - c_{11} - c_{22})/4$$
$$a_4 = (1 + c_{11} + c_{22} + c_{33})/4$$

2: Find i^* such that $a_{i^*} = \max(a_1, a_2, a_3, a_4)$ and calculate $q_{i^*} = \sqrt{a_{i^*}}$

3: **if** i^* is equal to 1 **then**

$$q_2 = (c_{12} + c_{21})/(4q_1)$$
$$q_3 = (c_{13} + c_{31})/(4q_1)$$
$$q_4 = (c_{23} - c_{32})/(4q_1)$$

4: **else if** i^* is equal to 2 **then**

$$q_1 = (c_{12} + c_{21})/(4q_2)$$
$$q_3 = (c_{23} + c_{32})/(4q_2)$$
$$q_4 = (c_{31} - c_{13})/(4q_2)$$

5: **else if** i^* is equal to 3 **then**

$$q_1 = (c_{13} + c_{31})/(4q_3)$$
$$q_2 = (c_{23} + c_{32})/(4q_3)$$
$$q_4 = (c_{12} - c_{21})/(4q_3)$$

6: **else if** i^* is equal to 4 **then**

$$q_1 = (c_{23} - c_{32})/(4q_4)$$
$$q_2 = (c_{31} - c_{13})/(4q_4)$$
$$q_3 = (c_{12} - c_{21})/(4q_4)$$

7: **end if**

8: **if** q_4 is negative **then**

$$q_1 \leftarrow -q_1,\ q_2 \leftarrow -q_2,\ q_3 \leftarrow -q_3,\ q_4 \leftarrow -q_4$$

9: **end if**

and σ_s is the standard deviation of the star sensor noise. The justification of this noise model is given in Shuster (1989). A more sophisticated star sensor noise modelling can be found in Fialho and Mortari (2019).

Consider the following two stars in the reference coordinates:

$$\mathbf{r}_R^1 = \begin{bmatrix} 0 & \frac{1}{\sqrt{2}} & -\frac{1}{\sqrt{2}} \end{bmatrix}_R^T \tag{2.42a}$$

$$\mathbf{r}_R^2 = \begin{bmatrix} \frac{1}{\sqrt{2}} & 0 & \frac{1}{\sqrt{2}} \end{bmatrix}_R^T \tag{2.42b}$$

Assume that $\mathbf{q}(t)$ represents the attitude of a satellite relative to the reference coordinates and the satellite equipped with multiple star sensors can see and identify the stars all the time. In reality, the stars would be in and out of the field of view of the star sensors. In Program 2.1 for MATLAB or 2.2 for Python, $\mathbf{q}(t)$ is given for every instance of time. For each instance, the corresponding body frame, where it is assumed that the sensor frames are aligned with the satellite body frame, representations of the stars are given by

$$\mathbf{r}_B^i = C_{BR}[\mathbf{q}(t)]\mathbf{r}_R^i \tag{2.43}$$

for $i = 1, 2, 3$, where $C_{BR}[\mathbf{q}(t)]$ is given by (2.37), and

$$\mathbf{r}_R^3 = \mathbf{r}_R^1 \times \mathbf{r}_R^2 \tag{2.44a}$$

$$\mathbf{r}_B^3 = \mathbf{r}_B^1 \times \mathbf{r}_B^2 \tag{2.44b}$$

Re-arrange (2.43) and (2.44) as follows:

$$\mathbf{r}_B^i = \begin{bmatrix} (\mathbf{r}_R^i)^T & 0_{1\times3} & 0_{1\times3} \\ 0_{1\times3} & (\mathbf{r}_R^i)^T & 0_{1\times3} \\ 0_{1\times3} & 0_{1\times3} & (\mathbf{r}_R^i)^T \end{bmatrix} \begin{bmatrix} c_{11} \\ c_{12} \\ c_{13} \\ c_{21} \\ c_{22} \\ c_{23} \\ c_{31} \\ c_{32} \\ c_{33} \end{bmatrix} = A^i \operatorname{vec}(C^T) \tag{2.45}$$

for $i = 1, 2, 3$, where A^i is a 3×9 matrix constructed using $\mathbf{r}_R^i$ and $\operatorname{vec}(\cdot)$ vectorizes the matrix in the column direction as follows:

$$A = \begin{bmatrix} 1 & 2 & 3 \\ 4 & 5 & 6 \\ 7 & 8 & 9 \end{bmatrix} \Rightarrow \operatorname{vec}(A) = \begin{bmatrix} 1 & 4 & 7 & 2 & 5 & 8 & 3 & 6 & 9 \end{bmatrix}^T \tag{2.46}$$

As MATLAB is the column-major language (The MathWorks, 2020), i.e. the elements in the array are indexed in the column direction first, the vectorization of the matrix is performed as

```
>> A=[1 2 3; 4 5 6; 7 8 9]
>> A(:)
```

Numpy array in Python, on the other hand, is the row-major, i.e. the elements in the array are indexed in the row direction first, and the following *flatten()* function returns $A = [1, 2, 3, 4, 5, 6, 7, 8, 9]$:

```
In [11]: A=np.array([[1, 2, 3], [4, 5, 6], [7, 8, 9]])
In [12]: A.flatten()
```

Hence, to obtain the same result for vec(·), the matrix is transposed first and then flattened as follows:

```
In [13]: A.transpose().flatten()
```

Using three matrices, A^1, A^2, and A^3,

$$\begin{bmatrix} \mathbf{r}_B^1 \\ \mathbf{r}_B^2 \\ \mathbf{r}_B^3 \end{bmatrix} = \begin{bmatrix} A^1 \\ A^2 \\ A^3 \end{bmatrix} \text{vec}(C^T) = A\text{vec}(C^T) \tag{2.47}$$

where A is not singular, which means the inversion exists, as long as $\mathbf{r}^1$ and $\mathbf{r}^2$ are not parallel to each other. Hence, the elements of the direction cosine matrix are simply determined by

$$\text{vec}(C^T) = A^{-1} \begin{bmatrix} \mathbf{r}_B^1 \\ \mathbf{r}_B^2 \\ \mathbf{r}_B^3 \end{bmatrix} \tag{2.48}$$

To construct an arbitrary attitude of the body relative to the reference frame, we generate the following four values in MATLAB using the uniform random number generator:

```
>> q_rand = 2*rand(4,1)-1;
>> q_rand = q_rand/norm(q_rand);
```

or in Python:

```
In [17]: q_rand=2*np.random.rand(4)-1
In [18]: q_rand=q_rand/np.linalg.norm(q_rand)
```

where they are normalized so that the random quaternion has the unit magnitude.

The corresponding direction cosine matrix to the random quaternion generated is calculated using (2.37) as follows in MATLAB:

```
>> q13=q_rand(1:3);
>> q4=q_rand(4);
>> q13x=[0 -q13(3) q13(2);
         q13(3) 0 -q13(1);
        -q13(2) q13(1) 0];
>> C_BR=(q4^2-q13'*q13)*eye(3)+2*q13*q13'-2*q4*q13x;
```

or in Python:

```
In [22]: q13=np.reshape(q_rand[0:3],(3,1))
In [23]: q4=q_rand[3]
In [24]: q13x = np.array([[0, -q13[2,0], q13[1,0]],[q13[2,0],0,-q13
    [0,0]],[-q13[1,0],q13[0,0],0]])
In [25]: C_BR = (q4**2-q13.transpose()@q13)*np.eye(3)+2*q13@q13.
    transpose()-2*q4*q13x
```

A simple check if the conversion to the direction cosine matrix is performed correctly is checking $C_{BR}^T C_{BR}$ equal to the identity matrix.

In Python, matrix multiplication is denoted by the '@' sign. Note that the result of multiplication of "*' and '@' multiplication in Python is different. For example,

```
In [80]: x=np.array([[1],[2],[3]])

In [81]: x.transpose()*x
Out[81]:
array([[1, 2, 3],
       [2, 4, 6],
       [3, 6, 9]])

In [82]: x.transpose()@x
Out[82]: array([[14]])
```

While 'x.transpose()@x' performs $\mathbf{x}^T\mathbf{x}$ operation, i.e. the dot product of the vectors, 'x.transpose()*x' performs

$$\begin{bmatrix} x_1\mathbf{x} & x_2\mathbf{x} & x_3\mathbf{x} \end{bmatrix} \tag{2.49}$$

where $\mathbf{x}^T = [x_1\ x_2\ x_3]$.

The measurements from the star sensor for the stars given in (2.42) corresponding to the random attitude generated are obtained by (2.43). The implementation in MATLAB is as follows:

```
>> r1R=[0 1/sqrt(2) -1/sqrt(2)]';
>> r2R=[1/sqrt(2) 0 1/sqrt(2)]';
>> r1B=C_BR*r1R;
>> r2B=C_BR*r2R;
>> r3R=cross(r1R,r2R);
>> r3B=C_BR*r3R;
```

and the implementation in Python is as follows:

```
In [86]: r1R=np.array([0, 1/np.sqrt(2), -1/np.sqrt(2)]).reshape
    ((3,1))
In [87]: r2R=np.array([1/np.sqrt(2), 0, 1/np.sqrt(2)]).reshape
    ((3,1))
In [88]: r1B=C_BR@r1R
In [89]: r2B=C_BR@r2R
In [90]: r3R=np.cross(r1R.flatten(),r2R.flatten()).reshape((3,1))
In [91]: r3B=C_BR@r3R
```

where 'r1R' and 'r2R' vectors in Python are shaped as 3×1 vectors. They are converted into one-dimensional arrays for *np.cross()* using *flatten()* function in numpy, and the result is reshaped as 3×1 vector.

2.2 Attitude Estimation Algorithm

2.2.1 A Simple Algorithm

For the noise-free perfect star sensor measurement case, i.e. $\mathbf{v}^i$ in (2.40) is zero for all i, C_{BR} is calculated using (2.48), where $\mathbf{r}^i_B$ is from the noise-free sensor, and $\mathbf{r}^i_R$ is from the star catalogue for $i = 1, 2, 3$, as follows:

```
>> A1=blkdiag(r1R(:)',r1R(:)',r1R(:)');
>> A2=blkdiag(r2R(:)',r2R(:)',r2R(:)');
>> A3=blkdiag(r3R(:)',r3R(:)',r3R(:)');
>> A=[A1;A2;A3];
>> vec_CT=A \ [r1B(:);r2B(:);r3B(:)];
>> C_BR_Cal=reshape(vec_CT,3,3)'
>> norm(C_BR-C_BR_Cal)
ans =

   1.6909e-16
```

In MATLAB, for calculating $A^{-1}\mathbf{x}$, the backslash operator '\' is preferred to the inverse function, i.e. 'A\x' instead of 'inv(A)*x'. The backslash operator is a lot faster and accurate than calculating the inversion matrix and performing the multiplication to x. In general, calculating the inverse matrix and performing some operations would cause more computing steps and produce larger numerical errors. The norm in the last line shows the difference between the true and the calculated direction cosine matrices. If the difference is not small enough, there would be some errors in the codes and/or the measurements would have some problem, e.g. two vector observations are too close and the matrix A would be close to being a singular matrix. The error is in the order of 10^{-16} that is close enough to zero.

Similarly, in Python,

```
In [148]: from scipy.sparse import block_diag
In [149]: A1=block_diag((r1R.transpose(),r1R.transpose(),r1R.
    transpose())).toarray()
In [150]: A2=block_diag((r2R.transpose(),r2R.transpose(),r2R.
    transpose())).toarray()
In [151]: A2=block_diag((r2R.transpose(),r2R.transpose(),r2R.
    transpose())).toarray()
In [152]: A3=block_diag((r3R.transpose(),r3R.transpose(),r3R.
    transpose())).toarray()

In [153]: A=np.vstack((A1,A2,A3))

In [154]: from scipy.linalg import solve
In [155]: vec_CT=solve(A,np.vstack((r1B,r2B,r3B)))
In [156]: C_BR_Cal = vec_CT.reshape(3,3)

In [158]: np.linalg.norm(C_BR-C_BR_Cal)
Out[158]: 2.3714374201337736e-16
```

numpy and scipy packages in Python have no backslash operator as in MATLAB. A similar efficient way of calculating $A^{-1}\mathbf{x}$ can be, however, achieved using the *solve* function in *scipy.linalg package*. The error is in the order of 10^{-16}, which is close enough to zero, and the calculated direction cosine matrix is close to the truth.

When there are four or more star vector observations, then (2.48) becomes an over-determined problem, and the minimum norm solution is calculated using the following formula:

$$\text{vec}(C^T) = \left(A^T A\right)^{-1} A^T \begin{bmatrix} \mathbf{r}_B^1 \\ \mathbf{r}_B^2 \\ \vdots \\ \mathbf{r}_B^k \end{bmatrix} \tag{2.50}$$

for $k \geq 4$.

2.2.2 QUEST Algorithm

Solving the following minimization problem determines the best direction cosine matrix, C_{BR}:

$$\underset{C_{BR}}{\text{Minimize}} \frac{1}{2} \sum_{i=1}^{k} a_i \| \mathbf{r}_B^i - C_{BR} \mathbf{r}_R^i \|^2 \tag{2.51}$$

where $\mathbf{r}_B^i$ is corrupted by stochastic noise with the known variance, which is found in the sensor specifications, a_i is the positive weight for each observation, which is typically set to the inverse of the variance of each observation, and this is known as Wahba's problem (Wahba, 1965).

Shuster and Oh (1981) presents the QUEST (quaternion estimation) algorithm,[2] which calculates the optimal solution for estimating the best quaternion solving Wahba's problem, where $C_{BR} = C_{BR}(\mathbf{q})$ given in (2.37). A pseudo-code for the QUEST is given in Algorithm 2.3, and the implementation of the QUEST algorithm is left as an exercise.

The QUEST algorithm provides the optimal quaternion estimation based on the current vector measurements. It does not use, however, any dynamic model. It purely solves the optimization problem for a fixed instant of time to estimate the quaternion.

2.2.3 Kalman Filter

The Kalman filter is originally developed for the linear systems, which are written as follows (Kalman, 1960):

$$\mathbf{x}_k = A\mathbf{x}_{k-1} + \mathbf{w}_k \tag{2.52a}$$

$$\mathbf{z}_k = H\mathbf{x}_k + \mathbf{v}_k \tag{2.52b}$$

where the system noise, $\mathbf{w}_k$, and the measurement noise, $\mathbf{v}_k$, are the zero-mean Gaussian white noise, their covariances are known as Q_k and R_k, respectively, and A and H are the matrices with appropriate dimensions. In practice, it is frequently that the covariance matrices are constant for all k. Note that the notation for the discrete-time instances, t_k and t_{k-1}, are simply written as the subscripts, k and $k-1$, respectively. Both are used interchangeably whenever it is convenient, e.g. $\mathbf{x}(t_k) = \mathbf{x}_k$.

The Kalman filter solves the following optimization problem:

$$\underset{K_k}{\text{Minimize}} \ \text{trace}\left[E\left(\Delta\mathbf{x}_k \ \Delta\mathbf{x}_k^T\right)\right] \tag{2.53}$$

where $\Delta\mathbf{x}_k$ is the estimation error, i.e. the difference between the true state and the estimated state, $\mathbf{x}_k - \hat{\mathbf{x}}_k$, and K_k is the Kalman gain to be designed. The pseudo-code for the Kalman filter is given in Algorithm 2.4. The optimal estimated state, $\hat{\mathbf{x}}_k^+$, is obtained by combining the predicted state, $\hat{\mathbf{x}}_k^-$, from the system model and the measurements, $\mathbf{z}_k$, from the sensor, using the Kalman gain as follows:

$$\hat{\mathbf{x}}_k^+ = \hat{\mathbf{x}}_k^- + K_k\left(\mathbf{z}_k - H\hat{\mathbf{x}}_k^-\right) \tag{2.54}$$

When the simulator is implemented, confusion occurs frequently between what physical object we simulate and what algorithm we implement. Both components

2 'QUEST is better than rest' quoted by John after dinner with Malcolm, John, Jinho, and Jongrae in Maryland, USA in May 1998.

Algorithm 2.3 QUEST (quaternion estimation) algorithm

1: Construct B using $\mathbf{r}_B^i$ and $\mathbf{r}_R^i$ for $i = 1, 2, \ldots, k$ as follows:

$$B = \sum_{i=1}^{k} a_i \mathbf{r}_B^i \left(\mathbf{r}_R^i\right)^T$$

where a_i is equal to the inverse ofthe variance of the i-th observation.

2: Calculate S, σ, δ, and κ as follows:

$$S = B + B^T,\ \sigma = \mathrm{trace}(B),\ \delta = \det(S) = |S|,\ \kappa = \mathrm{trace}\left[\mathrm{adj}(S)\right]$$

where $\det(\cdot) = |\cdot|$ is the determinant of the matrix and $\mathrm{adj}(\cdot)$ is the adjugate of the matrix. For the 3×3 matrix S, κ is given by

$$\kappa = (s_{22}s_{33} - s_{23}^2) + (s_{11}s_{33} - s_{13}^2) + (s_{11}s_{22} - s_{12}^2)$$

where s_{ij} is the i-th row and the j-th column element of S.

3: Construct $\mathbf{z}$ as follows:

$$\mathbf{z} = \sum_{i=1}^{k} a_i \mathbf{r}_B^i \times \mathbf{r}_R^i$$

4: Calculate the coefficients of the following fourth order polynomial in λ:

$$f(\lambda) = \lambda^4 - (a + b)\lambda^2 - c\lambda + (ab + c\sigma - d)$$

where $a = \sigma^2 - \kappa, b = \sigma^2 + \mathbf{z}^T\mathbf{z}, c = \delta + \mathbf{z}^T S\mathbf{z}, d = \mathbf{z}^T S^2 \mathbf{z}$.

5: Set the initial guess of λ^* equal to 10 as the maximum λ for $f(\lambda) = 0$ is known to be around 1.

6: Set the tolerance, ε, equal to a small positive number, e.g. 10^{-6}, and $\Delta\lambda$ equal to a positive number greater than ε, e.g. 1000.

7: Find the λ satisfying $f(\lambda) = 0$ using the Newton–Raphson method (Press et al., 2007) as follows:

8: **while** $\Delta\lambda > \varepsilon$ **do**

9: $\quad df(\lambda^*)/d\lambda \leftarrow 4(\lambda^*)^3 - 2(a + b)\lambda^* - c$

10: $\quad \lambda_{\mathrm{new}} \leftarrow \lambda^* - f(\lambda^*)/[df(\lambda^*)/d\lambda)]$

11: $\quad \Delta\lambda \leftarrow |\lambda_{\mathrm{new}} - \lambda^*|$

12: $\quad \lambda^* \leftarrow \lambda_{\mathrm{new}}$

13: **end while**

14: $\mathbf{y}^* \leftarrow \left[(\sigma + \lambda^*)I_3 - S\right]^{-1} \mathbf{z}$

15: $q^* \leftarrow \dfrac{1}{\sqrt{1 + \|\mathbf{y}^*\|^2}} \begin{bmatrix} \mathbf{y}^* \\ 1 \end{bmatrix}$

Algorithm 2.4 The Kalman filter for linear systems

1: Initialize

$$\hat{\mathbf{x}}_0^+,\ P_0^+ = E\left(\Delta\mathbf{x}_0\,\Delta\mathbf{x}_0^T\right),\ Q = E\left(\mathbf{w}_k\,\mathbf{w}_k^T\right),\ R = E\left(\mathbf{v}_k\,\mathbf{v}_k^T\right)$$

where Q and R are assumed to be constant for all k.

2: **for** $k = 1, 2, \ldots$ **do**

3: **Prediction:** from t_{k-1} to t_k

$$\hat{\mathbf{x}}_k^- = A\hat{\mathbf{x}}_{k-1}^+$$
$$P_k^- = AP_{k-1}^+A^T + Q$$

4: **Update:** the measurement, $\mathbf{z}_k$, is available at t_k

$$K_k = P_k^-H^T\left(HP_k^-H^T + R\right)^{-1}$$
$$\hat{\mathbf{x}}_k^+ = \hat{\mathbf{x}}_k^- + K_k\left(\mathbf{z}_k - H\hat{\mathbf{x}}_k^-\right)$$
$$P_k^+ = \left(I - K_kH\right)P_k^-$$

5: **Substitute:** No measurement, $\mathbf{z}_k$, is available at t_k

$$\mathbf{x}_k^+ = \mathbf{x}_k^-$$
$$P_k^+ = P_k^-$$

6: **end for**

of the simulator are implemented as parts of the MATLAB or Python program. Distinguishing clearly between what is simulated in the simulator and what algorithm is tested in the simulator reduces any conceptual confusion and leads to a clearer simulator structure.

Consider the following mass-spring-damper system:

$$\ddot{x} = -\frac{k}{m}x - \frac{c}{m}\dot{x} + w \tag{2.55}$$

where $m = 1$ kg, $k = 0.5$ N/m, $c = 0.1$ N/(m/s), and w is the process noise, which is the zero-mean Gaussian random noise. The standard deviation of process noise is usually identified experimentally.

The stochastic differential equation, (2.55), however, has a mathematical ambiguity as the right-hand side of the equation is discontinuous everywhere because of the random noise, w. It can be written in a mathematically preferred form called the Itô equation as follows (Van Kampen, 2007):

$$d\dot{x} = -\frac{k}{m}xdt - \frac{c}{m}\dot{x}dt + d\beta \tag{2.56}$$

where the left-hand side of the equation is the velocity increment, and $d\beta = wdt$ in the right-hand side of the equation is a random increment, which is the same as (2.29). From an engineering point of view, these two ways of expression, (2.55) and (2.56), do not make any significant difference as the random perturbation is not infinitely fast, and these are used interchangeably.

From experiments, the velocity increments, $d\dot{x}$, would be observed to diverge from the trajectory expected by the deterministic parts of the model with the speed corresponding to the variance of the process noise, w, as follows:

$$E\{d\beta\}^2 \approx \sigma_\beta^2 \Delta t \tag{2.57}$$

where the measurements are sampled at every Δt time interval. Suppose that the estimated value of σ_β is $\sqrt{0.5}$ m/s and Δt is equal to 0.01 seconds.

The next question is how to integrate the stochastic differential equation, (2.56). Solving (2.56) is the simulation of the physical object. For each time interval, $[t_k, t_k + \Delta t)$, w_k is fixed to a constant and (2.58) becomes simply an ordinary differential equation (ODE). Hence, it can be solved using the ODE solver. In the computer simulation, w is replaced by the sampled random noise, w_k, as follows:

$$\ddot{x} = -\frac{k}{m}x - \frac{c}{m}\dot{x} + w_k \tag{2.58}$$

where w_k is a constant between $t \in [t_k, t_k + \Delta t)$ for $k = 1, 2, \ldots$, and its variance is given by

$$E\{d\beta\}^2 = E\{w dt\}^2 \rightarrow \sigma_\beta^2 \Delta t = \sigma_w^2 (\Delta t)^2 \rightarrow \sigma_w = \frac{\sigma_\beta}{\sqrt{\Delta t}} \tag{2.59}$$

As long as Δt is shorter than the mass-spring-damper system response speed corresponding to the bandwidth of the system, the sampled noise simulates the white noise to the system reasonably close. w_k is sampled from the normal distribution with the mean zero and the standard deviation, σ_w, equal to $\sqrt{0.5}/\sqrt{0.01}$ m/s$^{3/2}$. Once the integration is complete, save the solution, reset the initial condition equal to the final value of the solution, and solve the differential equation for the next time interval. Repeat this until the simulation time reaches the final time. These steps are implemented in Programs 2.15 and 2.16 in MATLAB and Python, respectively. The stochastic realization of the position and the velocity histories is shown in Figure 2.18. Aware that we expect the trajectory to be different for every simulation as w_k changes randomly for each simulation.

```
1 clear;
2
3 m_mass = 1.0; %[kg]
4 k_spring = 0.5; %[N/m]
5 c_damper = 0.01; %[N/(m/s)]
6 msd_const = [m_mass k_spring c_damper];
7
8 init_pos = 0.0; %[m]
9 init_vel = 0.0; %[m/s]
```

```
10
11 init_time = 0; %[s]
12 final_time = 60; %[s]
13
14 Delta_t = 0.01; %[s]
15
16 time_interval = [init_time final_time];
17
18 num_w = floor((final_time-init_time)/Delta_t)+1;
19 sigma_beta = sqrt(0.5);
20 sigma_w =sigma_beta/sqrt(Delta_t);
21 wk_noise = sigma_w*(randn(num_w,1));
22
23 x0 = [init_pos init_vel];
24 t0 = init_time;
25 tf = t0 + Delta_t;
26
27 tout_all = zeros(num_w,1);
28 xout_all = zeros(num_w,2);
29
30 tout_all(1) = t0;
31 xout_all(1,:) = x0;
32
33 for idx=2:num_w
34
35     wk = wk_noise(idx);
36
37     [tout,xout] = ode45( ...
38         @(time,state)msd_noisy(time,state,wk,msd_const),...
39         [t0 tf],x0);
40
41     tout_all(idx) = tout(end);
42     xout_all(idx,:) = xout(end,:);
43
44     x0 = xout(end,:);
45
46     % time interval update
47     t0 = tf;
48     tf = t0 + Delta_t;
49
50 end
51
52 figure(1);
53 subplot(211);
54 plot(tout_all,xout_all(:,1));
55 hold on;
56 axis([init_time final_time -10 10]);
57 set(gca,'FontSize',12);
58 ylabel('position [m]');
59 xlabel('time [s]');
60 subplot(212);
61 plot(tout_all,xout_all(:,2));
62 hold on;
```

```
63 axis([init_time final_time -10 10]);
64 set(gca,'FontSize',12);
65 ylabel('velocity [m/s]');
66 xlabel('time [s]');
67
68 function dxdt = msd_noisy(time,state,wk,msd_const)
69     x1 = state(1);
70     x2 = state(2);
71     m = msd_const(1);
72     k = msd_const(2);
73     c = msd_const(3);
74
75     dxdt = zeros(2,1);
76     dxdt(1) = x2;
77     dxdt(2) = -(k/m)*x1 - (c/m)*x2 + wk;
78 end
```

Program 2.15 (MATLAB) Solve the stochastic mass-spring-damper system using the ODE solver

```
1 import numpy as np
2 from scipy.integrate import solve_ivp
3
4 m_mass = 1.0 #[kg]
5 k_spring = 0.5 #[N/m]
6 c_damper = 0.1 #[N/(m/s)]
7 msd_const = [m_mass, k_spring, c_damper]
8
9 init_pos = 0.0 #[m]
10 init_vel = 0.0 #[m/s]
11
12 init_time = 0 #[s]
13 final_time = 60 #[s]
14
15 Delta_t = 0.01 #[s]
16
17 time_interval = [init_time, final_time]
18
19 num_w = int((final_time-init_time)/Delta_t)+1
20 sigma_beta = np.sqrt(0.5)
21 sigma_w =sigma_beta/np.sqrt(Delta_t)
22 wk_noise = sigma_w*(np.random.randn(num_w))
23
24 x0 = [init_pos, init_vel]
25 t0 = init_time
26 tf = t0 + Delta_t
27
28 tout_all = np.zeros(num_w)
29 xout_all = np.zeros((num_w,2))
30
31 tout_all[0] = t0
32 xout_all[0] = x0
```

```
33
34
35 def msd_noisy(time, state,wk,msd_const):
36     x1, x2 = state
37     m, k, c = msd_const
38
39     dxdt = [x2,
40             -(k/m)*x1 - (c/m)*x2 + wk]
41     return dxdt
42
43 for idx in range(1,num_w):
44
45     wk = wk_noise[idx]
46
47     # RK45
48     sol = solve_ivp(msd_noisy,(t0, tf),x0,args=(wk,msd_const))
49     xout = sol.y.transpose()
50
51     tout_all[idx] = sol.t[-1]
52     xout_all[idx] = xout[-1]
53
54     x0 = xout[-1];
55
56     # time interval update
57     t0 = tf
58     tf = t0 + Delta_t
59
60 import matplotlib.pyplot as plt
61
62 fig_ms, (ax_ms_0, ax_ms_1) = plt.subplots(nrows=2,ncols=1)
63 ax_ms_0.plot(tout_all, xout_all[:,0])
64 ax_ms_0.set_ylabel(r'$x(t)$',fontsize=14)
65 ax_ms_0.set(xlim=(0, final_time),ylim=(-10,10))
66
67 ax_ms_1.plot(tout_all, xout_all[:,1])
68 ax_ms_1.set_ylabel(r'$\dot{x}(t)$',fontsize=14)
69 ax_ms_1.set_xlabel('time [s]',fontsize=14)
70 ax_ms_1.set(xlim=(0, final_time),ylim=(-10,10))
```

Program 2.16 (Python) Solve the stochastic mass-spring-damper system using the ODE solver

To implement the Kalman filter given in Algorithm 2.4, the physical system, (2.58), is seen as the discrete form given by (2.52). Integrate (2.58) from t_k to t_{k+1}

$$\int_{t=t_k}^{t=t_{k+1}} d\dot{x} = -\frac{1}{m}\int_{t=t_k}^{t=t_{k+1}} (kx + c\dot{x})\,dt + \int_{t=t_k}^{t=t_{k+1}} w_k\,dt \tag{2.60}$$

where assume that $\Delta t_k^{\mathrm{KF}} = t_{k+1} - t_k$ is sufficiently small so that $x(t)$ and $\dot{x}(t)$ remain constants during $t \in [t_k, t_{k+1})$.

$$\dot{x}(t_{k+1}) - \dot{x}(t_k) = -\frac{1}{m}\left[kx(t_k) + c\dot{x}(t_k)\right]\Delta t_k^{\mathrm{KF}} + w_k \Delta t_k^{\mathrm{KF}} \tag{2.61}$$

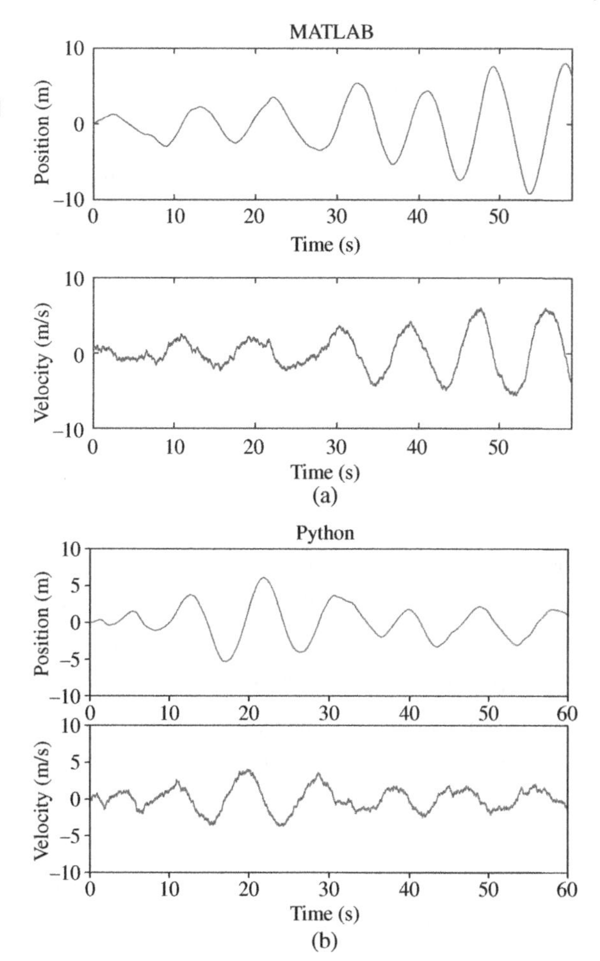

Figure 2.18 Stochastic simulations using the ODE solvers for $\Delta t = 0.01$ seconds in (a) MATLAB and (b) Python.

It is important to be aware of the difference between Δt in (2.57) and Δt_k^{KF} in the Kalman filter. Δt is to simulate the stochastic mass-spring-damper system, (2.56), in the computer, while Δt_{KF}^k is the time interval that the Kalman filter runs in the on-board computer, which is possibly attached to the mass-spring-damper system.

The discrete form for the velocity is given by

$$\dot{x}_{k+1} = -\frac{k\ \Delta t_k^{\text{KF}}}{m} x_k + \left(1 - \frac{c\ \Delta t_k^{\text{KF}}}{m}\right) \dot{x}_k + w_k\ \Delta t_k^{\text{KF}} \tag{2.62}$$

Similarly,

$$x_{k+1} = x_k + \dot{x}_k\ \Delta t_k^{\text{KF}} \tag{2.63}$$

In the state-space form,

$$\begin{bmatrix} x_{k+1} \\ \dot{x}_{k+1} \end{bmatrix} = \begin{bmatrix} 1 & \Delta t_k^{\mathrm{KF}} \\ -(k\ \Delta t_k^{\mathrm{KF}})/m & 1-(c\ \Delta t_k^{\mathrm{KF}})/m \end{bmatrix} \begin{bmatrix} x_k \\ \dot{x}_k \end{bmatrix} + \begin{bmatrix} 0 \\ \Delta t_k^{\mathrm{KF}} \end{bmatrix} w_k \tag{2.64}$$

In the Kalman filter algorithm design, the real system given by (2.56) *is viewed as* (2.64).

Comparing (2.64) with (2.52), the following equation is identified:

$$\mathbf{w}_k = \begin{bmatrix} 0 \\ \Delta t_k^{\mathrm{KF}} \end{bmatrix} w_k$$

The covariance matrix of the system as seen in the Kalman filter is

$$Q = E\left\{\mathbf{w}_k \mathbf{w}_k^T\right\} = \begin{bmatrix} 0 & 0 \\ 0 & (\Delta t_k^{\mathrm{KF}})^2 E\{w_k^2\} \end{bmatrix} = \begin{bmatrix} 0 & 0 \\ 0 & (\Delta t_k^{\mathrm{KF}})^2 \sigma_w^2 \end{bmatrix} \tag{2.65}$$

where Δt_k^{KF} is assumed to be constant, equal to 0.2 seconds, for all $k \in [1, \infty)$. Q is used to propagate the error covariance from P_{k-1}^+ to P_k^- in the prediction parts of the Kalman filter in Algorithm 2.4.

The prediction of the states, x_{k+1} and $\dot{x}_{k+1}$, in Algorithm 2.4, is performed by

$$\begin{bmatrix} x_{k+1} \\ \dot{x}_{k+1} \end{bmatrix} = \begin{bmatrix} 1 & \Delta t_k \\ -(k\ \Delta t_k)/m & 1-(c\ \Delta t_k)/m \end{bmatrix} \tag{2.66}$$

One last component to implement the Kalman filter is the sensor. Assume a sensor to measure the position, x_k, with the sampling frequency, Δt_k^{KF}, exists as follows:

$$z_k = x_k + v_k \tag{2.67}$$

where the noise characteristic of the sensor, i.e. the standard deviation of v_k, σ_v, is assumed to be equal to 0.75 m, which would be found in the sensor specifications. Comparing (2.67) with (2.52), the following values are identified:

$$H = \begin{bmatrix} 1 & 0 \end{bmatrix},\ R = \sigma_v^2$$

Figure 2.19 shows a time history example of the simulation scenario. The estimated states for the position and the velocity follow the true states reasonably close given that the noisy position measurements are indicated as the dots in the figure.

There is another important figure to be drawn for any Kalman filter simulations. Recall (2.53), where the Kalman gain, K_k, is to minimize the sum of the diagonal terms of the error covariance, P_k, which is equal to

$$P_k = E\{\Delta\mathbf{x}_k \Delta\mathbf{x}_k^T\} \tag{2.68}$$

The first and the second diagonal terms of P_k are $p_{11} = E\{(\Delta x_k)^2\}$ and $p_{12} = E\{(\Delta \dot{x}_k)^2\}$, respectively. $\pm 3\sqrt{p_{11}}$ and $\pm 3\sqrt{p_{22}}$ give the 3σ bounds about the error for the position and the velocity estimations, respectively. Given that

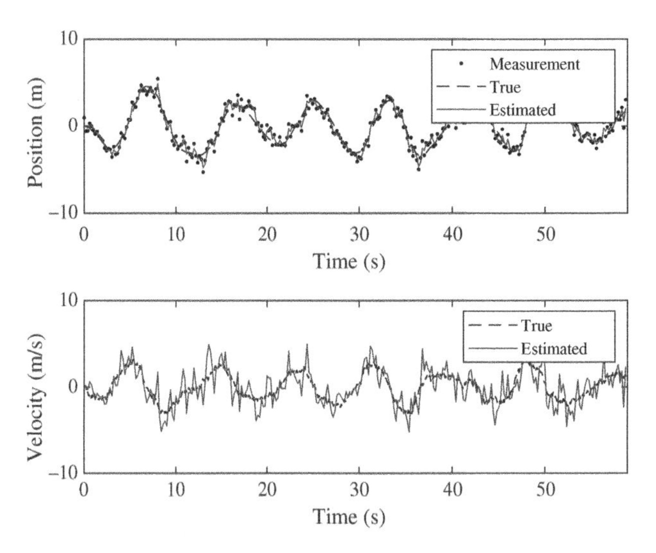

Figure 2.19 The Kalman filter for the mass-spring-damper system.

the error distribution is also Gaussian, the 3σ bounds provide a probabilistic guarantee that the error has a 99.7% chance of staying within the boundary.

In practice, the true states are unknown, and the error is also unknown. In simulations, the true states are accessible, and an accurate evaluation of the filter performance is made. As shown in Figure 2.20, the position error stays inside the bounds most of the time, and the bounds are reasonably tight to the actual error history. The velocity error is inside the bounds all the time, while the bounds are rather wide for the given time history.

2.2.4 Extended Kalman Filter

Most autonomous vehicles use gyroscopes to provide angular velocity measurements and optical sensors to provide absolute attitude measurements. The gyroscope measurement is simulated with the white noise and the bias noise in Figure 2.14. The main purpose of the Kalman filter in the attitude estimation is to estimate the bias error, $\boldsymbol{\beta}$, in the gyro measurements, (2.11), using optical sensors and a dynamic model. The angular velocity measurement is corrected by subtracting the bias error from the raw gyro measurement as follows:

$$\hat{\boldsymbol{\omega}}(t_k) = \tilde{\boldsymbol{\omega}}(t_k) - \hat{\boldsymbol{\beta}}(t_k) \tag{2.69}$$

where $\hat{\boldsymbol{\omega}}(t_k)$ is the estimated angular velocity at t_k, and $\hat{\boldsymbol{\beta}}(t_k)$ is the estimated bias at t_k from the Kalman filter to be designed. For the full detailed discussions of the attitude estimation Kalman filter to be introduced in the following

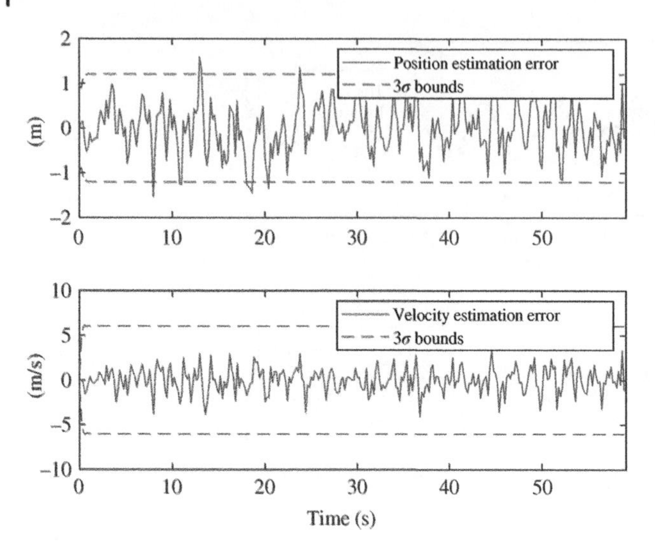

Figure 2.20 The Kalman filter states error and 3σ bounds for the mass-spring-damper system.

derivations, we refer the interested readers to Lefferts et al. (1982) and Crassidis and Junkins (2011).

2.2.4.1 Error Dynamics

We obtain the governing differential equation for the error dynamics of attitude. Define the error quaternion, $\delta\mathbf{q}$, equal to the quaternion between the estimated and actual attitude. The dynamics of the error quaternion is given by Bani Younes and Mortari (2019)

$$\delta\dot{\mathbf{q}}_{13} = -\frac{1}{2}[\delta\boldsymbol{\omega}\times]\delta\mathbf{q}_{13} - [\hat{\boldsymbol{\omega}}\times]\delta\mathbf{q}_{13} + \frac{1}{2}\delta q_4\delta\boldsymbol{\omega} \tag{2.70a}$$

$$\delta\dot{q}_4 = -\frac{1}{2}\delta\boldsymbol{\omega}^T\delta\mathbf{q}_{13} \tag{2.70b}$$

where $\delta\boldsymbol{\omega} = \boldsymbol{\omega} - \hat{\boldsymbol{\omega}}$. The error dynamics is non-linear and the original Kalman filter for linear systems cannot be applied directly. To use the Kalman filter for non-linear problems, non-linear dynamics is linearized. The technique is initially used to estimate spacecraft orbit trajectories, and it is called *the extended Kalman Filter* (EKF) (Grewal and Andrews, 2010).

For the linearization, assume that the magnitude of the error angle, $|\delta\theta|$, is small. The approximation of error quaternion is as follows:

$$\delta\mathbf{q}_{13} = \mathbf{e}\sin\frac{\delta\theta}{2} \approx \mathbf{e}\frac{\delta\theta}{2} \tag{2.71a}$$

$$\delta q_4 = \cos\frac{\delta\theta}{2} \approx 1 \tag{2.71b}$$

Also, the magnitude of error angular velocity, $\|\delta\boldsymbol{\omega}\|$, is assumed to be small. The non-linear error dynamics, (2.70), is linearized as follows:

$$\delta\dot{\mathbf{q}}_{13} \approx -[\hat{\boldsymbol{\omega}}\times]\delta\mathbf{q}_{13} + \frac{1}{2}\delta\boldsymbol{\omega} \tag{2.72a}$$

$$\delta\dot{q}_4 \approx 0 \tag{2.72b}$$

where the higher order terms are ignored. The error angular velocity can be expressed as

$$\delta\boldsymbol{\omega} = \boldsymbol{\omega} - \hat{\boldsymbol{\omega}} = \left(\tilde{\boldsymbol{\omega}} - \boldsymbol{\beta} - \boldsymbol{\eta}_v\right) - \left(\tilde{\boldsymbol{\omega}} - \hat{\boldsymbol{\beta}}\right) = -\delta\boldsymbol{\beta} - \boldsymbol{\eta}_v$$

where $\delta\boldsymbol{\beta} = \boldsymbol{\beta} - \hat{\boldsymbol{\beta}}$, and the gyro measurement model, (2.11), is used. The error dynamics, (2.72a), is re-written as

$$\delta\dot{\mathbf{q}}_{13} \approx -[\hat{\boldsymbol{\omega}}\times]\delta\mathbf{q}_{13} - \frac{1}{2}\delta\boldsymbol{\beta} - \frac{1}{2}\boldsymbol{\eta}_v \tag{2.73}$$

or in the Itô's form,

$$d\left(\delta\mathbf{q}_{13}\right) = -[\hat{\boldsymbol{\omega}}\times]\delta\mathbf{q}_{13}dt - \frac{1}{2}\delta\boldsymbol{\beta}dt - \frac{1}{2}\boldsymbol{\eta}_v dt$$

2.2.4.2 Bias Noise

While the bias is assumed to be constant in the estimated

$$\hat{\boldsymbol{\beta}}(t_{k+1}) = \hat{\boldsymbol{\beta}}(t_k) \tag{2.74}$$

the true bias follows (2.29). Subtract (2.74) from (2.29)

$$\boldsymbol{\beta}(t_{k+1}) - \hat{\boldsymbol{\beta}}(t_{k+1}) = \boldsymbol{\beta}(t_k) - \hat{\boldsymbol{\beta}}(t_k) + \boldsymbol{\eta}_u(t_{k+1})\Delta t_k$$

The bias error dynamics is written using the definition of $\delta\boldsymbol{\beta}$ as follows:

$$\delta\boldsymbol{\beta}(t_{k+1}) = \delta\boldsymbol{\beta}(t_k) + \boldsymbol{\eta}_u(t_{k+1})\Delta t_k$$

and in the Itô's form,

$$d\left(\delta\boldsymbol{\beta}\right) = \boldsymbol{\eta}_u\, dt$$

or in the common form frequently used,

$$\delta\dot{\boldsymbol{\beta}} = \boldsymbol{\eta}_u$$

To avoid frequent division by 2 in the equations in the following derivations, define

$$\delta\boldsymbol{\alpha} = 2\delta\mathbf{q}_{13} = \mathbf{e}\,\delta\theta = \begin{bmatrix}\delta\theta_1 & \delta\theta_2 & \delta\theta_3\end{bmatrix}^T$$

which is equal to the small-angle rotation for each body axis, and (2.73) becomes

$$\delta\dot{\boldsymbol{\alpha}} = -[\hat{\boldsymbol{\omega}}\times]\delta\boldsymbol{\alpha} - \delta\boldsymbol{\beta} - \boldsymbol{\eta}_v$$

Finally, the governing differential equation for the EKF for QUEST is summarized as follows:

$$\frac{d}{dt}\Delta\mathbf{x} = F\Delta\mathbf{x} + G\mathbf{w}_c \tag{2.75}$$

where

$$\Delta\mathbf{x} = \begin{bmatrix}\delta\alpha\\ \delta\beta\end{bmatrix}, \quad \mathbf{w}_c = \begin{bmatrix}\boldsymbol{\eta}_v\\ \boldsymbol{\eta}_u\end{bmatrix}, \quad F = \begin{bmatrix}-[\hat{\omega}\times] & -I_3\\ 0_3 & 0_3\end{bmatrix}, \quad G = \begin{bmatrix}-I_3 & 0_3\\ 0_3 & I_3\end{bmatrix},$$

and 0_3 is the 3×3 zero-matrix whose elements are all zero. The stochastic noise vectors, $\boldsymbol{\eta}_u$ and $\boldsymbol{\eta}_v$, are the zero-mean Gaussian, and the covariances are given by (2.26), $\sigma_v^2 I_3$, and (2.32), $\sigma_u^2 I_3$, respectively, i.e.

$$E(\boldsymbol{\eta}_v\boldsymbol{\eta}_v^T) = \sigma_v^2 I_3, \; E(\boldsymbol{\eta}_u\boldsymbol{\eta}_u^T) = \sigma_u^2 I_3$$

where the sensors towards three body frame directions have the same noise characteristics, and the two noises are not correlated, hence,

$$E(\boldsymbol{\eta}_v\boldsymbol{\eta}_u^T) = 0_3$$

2.2.4.3 Noise Propagation in Error Dynamics

For the discrete version of the Kalman filter, the governing equation, (2.75), is transformed into a discrete equation. Integrate (2.75) from t_k to t_{k+1} (Chen, 2009)

$$\Delta\mathbf{x}(t_{k+1}) = e^{F\Delta t_k}\Delta\mathbf{x}(t_k) + \int_{t=t_k}^{t=t_{k+1}} e^{F\Delta t} G\mathbf{w}_c(t)dt \tag{2.76}$$

where $\Delta t_k = t_{k+1} - t_k$ and $\Delta t = t_{k+1} - t$ for $t \in [t_k, t_{k+1})$. The Taylor series expansion of the exponential matrix up to the third order in Δt is given by

$$e^{F\Delta t} = I_6 + F\Delta t + \frac{\Delta t^2}{2}F^2 + \frac{\Delta t^3}{6}F^3 + \cdots = \begin{bmatrix}\Phi_1(t,\hat{\omega}) & \Phi_2(t,\hat{\omega})\\ 0_3 & I_3\end{bmatrix} \tag{2.77}$$

where

$$\Phi_1(t,\hat{\omega}) = I_3 - \Delta t[\hat{\omega}\times] + \frac{\Delta t^2}{2}[\hat{\omega}\times]^2 - \frac{\Delta t^3}{6}[\hat{\omega}\times]^3 + \cdots \tag{2.78a}$$

$$\Phi_2(t,\hat{\omega}) = \Delta t I_3 - \frac{\Delta t^2}{2}[\hat{\omega}\times] + \frac{\Delta t^3}{6}[\hat{\omega}\times]^2 - \cdots \tag{2.78b}$$

Perform the integration in (2.76)

$$\int_{t=t_k}^{t=t_{k+1}} e^{F\Delta t} G\mathbf{w}_c \, dt = \int_{t=t_k}^{t=t_{k+1}} \begin{bmatrix}-\Phi_1(t,\hat{\omega})\boldsymbol{\eta}_v + \Phi_2(t,\hat{\omega})\boldsymbol{\eta}_u\\ \boldsymbol{\eta}_u\end{bmatrix} dt = \mathbf{w}_d$$

where

$$\mathbf{w}_d = \begin{bmatrix}\int_{t=t_k}^{t=t_{k+1}} \Phi_1(t,\hat{\omega})\boldsymbol{\eta}_v \, dt - \int_{t=t_k}^{t=t_{k+1}} \Phi_2(t,\hat{\omega})\boldsymbol{\eta}_u\\ \int_{t=t_k}^{t=t_{k+1}} \boldsymbol{\eta}_u \, dt\end{bmatrix}$$

In a compact form similar to (2.64),

$$\Delta\mathbf{x}_{k+1} = \Phi\Delta\mathbf{x}_k + \mathbf{w}_d \tag{2.79}$$

where $\Phi = e^{F\Delta t_k}$. *This is how the error dynamics,* (2.79), *is viewed in the Kalman filter design.* Unlike the state prediction for the linear Kalman filter, e.g. (2.66), it would be surprising to see that the discrete error dynamics, (2.79), is not used to propagate the states in the EKF for the QUEST. Two main purposes of the discrete model are to obtain the state transition matrix, Φ, which is performed already and find stochastic properties of the process noise, $\mathbf{w}_d$. Specifically, we are to identify the mean and the covariance of the noise. It is easily shown that $E(\mathbf{w}_d) = \mathbf{0}$.

The Kalman filter relies on the knowledge of the process noise covariance matrix, i.e. $E(\mathbf{w}_d\mathbf{w}_d^T)$. Calculating the covariance matrix by hand following the formula shown above is tedious and prone to errors. Symbolic calculation in the computer is a powerful method for performing this type of long algebraic operation. For a single-axis rotation case, the covariance is derived in Farrenkopf (1978). The following procedures extend the derivation to arbitrary rotational motions in a three-dimensional space.

Symbolic math toolbox in MATLAB and Sympy in Python provide symbolic math operation capabilities. Although the symbolic operation capabilities are yet to be fully automatic, they could save time and minimize mistakes in derivations when we use them properly. Program 2.17 is the MATLAB m-script to define all symbols and the matrix for F, G, $\mathbf{w}_d$, and $e^{F\Delta t}$, where $e^{F\Delta t}$ is expanded up to the fourth order in Δt.

```
1  clear;
2
3  % symbols for omega, noise and the variances
4  syms w1 w2 w3 Dt nv1 nv2 nv3 nu1 nu2 nu3 real;
5  syms sgm2_u sgm2_v real; % these are variance, i.e. sigma-squared
6
7  wx=[ 0 -w3 w2; w3 0 -w1; -w2 w1 0];
8  nv=[nv1;nv2;nv3];
9  nu=[nu1;nu2;nu3];
10 wc=[nv;nu];
11
12 % F & G Matrices
13 F = [-wx -eye(3); zeros(3,6)];
14 G = [-eye(3) zeros(3); zeros(3) eye(3)];
15
16 % e^{Ft}
17 Phi = eye(6) + F*Dt + (1/2)*(F^2)*Dt^2 + (1/6)*(F^3)*Dt^3 + (1/24)
       *(F^4)*Dt^4;
18
19 % wd before integral
20 wd = Phi*wc;
21
22 % E(wd wd^T)
```

```
23 cov_wd = simplify(expand(wd*wd'));
24 Q_cov = sym(zeros(6));
25
26 eqn2=sgm2_u==nu1^2;
27 eqn3=sgm2_u==nu2^2;
28 eqn4=sgm2_u==nu3^2;
29 eqn5=sgm2_v==nv1^2;
30 eqn6=sgm2_v==nv2^2;
31 eqn7=sgm2_v==nv3^2;
```

Program 2.17 (MATLAB) Process noise covariance Q derivation using symbolic manipulations: Define variables

In line 23, *simplify* and *expand* commands in the symbolic math toolbox are used. These two functions should be used frequently to help symbolic computing in the computer. As the capability of symbolic computations is not perfect, it needs some help when these operations should be performed. Consider the following calculation:

```
>> syms x y real;
>> x*(y+1) - x*y
```

syms is the keyword in MATLAB to define symbols. Two variables, x and y, are defined and declared to be real variables. In the second line, we could expect x because $x(y+1) - xy = xy + x - xy = x$, but it is not. No automatic cancellation occurs. It performs the simplification only when either

```
>> expand(x*(y+1) - x*y)
```

or

```
>> simplify(x*(y+1) - x*y)
```

is explicitly called.

There is a difference between *expand* and *simplify*, for example,

```
>> expand(cos(x)^2+sin(x)^2)
```

returns the original expression without any further simplification, and

```
>> simplify(cos(x)^2+sin(x)^2)
```

returns 1, which is preferred. Without considering when and which one executes, we apply these two commands whenever we perform some algebraic manipulation so that they have their simplest form before the next symbolic operation.

In line 24, the symbolic 6×6 zero-matrix to store the covariance calculation result is declared using the *sym()* command. From line 26, define $\sigma_i^2 = E(\eta_{ij}^2)$ for $i = u, v$ and $j = x, y, z$ using '==' notation, where the noise characteristics of the sensor for each body axis are assumed to be equal to each other. 'eqn2' defines $\sigma_v^2 = E(v_{ux}^2)$. These definitions are later used in symbolic calculations to substitute v_{ij}^2 for σ_{ij}^2.

Program 2.18 continues Program 2.17. Define *eqn_1* equal to the first column and the first low element of the inside the integration of $\mathbf{w}_d\mathbf{w}_d^T$ and apply *expand* to simplify the expression. As it is a polynomial, no further simplification occurs by applying the *simplify* command. No harm to be done, of course, by calling the simplification command apart from the additional computation done in the computer giving the same result. In line 7, all v_{ij}^2 is replaced by σ_{ij}^2 using the equations defined in Program 2.17, where 'rhs(eqn2)' means the right-hand side of 'eqn2', which is equal to σ_{ux}^2, and 'lhs(eqn2)' means the left-hand side of 'eqn2'. *subs* command substitutes one symbol by another one. For example, the following lines

```
>> syms x y x2 y2 real;
>> eqn1 = x^2==x2
>> eqn2 = x^2 + y^2 + x*y + x*x*y
>> subs(eqn2,{lhs(eqn1)},{rhs(eqn1)})
```

replace x^2 in 'eqn1' with x_2, and the answer becomes $x_2 + y^2 + xy + x_2y$.

```
1
2  % (continue from Program 2.17)
3  syms q11 real;
4
5  % symbolic calculation of the inside integral
6  eqn_1=q11==expand(cov_wd(1,1));
7  PPT_11 = subs(eqn_1,{rhs(eqn2),rhs(eqn3),rhs(eqn4),rhs(eqn5),rhs(
       eqn6),rhs(eqn7)},{lhs(eqn2),lhs(eqn3),lhs(eqn4),lhs(eqn5),lhs(
       eqn6),lhs(eqn7)});
8  PPT_11 = subs(rhs(PPT_11),{nv1,nv2,nv3,nu1,nu2,nu3},{0,0,0,0,0,0});
9
10 % integral from t_k to t_k + Delta t
11 eqn_1=q11==expand(int(PPT_11,Dt,[0 Dt]));
12
13 % ignore higher order terms
14 Q_cov(1,1) = subs(rhs(eqn_1),{Dt^9,Dt^8,Dt^7,Dt^6,Dt^5,Dt
       ^4},{0,0,0,0,0,0});
```

Program 2.18 (MATLAB) Process noise covariance Q derivation using symbolic manipulations: Substitutions and Integration

Multiple substitutions are performed with additional symbols to be replaced provided as in line 7, Program 2.18. In the two lines of the substitutions, the expectations are applied by replacing η_{ij}^2 and η_{ij} with σ_i^2 and 0, respectively. In line 11, the symbolic integration by 'Dt', which is equal to Δt, is performed. As $t_k = 0$, $t_{k+1} = \Delta t_k$, $\Delta t = \Delta t_k - t$, and $d(\Delta t) = -dt$, the integration term in (2.76) becomes

$$\mathbf{w}_d = \int_{t=t_k}^{t=t_{k+1}} e^{F\Delta t} G\mathbf{w}_c(t)dt = \int_{\Delta t=\Delta t_k}^{\Delta t=0} e^{F\Delta t} G\mathbf{w}_c(\Delta t_k - \Delta t)\left[-d(\Delta t)\right]$$

$$= \int_{\Delta t=0}^{\Delta t=\Delta t_k} e^{F\Delta t} G\mathbf{w}_c(\Delta t_k - \Delta t)d(\Delta t)$$

Finally, the result is stored in the symbolic matrix defined earlier.

The same procedure is repeated for the rest of the eight elements of the covariance matrix. We obtain the following result:

$$E(\mathbf{w}_d\mathbf{w}_d^T) = \begin{bmatrix} \left(\sigma_v^2 \Delta t_k + \dfrac{\Delta t_k^3}{3}\sigma_u^2\right) I_3 & -\dfrac{\Delta t_k^2}{2}\sigma_u^2 I_3 - \dfrac{\Delta t_k^3}{6}\sigma_u^2[\hat{\omega}\times] \\ -\dfrac{\Delta t_k^2}{2}\sigma_u^2 I_3 - \dfrac{\Delta t_k^3}{6}\sigma_u^2[\hat{\omega}\times] & \sigma_u^2 \Delta t_k I_3 \end{bmatrix} \tag{2.80}$$

Introducing the assumption that the angular velocity, $\|\hat{\omega}\|$, is small enough such that $\Delta t^3[\hat{\omega}\times]$ is negligible compared to the other terms, then the covariance matrix becomes

$$E(\mathbf{w}_d\mathbf{w}_d^T) = \begin{bmatrix} \left(\sigma_v^2 \Delta t_k + \dfrac{\Delta t_k^3}{3}\sigma_u^2\right) I_3 & -\dfrac{\Delta t_k^2}{2}\sigma_u^2 I_3 \\ -\dfrac{\Delta t_k^2}{2}\sigma_u^2 I_3 & \sigma_u^2 \Delta t_k I_3 \end{bmatrix} \tag{2.81}$$

In the covariance matrix, (2.81), the process noise in each direction of the body frame is completely decoupled as the angular velocity is slow, while one in (2.80) has the coupling terms through $[\hat{\omega}\times]$. For Δt_k and $\|\hat{\omega}\|$ being sufficiently small, where the higher order terms are all negligible, the covariance simply becomes

$$E(\mathbf{w}_d\mathbf{w}_d^T) \approx \begin{bmatrix} \sigma_v^2 \Delta t_k I_3 & 0_3 \\ 0_3 & \sigma_u^2 \Delta t_k I_3 \end{bmatrix} \tag{2.82}$$

Now, each axis is decoupled, and the effect of $\boldsymbol{\eta}_u$ on $\delta\boldsymbol{\alpha}$ disappeared.

The Python program obtaining the first-row and the first-column element of (2.80) is shown in Program 2.19. *sympy* is the mathematical symbolic calculation module. We import the five specific functions. *symbols* are the function to define symbols, *Matrix* is to define symbolic matrices, *simplify* and *expand* are the same roles as in MATLAB symbolic toolbox, and *integrate* is the symbolic function integrator. Unlike the MATLAB program, line 23 in 2.17, the Python program, line 24 in 2.19, is not expanded and simplified. We only expand and simplify the first-row and the first-column element individually in line 28. As we observed, the expansion and the simplification for the whole matrix takes a longer computation time compared to the MATLAB commands, and it is not required to apply the operations for the whole elements. The same applies to the MATLAB operations to prevent long computation time for a single line, which requires a large memory to complete the operations. The substitution is not an independent function in sympy but a method in the symbolic equation. 'cov_wd_11' defined in line 28 has the substitution method. 'cov_wd_11.sub()' in line 29 performs the substitution. Finally, *integrate* at line 32 integrates 'cov_wd_11' by 'Dt' from 0 to 'Dt' indicated by the tuple, '(Dt,0,Dt)'.

```
1
2  from sympy import symbols, Matrix, simplify, expand, integrate
3
4  w1, w2, w3, Dt, nv1, nv2, nv3, nu1, nu2, nu3 = symbols('w1 w2 w3 Dt
       nv1 nv2 nv3 nu1 nu2 nu3')
5  sgm2_u, sgm2_v = symbols('sgm2_u sgm2_v') # these are variance, i.e
       . sigma-squared
6
7  wx = Matrix([[ 0, -w3, w2], [w3, 0, -w1], [-w2, w1, 0]])
8
9  nv = Matrix([[nv1],[nv2],[nv3]])
10 nu = Matrix([[nu1],[nu2],[nu3]])
11 wc = Matrix([nv,nu])
12
13 # F & G Matrices
14 F = Matrix([[-wx,-Matrix.eye(3)],[Matrix.zeros(3,6)]])
15 G = Matrix([[-Matrix.eye(3), Matrix.zeros(3)],[Matrix.zeros(3),
       Matrix.eye(3)]])
16
17 # e^{Ft}
18 Phi = Matrix.eye(6) + F*Dt + (1/2)*(F**2)*(Dt**2) + (1/6)*(F**3)*(
       Dt**3) + (1/24)*(F**4)*(Dt**4)
19
20 # wd before integral
21 wd = Phi@wc
22
23 # E(wd wd^T)
24 wd_wd_T = wd@wd.transpose()
25 Q_cov = Matrix.zeros(6)
26
27 # Q_11: integrate from 0 to Dt
28 cov_wd_11 = simplify(expand(wd_wd_T[0,0]))
29 cov_wd_11 = cov_wd_11.subs([[nu1**2,sgm2_u],[nu2**2,sgm2_u],[nu3
       **2,sgm2_u],[nv1**2,sgm2_v],[nv2**2,sgm2_v],[nv3**2,sgm2_v]])
30 cov_wd_11 = cov_wd_11.subs([[nu1,0],[nu2,0],[nu3**2,0],[nv1,0],[nv2
       ,0],[nv3,0]])
31
32 cov_wd_11 = integrate(cov_wd_11,(Dt,0,Dt))
33 cov_wd_11 = simplify(expand(cov_wd_11))
34 cov_wd_11 = cov_wd_11.subs([[Dt**4,0],[Dt**5,0],[Dt**6,0],[Dt
       **7,0],[Dt**8,0],[Dt**9,0]])
35 cov_wd_11 = expand(cov_wd_11)
36 Q_cov[0,0] = cov_wd_11
```

Program 2.19 (Python) Process noise covariance Q derivation using symbolic manipulations: Define variables and Integration

2.2.4.4 State Transition Matrix, Φ

The state transition matrices, Φ_1 and Φ_2, in (2.78) have the closed-form expressions. Using the following identity:

$$[\hat{\omega}\times]^3 = -\|\hat{\omega}\|^2[\hat{\omega}\times],$$

we show that the higher order terms satisfy the following equations:

$$\begin{aligned}
[\hat{\omega}\times]^4 &= -\|\hat{\omega}\|^2[\hat{\omega}\times]^2 \\
[\hat{\omega}\times]^5 &= -\|\hat{\omega}\|^2[\hat{\omega}\times]^3 = \|\hat{\omega}\|^4[\hat{\omega}\times] \\
[\hat{\omega}\times]^6 &= \|\hat{\omega}\|^4[\hat{\omega}\times]^2 \\
[\hat{\omega}\times]^7 &= \|\hat{\omega}\|^4[\hat{\omega}\times]^3 = -\|\hat{\omega}\|^6[\hat{\omega}\times] \\
&\vdots
\end{aligned}$$

We replace the terms in Φ_1 or Φ_2 with the above equations and collect the terms for $[\hat{\omega}\times]$ and $[\hat{\omega}\times]^2$, respectively. They are the Taylor series expansions of sin and cos functions. Hence, the transition matrices are written as (Markley and Crassidis, 2014)

$$\Phi_1 = I_3 - [\hat{\omega}\times]\left[\frac{\sin(\|\hat{\omega}\|\Delta t)}{\|\hat{\omega}\|}\right] + [\hat{\omega}\times]^2\left[\frac{1-\cos(\|\hat{\omega}\|\Delta t)}{\|\hat{\omega}\|^2}\right] \tag{2.83a}$$

$$\Phi_2 = -I_3\Delta t + [\hat{\omega}\times]\left[\frac{1-\cos(\|\hat{\omega}\|\Delta t)}{\|\hat{\omega}\|^2}\right] - [\hat{\omega}\times]^2\left[\frac{\|\hat{\omega}\|\Delta t - \sin(\|\hat{\omega}\|\Delta t)}{\|\hat{\omega}\|^3}\right] \tag{2.83b}$$

We carefully construct the transition matrix that performs the divisions by the power of the angular velocity magnitude, i.e. $\|\hat{\omega}\|^p$, *for* $p = 1, 2, 3$. These should be used only for $\|\hat{\omega}\|$ greater than ϵ, which is a small positive number, for example, 0.0001, and the following transition matrices are used for $\|\hat{\omega}\| \leq \epsilon$,

$$\Phi_1 = I_3 - [\hat{\omega}\times]\Delta t \tag{2.84a}$$

$$\Phi_2 = -I_3\Delta t \tag{2.84b}$$

Or, for Δt sufficiently small for any feasible magnitudes of the angular velocity, we use the simpler forms of the transition matrix above all the time.

2.2.4.5 Vector Measurements

Once in a while, a set of the vector measurements from optical sensors, e.g. star sensors, arrive as follows:

$$\begin{bmatrix} \tilde{\mathbf{r}}_B^1(t_k) \\ \tilde{\mathbf{r}}_B^2(t_k) \\ \vdots \\ \tilde{\mathbf{r}}_B^n(t_k) \end{bmatrix} = \begin{bmatrix} C_{BR}[\mathbf{q}(t_k)]\mathbf{r}_R^1 \\ C_{BR}[\mathbf{q}(t_k)]\mathbf{r}_R^2 \\ \vdots \\ C_{BR}[\mathbf{q}(t_k)]\mathbf{r}_R^n \end{bmatrix} + \begin{bmatrix} \mathbf{v}^1(t_k) \\ \mathbf{v}^2(t_k) \\ \vdots \\ \mathbf{v}^n(t_k) \end{bmatrix} \tag{2.85}$$

where the n-vector measurements, $\tilde{\mathbf{r}}_B^i$, are obtained from the sensor at time t_k, $\mathbf{r}_R^i$ is the direction vector towards the identified object, e.g. a star, stored in a database, and the direction cosine matrix, C_{BR}, in terms of the quaternion is given in (2.37). $\mathbf{q}(t_k)$ is the current true quaternion that is unknown.

The discrete non-linear measurement equation is given by

$$\mathbf{z}(t_k) = \mathbf{h}[\mathbf{q}(t_k)] + \mathbf{v}(t_k) \tag{2.86}$$

where

$$\mathbf{z}(t_k) = \begin{bmatrix} \tilde{\mathbf{r}}_B^1(t_k) \\ \tilde{\mathbf{r}}_B^2(t_k) \\ \vdots \\ \tilde{\mathbf{r}}_B^n(t_k) \end{bmatrix}, \quad \mathbf{v}(t_k) = \begin{bmatrix} \mathbf{v}^1(t_k) \\ \mathbf{v}^2(t_k) \\ \vdots \\ \mathbf{v}^n(t_k) \end{bmatrix}, \quad \mathbf{h}[\mathbf{q}(t_k)] = \begin{bmatrix} C_{BR}[\mathbf{q}(t_k)]\mathbf{r}_R^1 \\ C_{BR}[\mathbf{q}(t_k)]\mathbf{r}_R^2 \\ \vdots \\ C_{BR}[\mathbf{q}(t_k)]\mathbf{r}_R^n \end{bmatrix}$$

and the covariance of the vector measurement noise, R_k, is assumed to be known as

$$R_k = E\left[\mathbf{v}(t_k)\mathbf{v}^T(t_k)\right]$$

which is uncorrelated with the noises in the gyroscopes.

Unlike the measurement equation for linear Kalman filter, (2.52), the matrix H is not available, but the non-linear function $\mathbf{h}(\cdot)$ is given. The corresponding H in the non-linear measurement is obtained by the linearization procedures. Consider the following i-th vector measurement,

$$\mathbf{z}_i(t_k) = \tilde{\mathbf{r}}_B^i(t_k) = C_{BR}[\mathbf{q}(t_k)]\mathbf{r}_R^i + \mathbf{v}^i(t_k)$$

for $i = 1, 2, \dots, n$. The direction cosine matrix is written using the current estimated quaternion as follows:

$$C_{BR}[\mathbf{q}(t_k)] = C_{B\hat{B}}[\delta\mathbf{q}]C_{\hat{B}R}[\hat{\mathbf{q}}(t_k)] \tag{2.87}$$

where $\hat{B}$ is the estimated attitude, and $C_{B\hat{B}}$ is the direction cosine matrix between the estimated and actual attitudes, whose quaternion is given by $\delta\mathbf{q} = [\delta\mathbf{q}_{13}^T \ \delta q_4]^T$.

Apply the small attitude error assumption, (2.71), to the direction cosine matrix using the definition in (2.37)

$$C_{B\hat{B}}[\delta\mathbf{q}] \approx I_3 - 2[\delta\mathbf{q}_{13}\times] = I_3 - [\delta\boldsymbol{\alpha}\times] \tag{2.88}$$

where the higher order terms are neglected. Substitute (2.88) into (2.87)

$$C_{BR}[\mathbf{q}(t_k)] = C_{\hat{B}R}[\hat{\mathbf{q}}(t_k)] - [\delta\boldsymbol{\alpha}\times]C_{\hat{B}R}[\hat{\mathbf{q}}(t_k)]$$

Multiply both sides by $\mathbf{r}_R^i$, and it becomes

$$C_{BR}[\mathbf{q}(t_k)]\mathbf{r}_R^i - C_{\hat{B}R}[\hat{\mathbf{q}}(t_k)]\mathbf{r}_R^i = -[\delta\boldsymbol{\alpha}\times]\left\{C_{\hat{B}R}[\hat{\mathbf{q}}(t_k)]\mathbf{r}_R^i\right\}$$

Let $\mathbf{a} = C_{\hat{B}R}[\hat{\mathbf{q}}(t_k)]\mathbf{r}_R^i$, then

$$-[\delta\boldsymbol{\alpha}\times]\mathbf{a} = [\mathbf{a}\times]\delta\boldsymbol{\alpha} = \left[(C_{\hat{B}R}[\hat{\mathbf{q}}(t_k)]\mathbf{r}_R^i)\times\right]\delta\boldsymbol{\alpha}$$

Define $\Delta \mathbf{z}_i$ as follows:

$$\Delta \mathbf{z}_i(t_k) = C_{BR}\left[\mathbf{q}(t_k)\right]\mathbf{r}_R^i - C_{B'R}\left[\hat{\mathbf{q}}(t_k)\right]\mathbf{r}_R^i = \left[(C_{B'R}[\hat{\mathbf{q}}(t_k)]\mathbf{r}_R^i)\times\right]\delta\boldsymbol{\alpha},$$

for $i = 1, 2, \dots, n$. Therefore,

$$\Delta \mathbf{z}_k = \begin{bmatrix} \Delta \mathbf{z}_1(t_k) \\ \Delta \mathbf{z}_2(t_k) \\ \vdots \\ \Delta \mathbf{z}_n(t_k) \end{bmatrix} = \begin{bmatrix} [(C_{B'R}[\hat{\mathbf{q}}(t_k)]\mathbf{r}_R^1)\times] & 0_3 \\ [(C_{B'R}[\hat{\mathbf{q}}(t_k)]\mathbf{r}_R^2)\times] & 0_3 \\ \vdots & \vdots \\ [(C_{B'R}[\hat{\mathbf{q}}(t_k)]\mathbf{r}_R^n)\times] & 0_3 \end{bmatrix} \begin{bmatrix} \delta\boldsymbol{\alpha}(t_k) \\ \delta\boldsymbol{\beta}(t_k) \end{bmatrix} = H_k \Delta \mathbf{x}_k \tag{2.89}$$

and the expression for H_k is established.

2.2.4.6 Summary

The linearized state-space form, (2.79), and the linearized measurement equation, (2.89), are obtained as follows:

$$\begin{aligned} \Delta \mathbf{x}_{k+1} &= \Phi \Delta \mathbf{x}_k + \mathbf{w}_d \\ \Delta \mathbf{z}_k &= H_k \Delta \mathbf{x}_k \end{aligned}$$

where Φ is given by (2.83) and (2.84), H_k is given by (2.89), and $Q = E[\mathbf{w}_d \mathbf{w}_d^T]$ is given by (2.80).

2.2.4.7 Kalman Filter Update

When the vector measurements are available, update the Kalman gain, K_k, and the estimation error covariance matrix, $P_k = E[\Delta \mathbf{x}_k \Delta \mathbf{x}_k^T]$, as follows:

$$K_k = P_k^- H_k^T \left(H_k P_k^- H_k^T + R_k\right)^{-1} \tag{2.90a}$$

$$P_k^+ = \left(I_6 - K_k H_k\right) P_k^- \tag{2.90b}$$

$$\Delta \mathbf{x}_k = K_k \left[\mathbf{z}_k - \mathbf{h}\left(\hat{\mathbf{q}}_k^-\right)\right] \tag{2.90c}$$

In the standard EKF, the following state update equation is used

$$\mathbf{x}_k^+ = \mathbf{x}_k^- + \Delta \mathbf{x}_k, \tag{2.91}$$

which is the case for the bias estimation update as follows:

$$\hat{\boldsymbol{\beta}}_k^+ = \hat{\boldsymbol{\beta}}_k^- + \Delta \boldsymbol{\beta}_k = \hat{\boldsymbol{\beta}}_k^- + \begin{bmatrix} 0_3 & I_3 \end{bmatrix} \Delta \mathbf{x}_k \tag{2.92}$$

And the angular velocity is updated by (2.69).

The update equation, (2.91), is not, however, used for updating the quaternion. As the quaternion is attitude information, it does have little physical meaning in simple quaternion summation or subtraction. For the current quaternion estimate, $\hat{\mathbf{q}}_k^-$, and the error quaternion between the true and the current quaternions, $\delta \mathbf{q}_k$, then, $\hat{\mathbf{q}}_k^- + \delta \mathbf{q}_k$ does not have any clear physical interpretation to correct the error in the current quaternion estimate. Instead, acknowledge that the error quaternion itself is an attitude, hence, the update should be done such that rotating the current estimated quaternion with the amount of attitude indicated by the current error quaternion estimated. This is done by the quaternion algebra as follows that corresponds to the direction cosine matrix multiplication in (2.87) (Wie, 2008):

$$\hat{\mathbf{q}}_k^+ = \hat{\mathbf{q}}_k^- + \begin{bmatrix} \hat{q}_4^-(t_k)I_3 + [\hat{\mathbf{q}}_{13}^-(t_k)\times] \\ -\hat{\mathbf{q}}_{13}^-(t_k) \end{bmatrix} \delta\mathbf{q}_{13}(t_k) \tag{2.93}$$

where

$$\delta\mathbf{q}_{13}(t_k) = 2\,\delta\boldsymbol{\alpha}_k = 2\begin{bmatrix} I_3 & 0_3 \end{bmatrix}\Delta\mathbf{x}_k$$

2.2.4.8 Kalman Filter Propagation

The quaternion is propagated as follows:

$$\mathbf{q}_{k+1}^- = \begin{bmatrix} \cos\dfrac{\Delta\theta_k}{2}I_3 - \dfrac{\sin(\Delta\theta_k/2)}{\|\hat{\boldsymbol{\omega}}_k\|}[\hat{\boldsymbol{\omega}}_k\times] & \dfrac{\sin(\Delta\theta_k/2)}{\|\hat{\boldsymbol{\omega}}_k\|}\hat{\boldsymbol{\omega}}_k \\ -\dfrac{\sin(\Delta\theta_k/2)}{\|\hat{\boldsymbol{\omega}}_k\|}\hat{\boldsymbol{\omega}}_k^T & \cos\dfrac{\Delta\theta_k}{2} \end{bmatrix}\mathbf{q}_k^+ \tag{2.94}$$

which is the analytic solution of the quaternion kinematic equation with the constant angular velocity assumption, where $\Delta\theta_k = \|\hat{\boldsymbol{\omega}}_k\|\Delta t$. In addition, the bias is propagated by

$$\hat{\boldsymbol{\beta}}_{k+1}^- = \hat{\boldsymbol{\beta}}_k^+ \tag{2.95}$$

Finally, the error covariance, P, is propagated by

$$P_{k+1}^- = \Phi_k P_k^+ \Phi_k^T + Q \tag{2.96}$$

A summary of the quaternion and bias estimation Kalman filter is given in Algorithm 2.5.

We must be careful in the implementation of (2.94) for the sinusoidal terms. The sinusoidal terms divided by the magnitude of the angular velocity diverge to infinity when the magnitude is equal to zero or close to zero. To avoid this issue, robustly implement the sinusoidal terms considering the possibility to be divided by zero. For example,

$$\frac{\sin(\Delta\theta_k/2)}{\|\hat{\boldsymbol{\omega}}_k\|} = \begin{cases} \dfrac{\sin(\|\hat{\boldsymbol{\omega}}_k\|\Delta t/2)}{\|\hat{\boldsymbol{\omega}}_k\|} & \text{for } \|\hat{\boldsymbol{\omega}}_k\| \geq \varepsilon \\ \Delta t/2 & \text{for } \|\hat{\boldsymbol{\omega}}_k\| < \varepsilon \end{cases}$$

where ε is a small positive number to be chosen appropriately, and the small angle approximation, i.e. $\sin\theta \approx \theta$ for $\theta \approx 0$, is used.

Algorithm 2.5 Extended Kalman filter for quaternion estimation

1: Initialize

$$\hat{\mathbf{q}}_0^+,\ \hat{\beta}_0^+ = \mathbf{0},\ \hat{\omega}_0 = \tilde{\omega},\ P_0^+ = E\left(\Delta\mathbf{x}_0\, \Delta\mathbf{x}_0^T\right)$$

where, typically, the bias is set to zero, and the angular velocity is set to the gyro measurement.

2: **for** $k = 1, 2, \ldots$ **do**

3: Correct the gyro measurement using (2.69): $\hat{\omega}(t_k) = \tilde{\omega}(t_k) - \hat{\beta}(t_k)$

4: **Prediction:** from t_{k-1} to t_k

5: Propagate the quaternion using (2.94), $\hat{\mathbf{q}}_k^-$

6: Propagate the bias using (2.95), $\hat{\beta}_k^-$

7: Propagate the error covariance using (2.96), P_k^-

8: **Update:** when the measurement, $\mathbf{z}_k$, is available at t_k

9: Update K_k, P_k^+, and $\Delta\mathbf{x}_k$ using (2.90)

10: Update the bias using (2.92), β_k^+

11: Update the quaternion using (2.93), $\mathbf{q}_k^+$

12: **Substitute:** when no measurement, $\mathbf{z}_k$, is available at t_k

$$\mathbf{q}_k^+ = \mathbf{q}_k^-,\ \beta_k^+ = \beta_k^-,\ P_k^+ = P_k^-$$

13: **end for**

2.3 Attitude Dynamics and Control

2.3.1 Dynamics Equation of Motion

The angular velocity, ω, in the kinematic equation, (2.5), evolves governed by the attitude dynamic equation of motion from Newton's second law (N2L) of motion as follows:

$$\dot{\omega} = -J^{-1}\omega \times (J\omega) + J^{-1}\sum_i \mathbf{M}_i \tag{2.97}$$

where J is the moment of inertia of vehicle and $\mathbf{M}_i$ is the i-th external torque, which could also be the torque from attitude actuators. The moment of inertia is defined by

$$J = \begin{bmatrix} J_{11} & -J_{12} & -J_{13} \\ -J_{12} & J_{22} & -J_{23} \\ -J_{13} & -J_{23} & J_{33} \end{bmatrix}$$
$$= \begin{bmatrix} \int_m y^2 + z^2\, dm & -\int_m xy\, dm & -\int_m xz\, dm \\ -\int_m xy\, dm & \int_m x^2 + z^2\, dm & -\int_m yz\, dm \\ -\int_m xz\, dm & -\int_m yz\, dm & \int_m x^2 + y^2\, dm \end{bmatrix} \tag{2.98}$$

where x, y, and z are the coordinates of dm in the body frame. J is symmetric and positive definite, i.e. all eigenvalues are strictly greater than zero. The positive definiteness of J corresponds to the positiveness of mass. The definitions for the off-diagonal terms, J_{ij} for $i \neq j$, do not include the minus signs. It is also common to define the off-diagonal terms including the minus signs. *Hence, the off-diagonal term definitions must be checked when the moment of inertia matrix is provided by or to others.*

Unlike the mass in translational motions, however, the moment of inertia is a matrix. It has an evident difference compared to the mass. The direction of the vector multiplied by the mass remains the same, but the direction of the vector multiplied by the moment of inertia generally changes. For the translational velocity, $\mathbf{v}$, the linear momentum, $m\mathbf{v}$, remains the same direction as the direction of $\mathbf{v}$. For the rotational velocity, $\boldsymbol{\omega}$, the direction of the angular momentum (or the moment of momentum), $J\boldsymbol{\omega}$, is, in general, different from the direction of the angular velocity vector. Only for some special cases, for example, J is a sphere and its moment of inertia is equal to αI_3, where α is a positive constant, and the direction of $J\boldsymbol{\omega} = \alpha\boldsymbol{\omega}$ is the same as the angular velocity vector direction.

Another property of the inertia matrix is

$$J_{ii} < J_{jj} + J_{kk} \tag{2.99}$$

where (i, j, k) is (1,2,3), (2,1,3), or (3,1,2). The summation of any two diagonal terms must be greater than the other term. The following proves that for $(i, j, k) = (2, 1, 3)$

$$\begin{aligned} J_{11} + J_{33} &= \int_m y^2 + z^2 \, dm + \int_m x^2 + y^2 \, dm \\ &= \int_m x^2 + z^2 \, dm + 2\int_m y^2 \, dm = J_{22} + 2\int_m y^2 \, dm > J_{22} \end{aligned} \tag{2.100}$$

and a similar way proves for the other two cases.

The N2L of motion applies to the rotational motion exactly the same as the law to translational motion as indicated by the equivalency between them in Table 2.1, where $\mathbf{h}$ is the angular momentum or the moment of momentum and $\mathbf{M}$ is torque or moment. While N2L takes the derivative by time in the inertial coordinates, the angular momentum, $\mathbf{h}$, is written in the body coordinate, i.e. J is in the body coordinates as it is a property of satellite, and $\boldsymbol{\omega}$ is *expressed* in the body frame. The *transport theorem* applies to take the time derivative of a vector expressed in the body frame relative to the inertial frame as follows (Schaub and Junkins, 2018):

$$\frac{d(\cdot)^N}{dt} = \frac{d(\cdot)^B}{dt} + \boldsymbol{\omega}_B \times (\cdot)^B \tag{2.101}$$

Table 2.1 Translational and rotational motions.

Property	Translation	Rotation
Mass	$m = 2\,[\text{kg}]$	$J = \begin{bmatrix} 15 & -0.2 & -1.2 \\ -0.2 & 20.5 & 0.3 \\ -1.2 & 0.3 & 13 \end{bmatrix} [\text{kg m}^2]$
Velocity	$\mathbf{v} = \begin{bmatrix} 2 & -2.5 & 3 \end{bmatrix}^T$ [m/s]	$\boldsymbol{\omega} = \begin{bmatrix} -0.3 & 5 & 3 \end{bmatrix}^T$ [rad/s]
Momentum	$\mathbf{p} = m\mathbf{v}$	$\mathbf{h} = J\boldsymbol{\omega}$
Force	$\mathbf{F}$	$\mathbf{M}$
N2L	$\mathbf{F} = d\mathbf{p}/dt$	$\mathbf{M} = d\mathbf{h}/dt$

where $d(\cdot)^N/dt$ or $d(\cdot)^B/dt$ indicates that the time derivative is in the inertia frame or the body frame, respectively, and $\boldsymbol{\omega}_B$ is the angular velocity of B with respect to N *expressed in the body frame*. Most of the confusion in the rotational dynamics occurs in distinguishing the following differences:

- a vector, $\mathbf{x}$, *expressed* in N or B: $\mathbf{x}_N$ or $\mathbf{x}_B$
- a vector *differentiated* by time in N or B

All the following four combinations are possible:

$$\frac{d(\mathbf{x}_N)^N}{dt}, \frac{d(\mathbf{x}_N)^B}{dt}, \frac{d(\mathbf{x}_B)^N}{dt}, \text{or } \frac{d(\mathbf{x}_B)^B}{dt}$$

For example, a vector $\mathbf{x}$ could be expressed in B, and we want to calculate the derivative in the inertial frame, i.e. $d(\mathbf{x}_B)^N/dt$.

The time derivative in N2L must be in the inertial frame, i.e. $d(\cdot)^N/dt$, while the vector to be differentiated could be expressed in the inertial frame or the body frame. The derivative of the angular momentum expressed in the body frame must be in the inertial frame as follows:

$$\frac{d(\mathbf{h}_B)^N}{dt} \tag{2.102}$$

To obtain the derivative expressed in the body frame, apply the transport theorem to the derivative of the angular momentum, $\mathbf{h}$,

$$\frac{d(\mathbf{h}_B)^N}{dt} = \frac{d(\mathbf{h}_B)^B}{dt} + \boldsymbol{\omega}_B \times (\mathbf{h}_B)^B \tag{2.103}$$

In many engineering mechanics books, this derivative relationship is written in a compressed form as

$$\frac{d\mathbf{h}}{dt} = \dot{\mathbf{h}} + \boldsymbol{\omega} \times \mathbf{h} \tag{2.104}$$

where the confusion originates. We avoid using this notation.

Applying Newton's law, the following equation of motion is obtained:

$$\mathbf{M}_B = \frac{d(\mathbf{h}_B)^N}{dt} = \frac{d[(J\boldsymbol{\omega})_B]^N}{dt} + \boldsymbol{\omega}_B \times (J\boldsymbol{\omega})_B = J\dot{\boldsymbol{\omega}}_B + \boldsymbol{\omega}_B \times (J\boldsymbol{\omega}_B) \tag{2.105}$$

As it is clear that all vectors in the leftmost and the rightmost sides are in the body frame including $\boldsymbol{\omega}$ and its derivative, we drop the superscript, B, indicating the body frame expressions, and Euler's rigid-body rotational dynamic equation of motion is given by

$$J\dot{\boldsymbol{\omega}} = -\boldsymbol{\omega} \times (J\boldsymbol{\omega}) + \mathbf{M} \tag{2.106}$$

where $\mathbf{M}$ is from the environment or actuators.

The equation of motion given in (2.106) is solved with the quaternion kinematics, (2.5). We modify MATLAB Program 2.1 and Python Program 2.2 to include the rotational dynamics equation of motion. The angular velocity, $\boldsymbol{\omega}$, is obtained by solving (2.106). The moment of inertia matrix is given by

$$J = \begin{bmatrix} 0.005 & -0.001 & 0.004 \\ -0.001 & 0.006 & -0.002 \\ 0.004 & -0.002 & 0.004 \end{bmatrix} [\text{kg m}^2], \tag{2.107}$$

which are the approximate values for a quadcopter unmanned aerial vehicle (UAV) in Lee (2012). Let the initial angular velocity be zero and the initial quaternion equal to $[0, 0, 0, 1]^T$. Assume that the torque, $\mathbf{M}$, in the body coordinates is given by

$$\mathbf{M}(t) = \begin{bmatrix} 0.00001 + 0.0005 \sin 2t \\ -0.00002 + 0.0001 \cos 0.1t \\ -0.0001 \end{bmatrix} [\text{Nm}] \tag{2.108}$$

where t is in seconds.

2.3.1.1 MATLAB

The MATLAB program is given in 2.20. The differential equation includes $d\mathbf{q}/dt$ and $d\boldsymbol{\omega}/dt$. In line 18, the initial condition includes $\mathbf{q}(0)$ and $\boldsymbol{\omega}(0)$. In line 20, the max step size option is set to 0.01, which restricts the integration time interval smaller than 0.01. It would prevent sparse time resolution that might occur in some cases. The two functions implemented, $d\mathbf{q}/dt$ and $d\boldsymbol{\omega}/dt$, are separate, and they merge into one set of differential equations in *dqdt_dwdt* function, which is passed to *ode45*. The time histories of the quaternion and the angular velocity are shown in Figure 2.21.

```
1  clear;
2
3  init_time = 0; % [s]
4  final_time = 10.0; % [s]
5  time_interval = [init_time final_time];
6
7  J_inertia = [0.005 -0.001  0.004;
8               -0.001  0.006 -0.002;
9                0.004 -0.002  0.004];
10         % vehicle moment of inertia [kg m^2]
11 J_inv = inv(J_inertia);
12
13 J_inv_J_inertia = [J_inertia; J_inv];
14
15 q0 = [0 0 0 1]'; % initial quaternion
16 w0 = [0 0 0]'; % initial angular velocity
17
18 state_0 = [q0; w0]; % states including q0 and omega0
19
20 ode_options = odeset('RelTol',1e-6,'AbsTol',1e-9, 'MaxStep', 0.01);
21 [tout,state_out] = ode45(@(time,state) dqdt_dwdt(time,state,
       J_inv_J_inertia), ...
22     time_interval, state_0, ode_options);
23
24 qout = state_out(:,1:4);
25 wout = state_out(:,5:7);
26
27
28 % : (plot commands are left as an exercise)
29
30
31 function dstate_dt = dqdt_dwdt(time,state,J_inv_J_inertia)
32
33     q_current = state(1:4);
34     q_current = q_current(:)/norm(q_current);
35
36     w_current = state(5:7);
37     w_current = w_current(:);
38
39     J_inertia = J_inv_J_inertia(1:3,:);
40     inv_J = J_inv_J_inertia(4:6,:);
41
42     M_torque = [     0.00001+0.0005*sin(2*time);
43                     -0.00002+0.0001*cos(0.1*time);
44                     -0.0001]; % [Nm]
45
46     dqdt = dqdt_attitude_kinematics(q_current,w_current);
```

```
47      dwdt = dwdt_attitude_dynamics(w_current, J_inertia, inv_J,
            M_torque);
48
49      dstate_dt = [dqdt(:); dwdt(:)];
50
51  end
52
53  function dqdt = dqdt_attitude_kinematics(q_true,w_true)
54      q_true = q_true(:);
55      w_true = w_true(:);
56
57      wx = [  0             -w_true(3)   w_true(2);
58              w_true(3)     0            -w_true(1);
59              -w_true(2)    w_true(1)    0];
60
61      Omega = [   -wx          w_true;
62                  -w_true'     0];
63
64      dqdt = 0.5*Omega*q_true;
65  end
66
67  function dwdt = dwdt_attitude_dynamics(w_true, J_inertia,
        inv_J_inertia, M_torque)
68      w_true = w_true(:);
69      Jw = J_inertia*w_true;
70      Jw_dot = -cross(w_true,Jw) + M_torque(:);
71
72      dwdt = inv_J_inertia*Jw_dot;
73  end
```

Program 2.20 (MATLAB) Simulate rotational dynamics of quadcopter UAV

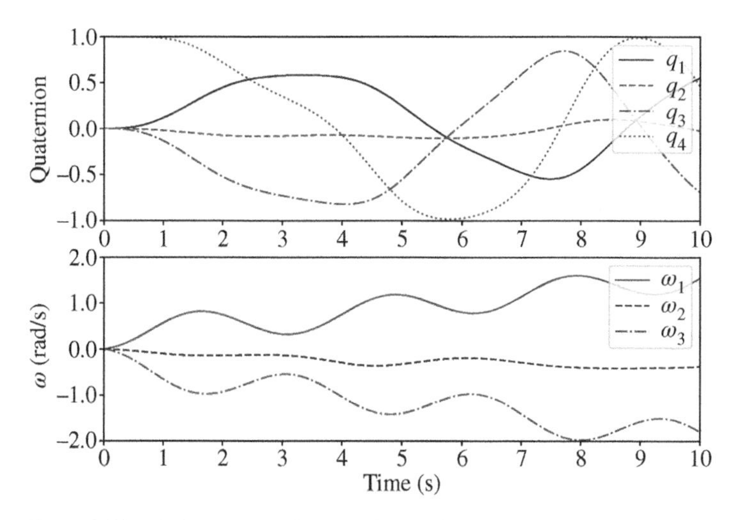

Figure 2.21 Attitude dynamics and kinematics solutions.

2.3.1.2 Python

While function definitions in MATLAB m-scripts must appear at the end of the scripts with no restrictions in the order of appearance, a Python function definition in Python scripts must appear before the function is used. In line 68 of Program 2.21, the relative tolerance, the absolute tolerance, and the maximum integration step size are set using *rtol*, *atol*, and *max_step*, respectively. 'J_inertia' and 'J_inv' for J and J^{-1} are stacked vertically using the numpy *vstack* command to make the 6×3 matrix, which is passed to the ODE solver using *args* argument in line 69.

```
1  import numpy as np
2  from numpy import linspace
3  from scipy.integrate import solve_ivp
4
5  init_time = 0 # [s]
6  final_time = 10.0 # [s]
7  num_data = 200
8  tout = linspace(init_time, final_time, num_data)
9
10 J_inertia = np.array([[0.005, -0.001, 0.004],
11                       [-0.001, 0.006, -0.002],
12                       [0.004, -0.002, 0.004]])
13 J_inv = np.linalg.inv(J_inertia)
14 J_inv_J_inertia = np.vstack((J_inertia,J_inv))
15
16 q0 = np.array([0,0,0,1])
17 w0 = np.array([0,0,0])
18
19 state_0 = np.hstack((q0,w0))
20
21 def dqdt_attitude_kinematics(q_true, w_true):
22     quat=q_true
23
24     wx=np.array([[0,            -w_true[2],       w_true[1]],
25                  [w_true[2],    0,               -w_true[0]],
26                  [-w_true[1],   w_true[0],        0]])
27
28     Omega_13 = np.hstack((-wx,np.resize(w_true,(3,1))))
29     Omega_4  = np.hstack((-w_true,0))
30     Omega = np.vstack((Omega_13, Omega_4))
31
32     dqdt = 0.5*(Omega@quat)
33
34     return dqdt
35
36
37 def dwdt_attitude_dynamics(w_true,J_inertia,inv_J_inertia, M_torque
       ):
38
39     Jw = J_inertia@w_true
40     Jw_dot = -np.cross(w_true,Jw) + M_torque
41
```

```
42        dwdt = inv_J_inertia@Jw_dot
43
44        return dwdt
45
46
47  def dqdt_dwdt(time, state, J_inv_J_inertia):
48
49        q_current = state[0:4]
50        q_current = q_current/np.linalg.norm(q_current)
51        w_current = state[4::]
52
53        J_inertia = J_inv_J_inertia[0:3,:]
54        J_inv = J_inv_J_inertia[3::,:]
55
56        M_torque = np.array([0.00001+0.0005*np.sin(2*time),
57                             -0.00002+0.0001*np.cos(0.75*time),
58                             -0.0001])
59
60        dqdt = dqdt_attitude_kinematics(q_current, w_current)
61        dwdt = dwdt_attitude_dynamics(w_current, J_inertia, J_inv,
              M_torque)
62
63        dstate_dt = np.hstack((dqdt,dwdt))
64        return dstate_dt
65
66  sol = solve_ivp(dqdt_dwdt, (init_time, final_time), state_0,
67        t_eval=tout,
68        r_tol=1e-6, atol=1e-9, max_step=0.01,
69        args=(J_inv_J_inertia,))
70
71  qout = sol.y[0:4,:]
72  wout = sol.y[4::,:]
73
74  #    ⋮ (plot commands are left as an exercise)
```

Program 2.21 (Python) Simulate rotational dynamics of quadcopter UAV

2.3.2 Actuator and Control Algorithm

Quadcopters have been widely used for many purposes typically equipped with four identical motors driving four propellers as shown in Figure 2.22. From the free-body diagram in Figure 2.22, the torques in $\mathbf{x}_B$ and $\mathbf{y}_B$ directions are given by Beard (2008)

$$M_1 = L\left(F_\ell - F_r\right) = L\,\Delta F_{\ell r} \tag{2.109a}$$

$$M_2 = L\left(F_f - F_b\right) = L\,\Delta F_{fb} \tag{2.109b}$$

where L is the length from the centre of the body frame to the centre of the propeller, F_f and F_b are the forces generated by the propellers in the forward and the backward of the quadcopter, respectively, and F_ℓ and F_r are the forces in the

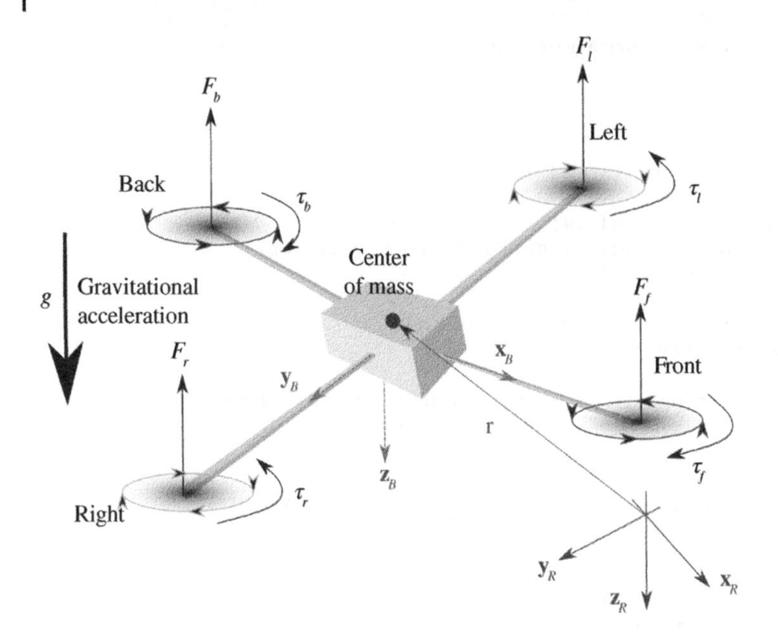

Figure 2.22 Quadcopter UAV with the four actuators, where the body frame and the reference frame are indicated by B and R, respectively, and the positive direction of $\mathbf{z}_R$ is the same as $\mathbf{z}_B$ so that they are aligned in the primary stabilized attitude.

left-hand side and the right-hand side of the quadcopter, respectively. The torque in $\mathbf{z}_B$ is produced by the reaction torque by the motor torque as follows:

$$M_3 = \tau_f + \tau_b - \tau_\ell - \tau_r = \Delta\tau \tag{2.110}$$

where τ_r, τ_ℓ, τ_f, and τ_b are the motor torque acting on the body of the quadcopter, whose direction is the opposite to the rotational direction of each propeller. The rotational directions of the front and the back propellers are the opposite of $\mathbf{z}_B$. The directions of the left and the right propellers are in the same direction as $\mathbf{z}_B$. The desired moment, M_1, M_2, and M_3, the desired force differences, $\Delta F_{\ell r}$ and ΔF_{bf}, and the desired torque difference, $\Delta\tau$, are to be determined by the control algorithm to be designed later. For the given desired forces and torques, the forces and the torques for the four motors are determined by

$$\begin{bmatrix} F \\ M_1 \\ M_2 \\ M_3 \end{bmatrix}_{\text{desired}} = \begin{bmatrix} -1 & -1 & -1 & -1 & 0 \\ 0 & 0 & L & -L & 0 \\ L & -L & 0 & 0 & 0 \\ 0 & 0 & 0 & 0 & 1 \end{bmatrix} \begin{bmatrix} F_f \\ F_b \\ F_\ell \\ F_r \\ \sum_{i=s} \tau_i \end{bmatrix} \tag{2.111}$$

where $s = \{f, b, \ell, r\}$. We concern only the attitude control here. There would be a desirable magnitude of the sum of the forces, F, to keep the same position.

The forces and the torques are proportional to the square of the rotational angular velocity of the propeller–motor as follows (Khodja et al., 2017):

$$F_s = C_T\, \omega_s^2 \tag{2.112a}$$

$$\tau_s = C_D\, \omega_s^2 \tag{2.112b}$$

where C_T is the thruster coefficient of the propeller, C_D is the drag coefficient of the propeller, and ω_s is the angular velocity of the motor relative to the quadcopter body frame for s, which would be measured by a rotary encoder or it would be calculated using the standard electrical motor equation providing the relationship between the electrical voltage signal to the motor and the motor torque. Substitute (2.112) into (2.111)

$$\begin{aligned}
\begin{bmatrix} F \\ M_1 \\ M_2 \\ M_3 \end{bmatrix}_{\text{desired}} &= \begin{bmatrix} -1 & -1 & -1 & -1 & 0 \\ 0 & 0 & L & -L & 0 \\ L & -L & 0 & 0 & 0 \\ 0 & 0 & 0 & 0 & 1 \end{bmatrix} \begin{bmatrix} C_T & 0 & 0 & 0 \\ 0 & C_T & 0 & 0 \\ 0 & 0 & C_T & 0 \\ 0 & 0 & 0 & C_T \\ C_D & C_D & -C_D & -C_D \end{bmatrix} \begin{bmatrix} \omega_f^2 \\ \omega_b^2 \\ \omega_\ell^2 \\ \omega_r^2 \end{bmatrix} \\
&= \begin{bmatrix} -C_T & -C_T & -C_T & -C_T \\ 0 & 0 & LC_T & -LC_T \\ LC_T & -LC_T & 0 & 0 \\ C_D & C_D & -C_D & -C_D \end{bmatrix} \begin{bmatrix} \omega_f^2 \\ \omega_b^2 \\ \omega_\ell^2 \\ \omega_r^2 \end{bmatrix}_{\text{desired}}
\end{aligned}$$

As the matrix is invertible, the desired squared-angular velocity is obtained by

$$\begin{bmatrix} \omega_f^2 \\ \omega_b^2 \\ \omega_\ell^2 \\ \omega_r^2 \end{bmatrix}_{\text{desired}} = \frac{1}{4} \begin{bmatrix} \frac{-1}{C_T} & 0 & \frac{2}{L\,C_T} & \frac{1}{C_D} \\ \frac{-1}{C_T} & 0 & \frac{-2}{L\,C_T} & \frac{1}{C_D} \\ \frac{-1}{C_T} & \frac{2}{L\,C_T} & 0 & \frac{-1}{C_D} \\ \frac{-1}{C_T} & \frac{-2}{L\,C_T} & 0 & \frac{-1}{C_D} \end{bmatrix} \begin{bmatrix} F \\ M_1 \\ M_2 \\ M_3 \end{bmatrix}_{\text{desired}} \tag{2.113}$$

The inversion can be easily obtained using the symbol manipulation functions in MATLAB or Python and is left as an exercise. Note that the multiplication of the inversion matrix with the desired force and torques does not guarantee to provide positive desired angular velocities. If negative values occur in ω_s^2, they are set to zero.

Set the propeller–motor parameters as follows: $C_T = 8.8 \times 10^{-7}$ [N/(rad/s)^2], $C_D = 11.3 \times 10^{-8}$ [Nm/(rad/s)^2], and $L = 0.127$ m (Khodja et al., 2017).

2.3.2.1 MATLAB Program

To include the motor model, Program 2.22 is updated from Program 2.20. Additional variables for motor characteristics are defined and included in 'quadcopter_uav', which is the variable to be passed to the differential equation solver. The variable includes different size variables, i.e. 3 × 3 inertia matrix and three scalar constants. The cell type, created using '{…}' brackets, can include different types and/or sizes of data. To access each element in the cell, for example, the third element is retrieved by 'quadcopter_uav{3}'. The variables defined in line 47 and below access each value in the cell.

The control algorithm for attitude stabilization, for example, has not been designed. It is important to consider all necessary components in simulation design and clearly distinguish what is simulated and what is control algorithm to be implemented. Inside the function, *dqdt_dwdt*, we declare the place to implement the controller. The functionality of the controller includes the following:

- calculate the desired force and torque
- convert them into the desired angular velocity of the motor
- send the desired angular velocity command to the motor

These are functionally implemented in the controller section. In the motor simulation, the motor receives the command angular velocity and converts it on line 94 to the motor angular velocity. An ideal motor is assumed such that the motor angular velocity is the same as the command. For a realistic motor model, we can implement a detailed motor model or simply implement a first-order model as follows:

$$\dot{\omega}_m = \frac{(-\omega_m + \omega_{\text{command}})}{\tau_m} \tag{2.114}$$

where ω_m and ω_{command} are the motor angular velocity and the command velocity, respectively, and τ_m is the motor time constant to be identified by some experiment.

> **Global variable**: Global variables should not be used for just immediate convenience. It should only be used when unavoidable or when other methods are too complex to implement.

In the simulation, we are mainly interested in the force and torques generated by the motor. However, extracting these values from the differential equations is not straightforward because they are not part of the state variables. There are several methods to extract these values from the integrator. Here, we use a *global* variable, which can be accessed from everywhere in the program. It is common practice to avoid using global variables as it is difficult to track changes in global variables. On the other hand, it is useful in this case to extract the inside values. Global variables should not be used for just immediate convenience, however. It is only used when unavoidable or when other methods are too complex. Line 13 in Program 2.22 defines the global variable to store time, force, and three torque values. To distinguish between global variables and local variables, global variable names start 'global_'. In MATLAB, global variables are created with the keyword, *global*. In the next line after the global variable is created, it is initialized. Also, in the function, 'dqdt_dwdt', where the global variable is used, the global variable is declared.

The four values are stored in each time instance. We do not know how many time instances are there between the initial simulation time and the final simulation time controlled by the integrator. To avoid excessive data size, the 'dt_save' variable to control the save time interval is defined as equal to 0.05 seconds. Inside the definition of the differential equation, shown in line 74, the data is only stored to the global variable when the current time is greater than the previous time by 0.05 seconds. After the simulation completes, the global variable values are transferred to local variables, and the global variable is deleted in line 32.

```
1
2    ⋮ (omitted)
3
4  C_T = 8.8e-7; %motor thruster coefficient [N/(rad/s)^2]
5  C_D = 11.3e-8;%motor drag coefficient [Nm/(rad/s)^2]
6  L_arm = 0.127;%length from centre of quadcopter to motor [m]
7
8  quadcopter_uav = {J_inertia, J_inv, C_T, C_D, L_arm};
9
10   ⋮ (omitted)
11
12 % use global variables only for saving values
13 global global_motor_time_FM_all;
14 global_motor_time_FM_all = [];
15
16 % minimum time interval for saving values to the global
17 dt_save = 0.05; %[s]
18
19 %% simulation
20 ode_options = odeset('RelTol',1e-6,'AbsTol',1e-9, 'MaxStep', 0.01);
21 [tout,state_out] = ode45(@(time,state) dqdt_dwdt(time,state,
       quadcopter_uav,dt_save), ...
22     time_interval, state_0, ode_options);
23
24 qout = state_out(:,1:4);
```

```
25 wout = state_out(:,5:7);
26
27 time_Motor = global_motor_time_FM_all(:,1);
28 Force_Motor = global_motor_time_FM_all(:,2);
29 Torque_Motor = global_motor_time_FM_all(:,3:5);
30
31 % clear all global variables
32 clearvars -global
33
34   ⋮ (omitted)
35
36 %% functions
37 function dstate_dt = dqdt_dwdt(time, state ,quadcopter_uav, dt_save)
38
39     global global_motor_time_FM_all;
40
41     q_current = state(1:4);
42     q_current = q_current(:)/norm(q_current);
43
44     w_current = state(5:7);
45     w_current = w_current(:);
46
47     J_inertia = quadcopter_uav{1};
48     inv_J = quadcopter_uav{2};
49     C_T = quadcopter_uav{3};
50     C_D = quadcopter_uav{4};
51     L_arm = quadcopter_uav{5};
52
53     %-----------------------------------
54     % Begin: this part is controller
55     %-----------------------------------
56     M_Desired = [      0.00001+0.0005*sin(2*time);
57                       -0.00002+0.0001*cos(0.1*time);
58                       -0.0001]; %[N]
59
60     mg = 10; %[N]
61     F_M_desired = [-mg; M_Desired];
62
63     w_motor_fblr_squared_desired = propeller_motor_FM2w_conversion(
          F_M_desired, ...
64        C_T, C_D, L_arm);
65     w_motor_fblr_desired = sqrt(w_motor_fblr_squared_desired);
66     %-----------------------------------
67     % End: this part is controller
68     %-----------------------------------
69
70     % Motor Force & Torque
71     FM_Motor = propeller_motor_actuator(C_T, C_D, L_arm,
          w_motor_fblr_desired);
72     M_torque = FM_Motor(2:4);
73
74     if time < 1e-200
75         global_motor_time_FM_all = [time FM_Motor(:)'];
```

```
76        elseif time > global_motor_time_FM_all(end,1)+dt_save
77            global_motor_time_FM_all = [global_motor_time_FM_all; time
                  FM_Motor(:)'];
78        end
79
80
81        % Kinematics & Dynamics
82        dqdt = dqdt_attitude_kinematics(q_current,w_current);
83        dwdt = dwdt_attitude_dynamics(w_current, J_inertia, inv_J,
              M_torque);
84
85        dstate_dt = [dqdt(:); dwdt(:)];
86
87  end
88
89    ⋮ (omitted)
90
91  function FM_Motor = propeller_motor_actuator(C_T,C_D,L_arm,
        w_command)
92
93        % assume perfect motor angular velocity control
94        w_motor = w_command(:);
95
96        F_fblr = C_T*(w_motor.^2);
97        tau_fblr = C_D*(w_motor.^2);
98
99        F_motor = sum(F_fblr);
100       M_motor = [ L_arm*(F_fblr(3)-F_fblr(4));
101                   L_arm*(F_fblr(2)-F_fblr(1));
102                   sum(tau_fblr(1:2))-sum(tau_fblr(3:4))];
103
104       FM_Motor = [F_motor; M_motor];
105 end
106
107   ⋮ (omitted)
```

Program 2.22 (MATLAB) Simulate rotational dynamics of quadcopter UAV with propeller–motor actuator model

2.3.2.2 Python

Program 2.23 is the Python program with the motor simulation model for quadcopter dynamics.

We define the global variable using the keyword, *global*. It must be defined also inside the function, 'dqdt_dwdt(⋅)', which uses the variable. In the lines from 64, the global variable values are transferred to the local variables, and the global variable is deleted using the command *del*. The simulation of the force and the torques from the four motors is shown in Figure 2.23.

```
1
2  : (omitted)
3
4  # use global variables only for saving values
5  global global_motor_time_FM_all
6
7  # minimum time interval for saving values to the global
8  dt_save = 0.05
9
10 : (omitted)
11
12 def dqdt_dwdt(time,state,J_inv_J_inertia,Motor_para,dt_save):
13
14     global global_motor_time_FM_all
15
16     q_current = state[0:4]
17     q_current = q_current/np.linalg.norm(q_current)
18     w_current = state[4::]
19
20     J_inertia = J_inv_J_inertia[0:3,:]
21     J_inv = J_inv_J_inertia[3::,:]
22     C_T = Motor_para[0]
23     C_D = Motor_para[1]
24     L_arm = Motor_para[2]
25
26     #-----------------------------------
27     # Begin: this part is controller
28     #-----------------------------------
29     M_Desired = np.array([0.00001+0.0005*np.sin(2*time),
30                          -0.00002+0.0001*np.cos(0.75*time),
31                          -0.0001])
32     mg = 10.0 #[N]
33     F_M_Desired = np.hstack([-mg, M_Desired])
34
35     w_motor_fblr_squared_desired = propeller_motor_FM2w_conversion(
           F_M_Desired, C_T, C_D, L_arm)
36     w_motor_fblr_desired = np.sqrt(w_motor_fblr_squared_desired)
37     #-----------------------------------
38     # End: this part is controller
39     #-----------------------------------
40
41     # Motor Force & Torque
42     FM_Motor = propeller_motor_actuator(C_T, C_D, L_arm,
           w_motor_fblr_desired)
43     M_torque = FM_Motor[1::]
44
45     current_data = np.hstack((time,FM_Motor,))
46     if time < 1e-200:
47         global_motor_time_FM_all = current_data.reshape(1,5)
48     elif time > global_motor_time_FM_all[-1,0]+dt_save:
49         global_motor_time_FM_all = np.vstack((
               global_motor_time_FM_all,current_data,))
50
```

```
51        dqdt = dqdt_attitude_kinematics(q_current, w_current)
52        dwdt = dwdt_attitude_dynamics(w_current, J_inertia, J_inv,
              M_torque)
53
54        dstate_dt = np.hstack((dqdt,dwdt))
55        return dstate_dt
56
57  # solve ode
58  sol = solve_ivp(dqdt_dwdt, (init_time, final_time), state_0,
59      t_eval=tout, atol=1e-9, rtol=1e-6, max_step=0.01,
60      args=(J_inv_J_inertia, Motor_para, dt_save,))
61  qout = sol.y[0:4,:]
62  wout = sol.y[4::,:]
63
64  time_Motor = global_motor_time_FM_all[:,0]
65  Force_Motor = global_motor_time_FM_all[:,1]
66  Torque_Motor = global_motor_time_FM_all[:,2::]
67  del global_motor_time_FM_all
68
69  : (omitted)
```

Program 2.23 (Python) Simulate rotational dynamics of quadcopter UAV with propeller–motor actuator model

2.3.2.3 Attitude Control Algorithm

There are a plethora of attitude control algorithms for aircraft, satellites, UAVs, etc. The quaternion feedback control is one of the commonly used attitude controllers, particularly for satellites (Wie, 2008). It has the general form that could

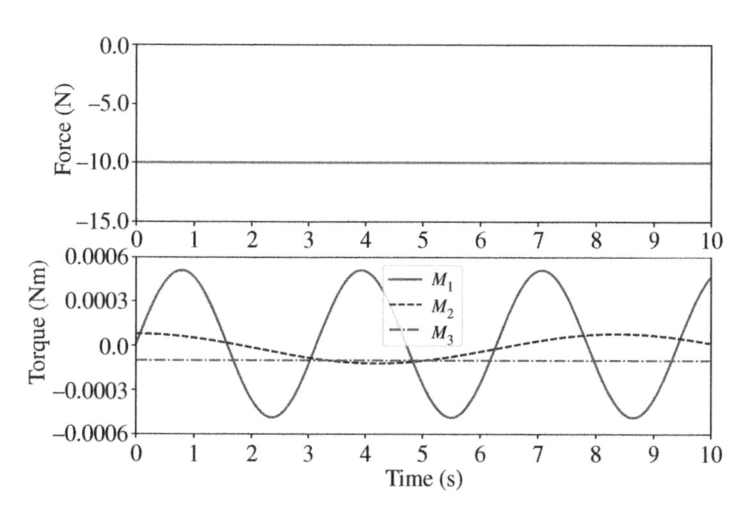

Figure 2.23 Quadcopter motor force and torques.

also be used for quadcopter attitude stabilization. The quaternion feedback control is given by

$$\mathbf{u} = -K\mathbf{q}_{13} - C\boldsymbol{\omega} - \boldsymbol{\omega} \times (J\boldsymbol{\omega}) \tag{2.115}$$

where the quaternion represents the attitude of quadcopter body frame, B, with respect to the reference frame, R, in Figure 2.22, K and C are the control gain positive definite matrices, i.e. symmetric and all eigenvalues are positive, and the last term in the right-hand side is to cancel the non-linear gyroscopic effect in the dynamics. The desired equilibrium point is

$$\mathbf{q}_{13}^{\text{eq}} = \begin{bmatrix} 0 & 0 & 0 \end{bmatrix}^T, \quad q_4^{\text{eq}} = 1, \quad \omega_1^{\text{eq}} = 0, \quad \omega_2^{\text{eq}} = 0, \quad \omega_3^{\text{eq}} = 0,$$

To provide stability conditions at the equilibrium point, firstly, define a candidate Lyapunov function, V, as a function of all states as follows:

$$V = V(\mathbf{q}_{13}, q_4, \boldsymbol{\omega}),$$

which must satisfy the following two conditions:

$$V(\mathbf{q}_{13}, q_4, \boldsymbol{\omega}) = 0, \text{if and only if } \mathbf{q}_{13} = \mathbf{q}_{13}^{\text{eq}},\ q_4 = q_4^{\text{eq}}, \boldsymbol{\omega} = \boldsymbol{\omega}^{\text{eq}} \tag{2.116a}$$

$$V(\mathbf{q}_{13}, q_4, \boldsymbol{\omega}) > 0, \text{otherwise} \tag{2.116b}$$

In dynamical systems, the total energy-like function satisfies the above conditions. Let

$$V = \frac{1}{2}\boldsymbol{\omega}^T K^{-1} J\boldsymbol{\omega} + \mathbf{q}_{13}^T\mathbf{q}_{13} + (q_4 - 1)^2 \tag{2.117}$$

where K is symmetric, and K^{-1} is also symmetric. Take the derivative V with respect to time

$$\frac{dV}{dt} = \boldsymbol{\omega}^T K^{-1} J\dot{\boldsymbol{\omega}} + 2\mathbf{q}_{13}^T\dot{\mathbf{q}}_{13} + 2(q_4 - 1)\dot{q}_4 \tag{2.118}$$

Substitute

$$J\dot{\boldsymbol{\omega}} = -\boldsymbol{\omega} \times (J\boldsymbol{\omega}) + \mathbf{u} = -K\mathbf{q}_{13} - C\boldsymbol{\omega} \tag{2.119a}$$

$$\dot{\mathbf{q}}_{13} = \frac{1}{2}\left(-[\boldsymbol{\omega}\times]\mathbf{q}_{13} + \boldsymbol{\omega}q_4\right) \tag{2.119b}$$

$$\dot{q}_4 = -\frac{1}{2}\boldsymbol{\omega}^T\mathbf{q}_{13} \tag{2.119c}$$

into dV/dt

$$\begin{aligned} \frac{dV}{dt} &= \boldsymbol{\omega}^T K^{-1}\left[-K\mathbf{q}_{13} - C\boldsymbol{\omega}\right] + \mathbf{q}_{13}^T\left(-[\boldsymbol{\omega}\times]\,\mathbf{q}_{13} + \boldsymbol{\omega}q_4\right) - (q_4 - 1)\,\boldsymbol{\omega}^T\mathbf{q}_{13} \\ &= \cancel{-\boldsymbol{\omega}^T\mathbf{q}_{13}} - \boldsymbol{\omega}^T K^{-1} C\boldsymbol{\omega} - \underbrace{\cancel{\mathbf{q}_{13}^T[\boldsymbol{\omega}\times]\mathbf{q}_{13}}}_{=0} + \cancel{\mathbf{q}_{13}^T\boldsymbol{\omega}q_4} - \cancel{q_4\boldsymbol{\omega}^T\mathbf{q}_{13}} + \cancel{\boldsymbol{\omega}^T\mathbf{q}_{13}} \\ &= -\boldsymbol{\omega}^T K^{-1} C\boldsymbol{\omega} \end{aligned} \tag{2.120}$$

where the fact that $\mathbf{a}$ is perpendicular to $\mathbf{b} \times \mathbf{a}$ and $\mathbf{a} \cdot (\mathbf{b} \times \mathbf{a}) = 0$ is used for the term to be zero. Hence, the derivative of V is less than or equal to zero, i.e. $\dot{V} \leq 0$, as $K^{-1}C$ is positive definite.

If it should have proven that $\dot{V} < 0$, i.e. the negative definite, the stability proof is completed, and it concludes that the equilibrium point is asymptotically stable. The equality sign in $\dot{V} \leq 0$ implies that $\dot{V}$ could be zero for non-equilibrium $\mathbf{q}$ as $\dot{V}$ is not a function of $\mathbf{q}$. We have to consider how the subset, where $\dot{V} = 0$, in the state-space, would be defined. For $\dot{V}$ equal to zero, $\boldsymbol{\omega}$ must be zero by the equation derived for $\dot{V}$. Substitute $\boldsymbol{\omega} = 0$ and $\dot{\boldsymbol{\omega}} = 0$ into (2.119a), and we found that $\mathbf{q}_{13}$ must be zero. Because of the unit norm condition for the quaternion, q_4 must be 1. The subset called *the invariant set* is composed of only one point, i.e. the equilibrium point. *LaSalle's invariance theorem* says that all state trajectories converge to the invariant set (Slotine et al., 1991). Therefore, conclude that the equilibrium point is asymptotically stable.

The desired torque is set to equal to the control input calculated by the quaternion feedback controller as follows:

$$\begin{bmatrix} M_1 \\ M_2 \\ M_3 \end{bmatrix}_{\text{desired}} = \mathbf{u}$$

2.3.2.4 Altitude Control Algorithm

Although the main purpose of the control design in this section is for stabilizing the attitude, the translational motion is simulated using

$$\dot{\mathbf{r}} = \mathbf{v}$$

$$\dot{\mathbf{v}} = \begin{bmatrix} 0 \\ 0 \\ g \end{bmatrix} + \frac{1}{m} C_{BR}^T(\mathbf{q}) \begin{bmatrix} 0 \\ 0 \\ -\sum_{i \in S} F_i \end{bmatrix}$$

where $\mathbf{r}$ is 3×1 to represent the coordinates of the quadcopter position in the three-dimensional reference frame, $g = 9.81\,\text{m/s}^2$, the mass of quadcopter, m, is equal to 0.45 kg, and $C_{BR}(\mathbf{q})$ is calculated using (2.37). To achieve the desired altitude, $h_{\text{desired}} = -(z_R)_{\text{desired}}$, the desired force in $\mathbf{z}_R$-direction cancels the gravitational force and generates the feedback force proportional to the error in the altitude and the velocity as follows:

$$(f_z)_{\text{desired}} = \begin{bmatrix} 0 \\ 0 \\ -mg \end{bmatrix} + k_1 \left[(z_R)_{\text{desired}} - z_R \right] + k_2 \left[(\dot{z}_R)_{\text{desired}} - \dot{z}_R \right]$$

where k_1 and k_2 are the control gains to be designed. The two control gains are multiplied by the position difference and the velocity difference. This is the proportional derivative (PD) control, and it is one of the most common controllers in the industry. In addition, the desired force for the sum of four propellers is obtained by

$$F_{\text{desired}} = \begin{bmatrix} 0 & 0 & 1 \end{bmatrix} C_{BR}(\mathbf{q}) \begin{bmatrix} 0 \\ 0 \\ (f_z)_{\text{desired}} \end{bmatrix}$$

Designing the position tracking control for quadcopter is not the same control problem as this attitude stabilization control design. The desired attitude must be calculated to direct the propeller force in the desired direction. Some examples are found in Beard (2008), Yu et al. (2019), and Xie et al. (2021).

2.3.2.5 Simulation

Set the initial conditions as follows:

$$\mathbf{q}(0) = \frac{1}{\sqrt{4}} \begin{bmatrix} 1 & 1 & -1 & 1 \end{bmatrix}^T$$

$$\boldsymbol{\omega}(0) = \begin{bmatrix} 0.1 & -0.2 & 0.1 \end{bmatrix}^T \text{ [rad/s]}$$

$$\mathbf{r}(0) = \begin{bmatrix} 0 & 0 & -30 \end{bmatrix}^T \text{ [m]}$$

$$\mathbf{v}(0) = \begin{bmatrix} 0 & 0 & 0 \end{bmatrix}^T \text{ [m/s]}$$

and the desired altitude and the velocity in $\mathbf{z}_R$ direction are set to −30 m and 0 m/s, respectively.

From several trials and errors to achieve reasonable convergent speeds and control input magnitudes, the control gains are designed as

$$K = 0.01I_3, \quad C = 0.001I_3, \quad k_1 = 0.1, \quad k_2 = 0.5$$

Translational dynamics is coupled with attitude dynamics, and the above control design ignores the coupling. If k_1 and k_2 gains are too big, the coupling effect would be significant, and the quadcopter would be unstable. Figure 2.24a shows that the attitude converges to the desired equilibrium point in about 30 seconds. Figure 2.24b shows that the altitude is settled to 30 m in a similar time length. The velocity for $\mathbf{x}_R$ and $\mathbf{y}_R$ directions, v_1 and v_2, converge to non-zero values, and the quadcopter will fly away from the starting location. The force and the torques generated by the motors are shown in Figure 2.25a, and the corresponding angular velocity of the motor is shown in Figure 2.25b. If the motor angular velocities are too high, they could be adjusted by

- tuning the control gains: K, C, k_1, and k_2
- changing the propeller: C_T and C_D
- resizing the quadcopter: m and L

2.3.2.6 MATLAB

The controller is given in Program 2.24. Updating the simulation with the control and producing Figures 2.24 and 2.25 are left as an exercise.

```
1  function w_motor_fblr_desired =
       quaternion_feedback_and_altitude_control(q_current, ...
2      w_current, rv_current, J_inertia, C_T, C_D, L_arm, C_BR, ...
3      mass_quadcopter, grv_acce)
4
5      zR_desired = -30; %[m]
6      zdotR_desired = 0; %[m/s]
7      K_qf = 0.01*eye(3);
8      C_qf = 0.001*eye(3);
9      k1 = 0.1;
10     k2 = 0.5;
11
12     q_13 = q_current(1:3); q_13 =q_13(:);
13     w = w_current(:);
14
15     Fmg_R = grv_acce*mass_quadcopter; %[N]
16     Falt_R = k1*(zR_desired-rv_current(3))+k2*(zdotR_desired-
           rv_current(6));
17     F_desired_R = [0;0;-Fmg_R+Falt_R];
18     F_desired_B = C_BR*F_desired_R;
19
20     u_qf = -K_qf*q_13 - C_qf*w - cross(w,J_inertia*w);
21     M_Desired = u_qf;
22
23     F_M_desired = [F_desired_B(3); M_Desired];
24
25     w_motor_fblr_squared_desired = propeller_motor_FM2w_conversion(
           F_M_desired, ...
26         C_T, C_D, L_arm);
27
28     w_motor_fblr_desired = sqrt(w_motor_fblr_squared_desired);
29
30 end
```

Program 2.24 (MATLAB) Quaternion feedback control and PD altitude control

2.3.2.7 Robustness Analysis

Once the control design is completed, the next step is the robustness analysis. When the control gains are chosen or optimized, various physical parameters are assumed to be certain values. In reality, the values would be different from the assumed ones. Before the controller is finally implemented into the on-board computer of an autonomous vehicle, it needs to be checked if the performance is acceptable for some ranges of the uncertain variable changes. Consider the settling time, t_s, of the attitude as the performance measure of the quadcopter

$$\|\mathbf{q}_{13}(t)\|_2 \leq 0.01\,[\text{rad/s}], \quad \text{for } t \geq t_s \tag{2.121}$$

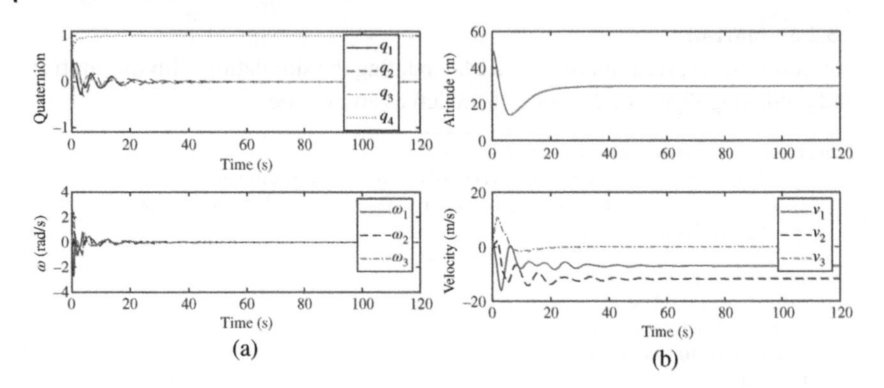

Figure 2.24 (a) Attitude stabilization using the quaternion feedback control and (b) altitude stabilization using the PD control, where the altitude is the opposite sign to $\mathbf{z}_R$.

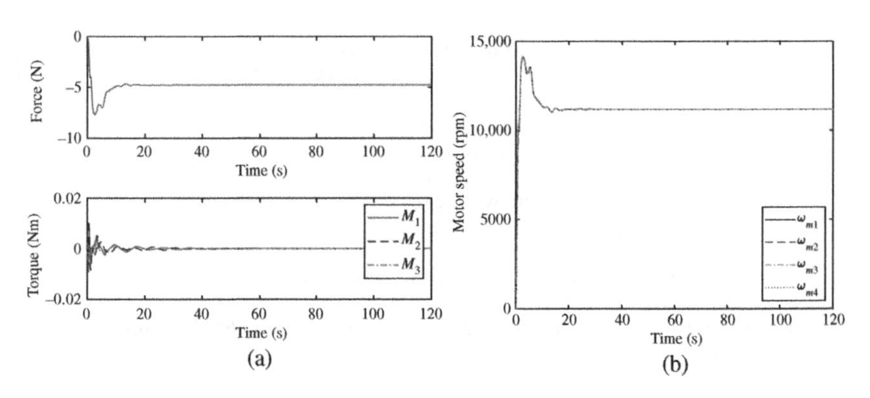

Figure 2.25 (a) Total propeller force and the torque for each direction in the body coordinates and (b) quadcopter motor angular velocity in rpm.

where $\|\cdot\|_2$ is the 2-norm, i.e. $(\mathbf{q}_{13}^T\mathbf{q}_{13})^{1/2}$. The settling time t_s is calculated in the following program:

```
q13 = qout(:,1:3);
q13_norm = sqrt(sum(q13.^2,2));
q13_ts = int32(q13_norm>0.01);
q13_ts = cumsum(q13_ts);
q13_ts = tout(q13_ts==q13_ts(end));
ts = q13_ts(1);
```

where $\|\mathbf{q}_{13}\|$ is calculated, *int32* changes true/false into the integer 1 or 0, *cumsum* returns the cumulative summation, for example,

$$\text{cumsum}([1\ 0\ 3-2\ 5]) \Rightarrow [1\ 1\ 4\ 2\ 7] \tag{2.122}$$

and the settling time index is the first index, whose corresponding cumulative summation is equal to the last value. The main assumption in this calculation is that the settling time exits in the simulation time interval. If the calculated settling time is close to the final time of the simulation, the final time must increase to confirm the calculated settling time as the trajectory might fluctuate out of the bound, 0.01, after the calculated settling time.

In this robustness analysis, the uncertainty is introduced to the moment of inertia matrix as follows:

$$J = \begin{bmatrix} 0.005 + \delta J_1 & -0.001 & 0.004 \\ -0.001 & 0.006 + \delta J_2 & -0.002 \\ 0.004 & -0.002 & 0.004 + \delta J_3 \end{bmatrix} \text{ [kg m}^2\text{]}, \tag{2.123}$$

and δJ_i for $i = 1, 2, 3$ is sampled from the normal distribution with the standard deviation equal to $0.002\,\text{kg}{\cdot}\text{m}^2$. As the moment of inertia matrix, J, must satisfy the positive definiteness and (2.99) inequality, these conditions must be checked for the perturbed J and reject the perturbation if the conditions are not met. The rejection rate would not be high as the random sample space is three dimensional. It becomes very high for high-dimension sample space, and we need better sampling methods. Random sampling methods for control design and analysis have been discussed thoroughly in Tempo et al. (2012). The pseudo-code of the random-sampling-based robustness analysis is given in Algorithm 2.6.

Figure 2.26 shows that the settling time is shorter than about 80 seconds in the worst case, where the uncertainty magnitude is up to $0.008\,\text{kg m}^2$, where the number of random samples is 10,000. Based on the results, probabilistic conclusions are drawn. The reader would find interesting concepts and methods in Tempo et al. (2012) such as the *Chernoff bound*.

Algorithm 2.6 Robustness analysis using random samplings

1: Set the nominal moment of inertia, $\bar{J}$
2: Initialize the rest of simulation parameters
3: **for** i **do**dx = 1, 2, ..., (Maximum Simulation Number)
4:     **while** the positive definiteness and (2.99) are not met **do**
5:         Generate random δJ_1, δJ_2, and δJ_3 from some distribution
6:         Set the moment of inertia, $J = \bar{J} + \text{diag}[\delta J_1, \delta J_2, \delta J_3]$
7:         Check J for the conditions
8:     **end while**
9:     Run the simulation
10:     Calculate the settling time
11: **end for**

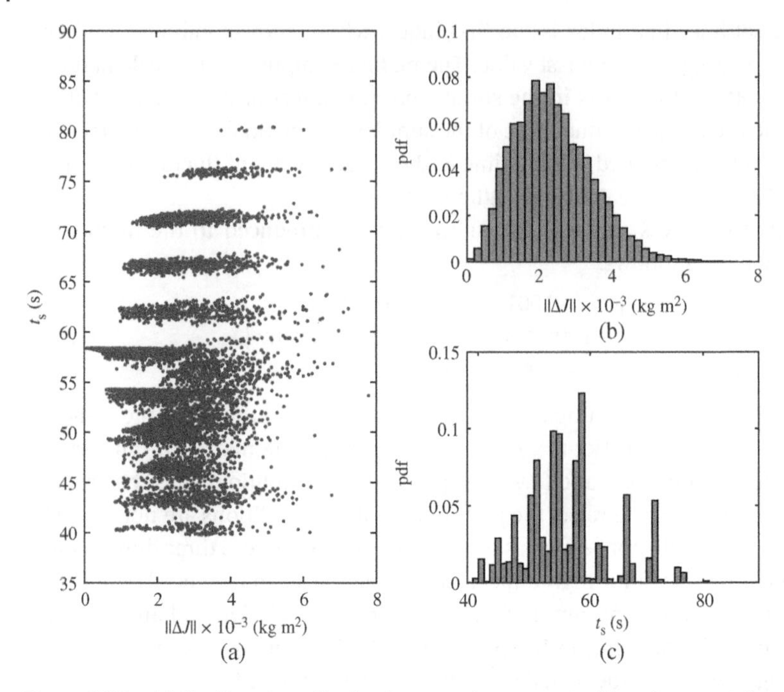

Figure 2.26 (a) Settling time distribution over the moment of inertia uncertainty distribution; (b) the moment of inertia distribution; (c) Settling time distribution.

2.3.2.8 Parallel Processing

As the modern Central Processing Units (CPU) has multiple cores, multiple simulations can be run in parallel, and the total computation time can be reduced. The random-sampling-based robustness analysis, i.e. Monte Carlo Simulations, is *embarrassingly parallel* as each simulation is completely independent of each other and no effort is required to make the simulations running in parallel. In the MATLAB parallel computing toolbox, the parallelization of the robustness analysis is done simply replacing *for* into *parfor* in the implementation of Algorithm 2.6. The number of cores used by *parfor* is set in the MATLAB preference configuration.

Parallel computation in Python is simple enough but needs some minor modifications from the non-parallel simulation program. Multiple executions of a function are performed using the *multiprocessing* library in Python as shown in Program 2.25. Firstly, it imports *Pool* from *multiprocessing*. Secondly, using *map* function executes *robustness_analysis_MC* function implemented in Program 2.26, which solves the quadcopter closed-loop dynamics for each perturbed moment of inertia. The first argument of *map* function is the name of the function to be executed in parallel, and the second argument is the list of values, where

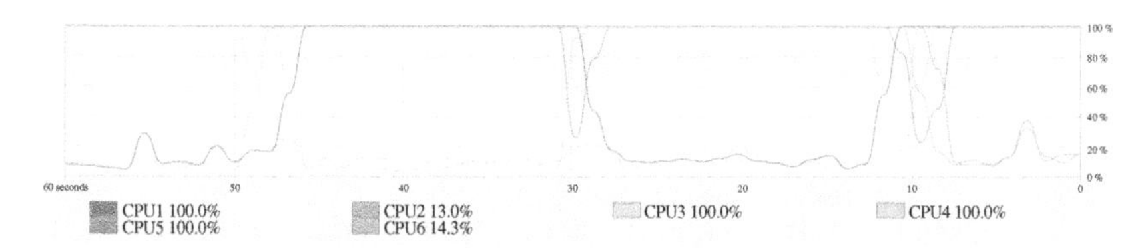

Figure 2.27 CPU loading shows four of them running in 100%.

the size of the list is the number of simulations equal to 3000 for this example. The number of CPU cores to be used is set to 4. Finally, the results are stored in the return variable, *result*. *result* is a list, whose element is tuple as (t_s, $\|\delta J\|$), which are the return values by the function, *robustness_analysis_MC*.

```
1 from multiprocessing import Pool
2 num_MC = 3000
3 num_core = 4
4 with Pool(num_core) as p:
5     result = p.map(robustness_analysis_MC , range(num_MC))
```

Program 2.25 (Python) Parallel processing using multiprocessing library

It takes about 93 minutes of the total computation time for the sequential simulation one by one for the total number of simulations equal to 3000. The same simulations using the parallel processing with 4 CPU cores takes about 25 minutes, which is only 27% of 93 minutes, slightly more than 25% of 93 minutes as expected. As shown in Figure 2.27, four CPUs are running 100%.

In Program 2.25, the Python keyword, *with*, is used to call *map* function. This is one of the Python programming methods frequently used. In a simple explanation, *with* makes the program finish properly even for some abnormal situations. For example, while *robustness_analysis_MC* executes in parallel, some calculations would cause *map* function to crash and hang. Even for the case, the *with* statement makes the function properly end so that there are no further memory usages from the ghost process, which would slow down the computer.

```
1 def robustness_analysis_MC(MC_id):
2
3     J_inertia = np.array([[0.005, -0.001, 0.004],
4                           [-0.001, 0.006, -0.002],
5                           [0.004, -0.002, 0.004]])
6
7     not_find_dJ = True
8
9     np.random.seed()
10
11     while not_find_dJ:
12
13         dJ = np.diag(0.002*np.random.randn(3))
```

```
14
15            J_inertia_perturbed = J_inertia + dJ
16
17            pd_cond = np.min(np.linalg.eig(J_inertia_perturbed)[0])>0
18            j3_cond = J_inertia_perturbed[0,0]+J_inertia_perturbed[1,1]
                  > J_inertia_perturbed[2,2]
19            j2_cond = J_inertia_perturbed[0,0]+J_inertia_perturbed[2,2]
                  > J_inertia_perturbed[1,1]
20            j1_cond = J_inertia_perturbed[1,1]+J_inertia_perturbed[2,2]
                  > J_inertia_perturbed[0,0]
21
22            if pd_cond and j1_cond and j2_cond and j3_cond:
23                not_find_dJ = False
24
25        dJ_norm = np.linalg.norm(dJ)
26        J_inv_perturbed = np.linalg.inv(J_inertia_perturbed)
27
28        quadcopter_uav=(J_inertia_perturbed, J_inv_perturbed, C_T, C_D,
              L_arm)
29
30        sol = solve_ivp(dqdt_dwdt_drvdt, (init_time, final_time),
              state_0, t_eval=tout,
31                        atol=1e-9, rtol=1e-6, max_step=0.01,
32                        args=(quadcopter_uav,))
33        qout = sol.y[0:4,:]
34
35        q13=qout[0:3,:]
36        q13_norm = np.sqrt((np.sum(q13**2,axis=0)))
37        q13_ts = (q13_norm>0.01)*np.ones(num_data)
38        q13_ts = np.cumsum(q13_ts)
39        q13_ts = tout[q13_ts==q13_ts[-1]]
40        ts = q13_ts[0]
41
42        print(f'#{MC_id}: {np.linalg.norm(dJ):6.5f}, {q13_ts:4.2f}\n')
43
44        return ts, dJ_norm
```

Program 2.26 (Python) Quadcopter closed-loop dynamics for perturbed moment of inertia

The random number generations inside the function to be executed in parallel using *map* must be done with the explicit random seed function call as line 9, Program 2.26. Otherwise, all 3000 executions use the same random seed, and the random perturbation in line 13 generates the same number all the time. To generate a different random number at each iteration, we call the seed function explicitly before each run.

The print command in line 42 is introduced from Python 3.6. This is very convenient to print formatted numbers, strings, etc. The number to be printed is simply placed inside the curly bracket, and the format is indicated next to the colon.

Exercises

Exercise 2.1 (MATLAB) Plot Figure 2.5 for the following three tolerance cases: (i) RelTol=1e-3, AbsTol=1e-6; (ii) RelTol=1e-6, AbsTol=1e-9; and (iii) RelTol=1e-9, AbsTol=1e-12. Also indicate and place the legend and the axis labels the same as shown in the figure. Hint: Use odeset function to set the tolerances and pass the option to ode45.

Exercise 2.2 (MATLAB) Draw Figure 2.11 using surf command and execute the following in the MATLAB command prompt to see what changes they make in the pdf plot: (i) shading flat, (ii) shading interp, and (iii) shading faceted.

Exercise 2.3 (Python) Change rstride and cstride values in line 38 Program 2.12 to positive integers larger than 1 and confirm the functionality of the two optional arguments in Figure 2.12. Also, try to change the colormap to one of these, {'plasma', 'inferno', 'magma'}, and check the effect for each colormap.

Exercise 2.4 (MATLAB/Python) Implement MATLAB or Python program for the bias noise generation using Algorithm 2.1 from 0 to 120 seconds, where $\sigma_{\beta x} = \sigma_{\beta y} = 0.01\ ^\circ/\sqrt{\text{s}}$, $\sigma_{\beta z} = 0.02\ ^\circ/\sqrt{\text{s}}$, and $\Delta t_k = 0.1$ seconds. The initial bias, $\beta(t_0 = 0)$, is taken from the uniform distribution between $-0.03\,^\circ/\text{s}$ and $+0.03\,^\circ/\text{s}$.

Exercise 2.5 Construct the simulation algorithm for the body coordinates and the sensor coordinates, which are shown in Figure 2.28. How $\mathbf{r}^1$ in the body coordinates could be obtained from $\mathbf{r}^1$ in the sensor coordinates?

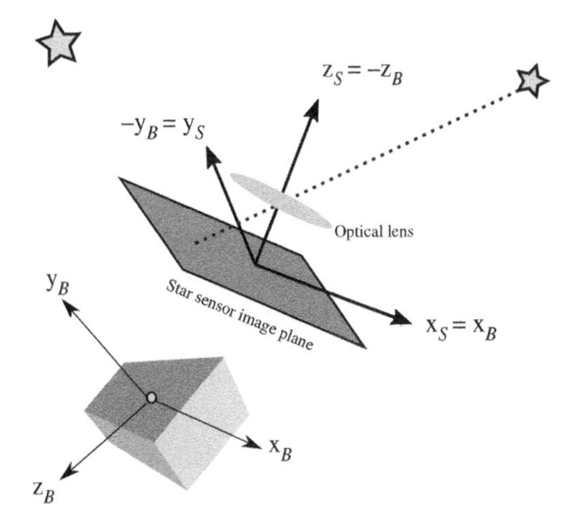

Figure 2.28 The star sensor frame and the body frame.

Exercise 2.6 (MATLAB/Python) Implement the functions to convert the direction cosine matrix to the quaternion and the quaternion to the direction cosine matrix using (2.37) and Algorithm 2.2.

Exercise 2.7 (MATLAB/Python) The following five stars in the star catalogue database are identified by the star sensor:

$$\mathbf{r}_R^1 = \begin{bmatrix}-0.6794 & -0.3237 & -0.6586\end{bmatrix}_R^T$$
$$\mathbf{r}_R^2 = \begin{bmatrix}-0.7296 & 0.5858 & 0.3528\end{bmatrix}_R^T$$
$$\mathbf{r}_R^3 = \begin{bmatrix}-0.2718 & 0.6690 & -0.6918\end{bmatrix}_R^T$$
$$\mathbf{r}_R^4 = \begin{bmatrix}-0.2062 & -0.3986 & 0.8936\end{bmatrix}_R^T$$
$$\mathbf{r}_R^5 = \begin{bmatrix}0.6858 & -0.7274 & -0.0238\end{bmatrix}_R^T$$

and the star measurements are given by

$$\mathbf{r}_B^1 = \begin{bmatrix}-0.2147 & -0.7985 & 0.5626\end{bmatrix}_B^T$$
$$\mathbf{r}_B^2 = \begin{bmatrix}-0.7658 & 0.4424 & 0.4667\end{bmatrix}_B^T$$
$$\mathbf{r}_B^3 = \begin{bmatrix}-0.8575 & -0.4610 & -0.2284\end{bmatrix}_B^T$$
$$\mathbf{r}_B^4 = \begin{bmatrix}0.4442 & 0.6863 & 0.5758\end{bmatrix}_B^T$$
$$\mathbf{r}_B^5 = \begin{bmatrix}0.9407 & -0.1845 & -0.2847\end{bmatrix}_B^T$$

Calculate C_{BR} using (2.50) to show

$$C_{BR} = \begin{bmatrix}0.4885 & -0.8403 & 0.2350 \\ 0.1096 & 0.3263 & 0.9389 \\ -0.8656 & -0.4329 & 0.2516\end{bmatrix}$$

Exercise 2.8 (MATLAB/Python) Implement QUEST shown in Algorithm 2.3 and run a test simulation introducing multiple stars in the reference and the body coordinates.

Exercise 2.9 The star sensor frame is given in Figure 2.28. Extend the simulation implemented in Exercise 2.8 to include the effect of the sensor field of view. The star sensor can only detect the stars when the angle between $\mathbf{z}_S$ and the star vector, which is from the origin of the sensor coordinates to each star, is less than or equal to 12°.

Exercise 2.10 (MATLAB/Python) Complete the MATLAB Program 2.15 or the Python Program 2.16 to implement the Kalman filter for (2.64) and (2.67) using Algorithm 2.4. Plot Figures 2.19 and 2.20.

Exercise 2.11 Derive the quaternion error dynamics given in (2.70).

Exercise 2.12 (MATLAB/Python) Complete the MATLAB Program starting from Programs 2.17 and 2.18 or the Python Program from Program 2.19 to calculate the covariance matrix given in (2.80).

Exercise 2.13 (MATLAB/Python) Implement the Kalman filter given in Algorithm 2.5 using MATLAB or Python. Use the vector measurements given in Exercise 2.7 for the optical sensor with the assumption that they are always visible from the sensor.

Exercise 2.14 (Attitude Dynamics) Complete Programs 2.20 and 2.21 to produce Figure 2.21.

Exercise 2.15 (MATLAB/Python) Obtain the matrix in (2.113) by the symbolic matrix inversions in MATLAB or Python.

Exercise 2.16 (MATLAB/Python) Update the simulator Program 2.22 or 2.23 to include the first-order motor model given in (2.114), where τ_m is equal to 0.01 seconds.

Exercise 2.17 (MATLAB/Python) Update the simulator Program 2.22 or 2.23 to plot the angular velocities of the four motors.

Exercise 2.18 (MATLAB) Using Program 2.24, update line 94 of Program 2.22 and produce Figures 2.24 and 2.25.

Exercise 2.19 (MATLAB/Python) With the uncertain moment of inertia given in (2.123), perform the robustness analysis described in the paragraph and produce Figure 2.26.

Bibliography

Ahmad Bani Younes and Daniele Mortari. Derivation of all attitude error governing equations for attitude filtering and control. *Sensors*, 19(21):4682, 2019. https://doi.org/10.3390/s19214682.

Randal W. Beard. Quadrotor dynamics and control. *Brigham Young University*, 19(3):46–56, 2008.

C. T. Chen. *Linear System Theory and Design, Third Edition, International Edition.* OUP USA, 2009. ISBN 9780195392074. https://books.google.co.uk/books?id=D9nXSAAACAAJ.

J. L. Crassidis. Angular velocity determination directly from star tracker measurements. *Journal of Guidance Control and Dynamics - J GUID CONTROL DYNAM*, 25:11, 2002. https://doi.org/10.2514/2.4999.

J. L. Crassidis and J. L. Junkins. *Optimal Estimation of Dynamic Systems*. Chapman & Hall/CRC Applied Mathematics & Nonlinear Science. CRC Press, 2011. ISBN 9781439839867.

R. L. Farrenkopf. Analytic steady-state accuracy solutions for two common spacecraft attitude estimators. *Journal of Guidance and Control*, 1(4):282–284, 1978. https://doi.org/10.2514/3.55779. https://arc.aiaa.org/doi/abs/10.2514/3.55779.

M. A. A. Fialho and D. Mortari. Theoretical limits of star sensor accuracy. *Sensors*, 19(24), 2019. https://doi.org/10.3390/s19245355.

M. S. Grewal and A. P. Andrews. Applications of kalman filtering in aerospace 1960 to the present [historical perspectives]. *IEEE Control Systems Magazine*, 30(3):69–78, 2010. https://doi.org/10.1109/MCS.2010.936465.

R. E. Kalman. A new approach to linear filtering and prediction problems. *Journal of Basic Engineering*, 82(1):35–45, 1960. ISSN 0021-9223. https://doi.org/10.1115/1.3662552.

M. A. Khodja, M. Tadjine, M. S. Boucherit, and M. Benzaoui. Experimental dynamics identification and control of a quadcopter. In *2017 6th International Conference on Systems and Control (ICSC)*, pages 498–502, May 2017. https://doi.org/10.1109/ICoSC.2017.7958668.

Taeyoung Lee. Robust adaptive attitude tracking on so(3) with an application to a quadrotor UAV. *IEEE Transactions on Control Systems Technology*, 21(5):1924–1930, 2012.

E. J. Lefferts, F. L. Markley, and M. D. Shuster. Kalman filtering for spacecraft attitude estimation. *Journal of Guidance, Control, and Dynamics*, 5(5):417–429, 1982. https://doi.org/10.2514/3.56190.

F. Landis Markley and John L. Crassidis. *Fundamentals of Spacecraft Attitude Determination and Control*. Springer, 2014.

W. H. Press, S. A. Teukolsky, W. T. Vetterling, and B. P. Flannery. *Numerical Recipes 3rd Edition: The Art of Scientific Computing*. Cambridge University Press, 2007. ISBN 9780521880688.

H. Schaub and J. L. Junkins. *Analytical Mechanics of Space Systems*. AIAA Education Series. American Institute of Aeronautics and Astronautics, 2003. ISBN 9781600860270.

Hanspeter Schaub and John L. Junkins. *Analytical Mechanics of Space Systems*. AIAA Education Series, Reston, VA, 4th edition, 2018. https://doi.org/10.2514/4.105210.

K. S. Shanmugan and A. M. Breipohl. *Random Signals: Detection, Estimation and Data Analysis*. John Wiley & Sons, 1988. ISBN 978-0471815556.

M. D. Shuster. Maximum likelihood estimation of spacecraft attitude. *Journal of the Astronautical Sciences*, 37:79–88, 1989. https://ci.nii.ac.jp/naid/20001198295/en/.

M. D. Shuster and S. D. Oh. Three-axis attitude determination from vector observations. *Journal of Guidance and Control*, 4(1):70–77, 1981. https://doi.org/10.2514/3.19717.

J. J. E. Slotine, J. J. E. Slotine, and W. Li. *Applied Nonlinear Control*. Prentice Hall, 1991. ISBN 9780130408907. https://books.google.co.uk/books?id=cwpRAAAAMAAJ.

Roberto Tempo, Giuseppe Calafiore, and Fabrizio Dabbene. *Randomized Algorithms for Analysis and Control of Uncertain Systems: With Applications*. Springer Science & Business Media, 2012.

The LaTeX Project Team. LaTeX - A document preparation system. https://www.latex-project.org/, 2020. Accessed: 2020-10-23.

The MathWorks. Row-major and column-major array layout. https://uk.mathworks.com/help/coder/ug/what-are-column-major-and-row-major-representation-1.html, 2020. Accessed: 2020-11-26.

N. G. Van Kampen. *Stochastic Processes in Physics and Chemistry*. North Holland, 2007.

Grace Wahba. A least squares estimate of satellite attitude. *SIAM Review*, 7(3):409–409, 1965. https://doi.org/10.1137/1007077.

B. Wie. *Space Vehicle Dynamics and Control*. AIAA Education Series. American Institute of Aeronautics and Astronautics, 2008. ISBN 9781563479533.

Oliver J. Woodman. An introduction to inertial navigation. Technical Report UCAM-CL-TR-696, University of Cambridge, Computer Laboratory, August 2007. https://www.cl.cam.ac.uk/techreports/UCAM-CL-TR-696.pdf.

W. Xie, G. Yu, D. Cabecinhas, R. Cunha, and C. Silvestre. Global saturated tracking control of a quadcopter with experimental validation. *IEEE Control Systems Letters*, 5(1):169–174, 2021. https://doi.org/10.1109/LCSYS.2020.3000561.

G. Yu, D. Cabecinhas, R. Cunha, and C. Silvestre. Nonlinear backstepping control of a quadrotor-slung load system. *IEEE/ASME Transactions on Mechatronics*, 24(5):2304–2315, 2019. https://doi.org/10.1109/TMECH.2019.2930211.

3

Autonomous Vehicle Mission Planning

Mission planning is an essential part of autonomous vehicles. Because of the wide practical applicabilities and the simplicity path planning problem, we consider the following simplified dynamics:

$$\dot{\mathbf{r}} = \mathbf{v} \tag{3.1a}$$

$$\dot{\mathbf{v}} = \mathbf{u} \tag{3.1b}$$

where $\mathbf{r}$ is the position vector, $\mathbf{v}$ is the velocity vector, and $\mathbf{u}$ is the command from the mission planner. Solving (3.1) with the mission planner command, $\mathbf{u}$, provides a trajectory, $\mathbf{r}(t)$, from the initial location, $\mathbf{r}(t_0)$, at the initial time, t_0, to the desired location, $\mathbf{r}(t_f)$, at the final time, t_f.

3.1 Path Planning

Many practical path planning problems are in the two-dimensional spatial space. $\mathbf{r}$ is given by the coordinates in $\mathbf{x}$ and $\mathbf{y}$ indicated in Figure 3.1. The vehicle is at the origin of the coordinates, and the desired position is where the attractive potential is, whose potential field is the lowest value at the desired destination, (x_d, y_d).

3.1.1 Potential Field Method

The potential field method introduces artificial potential functions to generate the force, $\mathbf{u}$, in the operation area. The idea is inspired by the physical forces generated by the potential fields. For example, the spring force, $-kx$, where k is the spring constant and x is the distance from the equilibrium, comes from the derivative of the potential function, i.e. $-dV/dh$, where the potential function, V, is equal to $kx^2/2$. The minus sign in $-dV/dx$ indicates that the force is in the direction of the potential energy decreasing.

Dynamic System Modelling and Analysis with MATLAB and Python: For Control Engineers,
First Edition. Jongrae Kim.

Companion Website: www.wiley.com/go/kim/dynamicmodeling

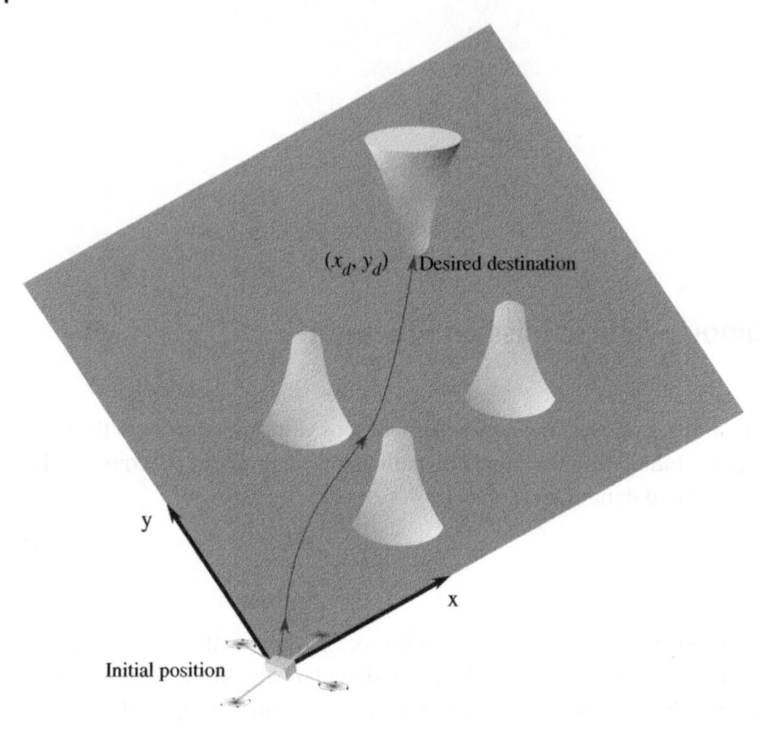

Figure 3.1 Three repulsive potential functions and one attractive potential function.

Figure 3.1 has three repulsive potential functions placed, where the obstacles or hazardous areas for the vehicle to avoid are, and one attractive potential function to pull the vehicle towards the desired destination. Many forms of the potential fields would perform the desired functions, and one of the most common forms of the potential functions is as follows:

$$U_a = \frac{1}{2} k_a \rho_a^2 \tag{3.2a}$$

$$U_r^i = \begin{cases} 0, & \text{for } \rho_r^i > \rho_0^i \\ \frac{1}{2} k_r \left(\frac{1}{\rho_r^i} - \frac{1}{\rho_0^i} \right), & \text{otherwise} \end{cases} \tag{3.2b}$$

where U_a is the attractive potential function, U_r^i is the i-th repulsive potential function for the i-th obstacle, k_a and k_f are the attractive and the repulsive potential strength, respectively, ρ_0^i is the radius of the i-th obstacle size,

$$\rho_a = \sqrt{(x - x_{\text{dst}})^2 + (y - y_{\text{dst}})^2} \tag{3.3a}$$

$$\rho_r^i = \sqrt{(x - x_{\text{ost}}^i)^2 + (y - y_{\text{ost}}^i)^2} \tag{3.3b}$$

(x_{dst}, y_{dst}) is the coordinates of the desired position of the vehicle, and (x^i_{ost}, y^i_{ost}) is the coordinates of the centre of the i-th obstacle.

The total sum of the forces from the potential functions are given by

$$\mathbf{F} = \begin{bmatrix} F_x \\ F_y \end{bmatrix} = -\begin{bmatrix} \dfrac{\partial U_a}{\partial x} + \displaystyle\sum_{i=1}^{N_{ost}} \dfrac{\partial U^i_r}{\partial x} \\ \dfrac{\partial U_a}{\partial y} + \displaystyle\sum_{i=1}^{N_{ost}} \dfrac{\partial U^i_r}{\partial y} \end{bmatrix} \tag{3.4}$$

where N_{ost} is the number of obstacles. The attractive force in the positive $\mathbf{x}$ direction is obtained by

$$-\frac{\partial U_a}{\partial x} = -\frac{\partial}{\partial x}\left(\frac{1}{2}k_a\rho_a^2\right) = -\frac{\partial}{\partial \rho_a}\left(\frac{1}{2}k_a\rho_a^2\right)\frac{\partial \rho_a}{\partial x} = -k_a(x - x_{dst})$$

and the i-th repulsive in the positive $\mathbf{x}$-direction for $\rho^i_r \leq \rho^i_o$

$$-\frac{\partial U^i_r}{\partial x} = -\frac{\partial}{\partial x}\left[\frac{1}{2}k_r\left(\frac{1}{\rho^i_r} - \frac{1}{\rho^i_o}\right)\right] = \frac{1}{2}k_r\left(\frac{1}{\rho^i_r}\right)^2\frac{\partial \rho^i_r}{\partial x} = \frac{k_r(x - x^i_{ost})}{(\rho^i_r)^3} \tag{3.5}$$

We obtain the expression for the forces in the $\mathbf{y}$-direction similarly, and it is left as an exercise in Exercise 3.1.

The attractive force, $k_a(x - x_{dst})$, becomes large when the vehicle is far from the desired location. Hence, we restrict the force magnitude to avoid having a large acceleration. The equation of motion for the $\mathbf{x}$-axis with the saturation is given by

$$\dot{v}_x = u_x = \frac{F_x}{m} = \begin{cases} \text{sat}[-k_a(x - x_{dst}), \overline{w}_x] \text{ for no contact to obstacle} \\ \text{sat}[-k_a(x - x_{dst}), \overline{w}_x] + \dfrac{k_r(x - x^{i^*}_{ost})}{(\rho^{i^*}_r)^3} \end{cases} \tag{3.6}$$

where m is the unit mass. The vehicle meets either no obstacle or only one obstacle at a time among $i^* \in [1, N_{obs}]$. $\overline{w}_x$ is a positive number indicating the possible maximum magnitude of the attractive force. The saturation function is defined by

$$\text{sat}(a, b) = \begin{cases} \text{sgn}(a)b \text{ for } |a| > b \\ a \text{ for } |a| \leq b \end{cases}$$

where b is positive, and the sign function is defined by

$$\text{sgn}(a) = \begin{cases} -1 \text{ for } a < 0 \\ 0 \text{ for } a = 0 \\ +1 \text{ for } a > 0 \end{cases}$$

The saturation function can be simply written as

$$\text{sat}(a, b) = \text{sgn}(a)\min(|a|, b)$$

and this is compact to implement.

For the no contact case, the equation of motion is the mass-spring system with no damping. It will oscillate at the destination point. For example, let $v(0) = 0$, $x_{\text{dst}} = 0$, and $k_a = 1$, and the equation of motion becomes

$$\dot{v}_x = \ddot{x} = -x$$

The solution is simply $x(t) = x(0) \sin t$, which never converges to the desired x. It is fixed by introducing a damping force such as

$$\mathbf{F}_d = -c_d \mathbf{v} \tag{3.7}$$

and the equation of motion becomes

$$\dot{\mathbf{v}} = \frac{1}{m} \left(\mathbf{F} + \mathbf{F}_d \right) \tag{3.8}$$

Consider the following scenario: initially, the vehicle is at $x = 0$, $y = 5$ m, the desired final destination, (x_d, y_d), is equal to (20 m, 5 m), and the four obstacles, whose radius is equal to 2.4 m, is at (5 m, 8 m), (10 m, 5 m), (15 m, 8 m), and (15 m, 2 m) (Chou et al., 2017), and the maximum magnitude of the attractive force is 10.

One important issue to consider in the potential function method is the local equilibrium point, where $\mathbf{F} + \mathbf{F}_d = 0$. Once the vehicle path is stuck in a local equilibrium point, it cannot escape as $\mathbf{v} = 0$. In addition, if the force is normal towards the obstacle surface, the path cannot escape from the obstacle as shown in Figure 3.2. In this scenario, the straight line from the initial position to the desired position shown in the figure aligns to the line from the contact point to the centre of the circular obstacle. There is no tangential component of the force at the contact point, and the velocity in $\mathbf{x}$ direction, v_x, vibrates with high frequencies converging to zero.

There are several improvements and many variations of the potential field method. Waydo and Murray (2003) and Chou et al. (2017) are presenting the streamline function method to remove the local-minima problem. We take an approach that adds a little noise that is relatively small in magnitude compared to attractive and repulsive forces. The equation of motion is finally given by

$$\dot{\mathbf{v}} = \frac{1}{m} \left(\mathbf{F} + \mathbf{F}_d + \mathbf{w}_k \right) \tag{3.9}$$

where $\mathbf{w}_k$ is the noise force.

3.1.1.1 MATLAB

In Program 3.1, the simulation parameters, whose sizes are different from each other, are packed into one cell array, 'sim_para'. The cell array is defined using the braces in line 24. Access the elements of cell is similar to accessing the elements of matrices but using the curly bracket, e.g. 'r_obs', is 'sim_para{2}'.

The integrator tolerances are carefully set using *odeset* so that the computation time is not too long. In many path planning scenarios, we do not need highly

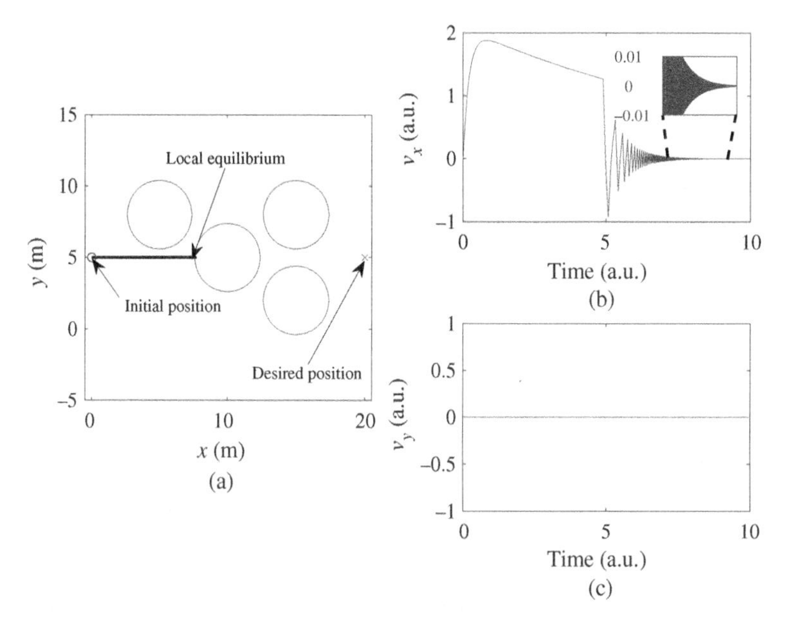

Figure 3.2 (a) A path converges to the local equilibrium point; (b) the horizontal direction velocity; (c) the vertical direction velocity.

accurate integration results as the missions could be achieved with more or less several centimetres of numerical errors in the path commands.

```
1  %% simulation parameters
2  r_vehicle = [0 5]; % initial vehicle position (x,y) [m]
3  v_vehicle = [0 0]; % initial vehicle velocity (vx,vy) [m/s]
4
5  r_desired = [20 5]; % desired vehicle position (xdst, ydst) [m]
6  r_obs = [5 8; 10 5; 15 8; 15 2]; % obstacles (xobs,yobs) [m]
7  rho_o_i = 2.4; % obstacle radius [m]
8
9  ka = 0.5;
10 kr = 100;
11 c_damping = 5;
12 F_attractive_max = 10;
13
14 time_interval = [0 50]; % [s]
15 state_0 = [r_vehicle v_vehicle];
16
17 % Perturbation Force
18 dt = 0.1;
19 wk_mag = 0.01;
20 N_noise = floor(diff(time_interval)/dt);
21 wk_time = linspace(time_interval(1),time_interval(2),N_noise);
22 wk_noise = wk_mag*(2*rand(N_noise,2)-1);
```

```
23
24 sim_para = {r_desired, r_obs, rho_o_i, ka, kr, c_damping,
       F_attractive_max, wk_time, wk_noise};
25
26 %% simulation
27 ode_options = odeset('RelTol',1e-2,'AbsTol',1e-3, 'MaxStep', 0.1);
28 [tout,state_out] = ode45(@(time,state) drvdt_potential_field(time,
       state,sim_para), ...
29     time_interval, state_0, ode_options);
```

Program 3.1 (MATLAB) Pass multiple variables to *ode45* using a cell array

The differential equation passed to the differential equation solver, *ode45*, is defined in Program 3.2. The cell array, 'sim_para', is used to define the simulation parameters. The noise force, $\mathbf{w}_k$, is sampled from the uniform random distribution between ± 0.01 every 0.1 seconds at line 22. In line 49, $\mathbf{w}_k$ is interpolated using the noise samples. The MATLAB function, *interp1*, is the one-dimensional interpolator, where each column of 'wk_noise' is interpolated with the default linear option. There are several other interpolation methods available in this function. As *ode45* would call this function for any time in the given integration time intervals, the noise value should be interpolated for the time instance, which is not generated a priori.

> **Random number in differential equation**: Direct inclusion of random number generator inside the differential equation function causes highly discontinuous behaviours. Every time the equation is called, it returns different values because of the random number. The numerical integrator would slow down significantly not be able to finish the integration.

```
1  function dstate_dt = drvdt_potential_field(time,state ,sim_para)
2
3      % states
4      x_vehicle = state(1);
5      y_vehicle = state(2);
6      v_current = state(3:4);
7
8      % simulation setting
9      xy_dst = sim_para{1};
10     xy_obs = sim_para{2};
11     rho_o_i = sim_para{3};
12     ka = sim_para{4};
13     kr = sim_para{5};
14     c_damping = sim_para{6};
15     Famax = sim_para{7}*ones(2,1);
16     wk_time = sim_para{8};
17     wk_noise = sim_para{9};
```

```
18
19        num_obs = size(xy_obs,1);
20
21        % desired position
22        x_dst = xy_dst(1);
23        y_dst = xy_dst(2);
24
25        % attaractive & damping force
26        Fa = -ka*[(x_vehicle-x_dst); (y_vehicle-y_dst)];
27        Fa = sign(Fa).*min([abs(Fa(:)) Famax(:)],[],2);
28        Fd = -c_damping*v_current(:);
29
30        % repulsive force
31        Fr = [0; 0];
32        for idx=1:num_obs
33
34            x_ost = xy_obs(idx,1);
35            y_ost = xy_obs(idx,2);
36            rho_r_i = sqrt((x_vehicle-x_ost)^2+(y_vehicle-y_ost)^2);
37            if rho_r_i > rho_o_i
38                Frx_idx = 0;
39                Fry_idx = 0;
40            else
41                Frx_idx = kr*(x_vehicle-x_ost)/(rho_r_i^3);
42                Fry_idx = kr*(y_vehicle-y_ost)/(rho_r_i^3);
43            end
44
45            Fr(1) = Fr(1) + Frx_idx;
46            Fr(2) = Fr(2) + Fry_idx;
47        end
48
49        wk = interp1(wk_time,wk_noise,time);
50
51        F_sum = Fa(:) + Fr(:) + Fd(:) + wk(:);
52
53        drdt = v_current(:);
54        dvdt = F_sum;
55        dstate_dt = [drdt; dvdt];
56
57    end
```

Program 3.2 (MATLAB) Differential equation using potential fields for the path planning

Figure 3.3 shows two example paths obtained for the same scenario. Because of the random force, the path could go either direction from the first contact to the obstacle. The obstacle configuration is symmetric along $y = 5$, and both paths are qualitatively identical. The same behaviour of the path planner is shown in Figure 3.4 when it determines the direction at the first contact point. The path going above and the path going down are not qualitatively equal as the obstacle configurations are not symmetric in this scenario. The lower path seems to

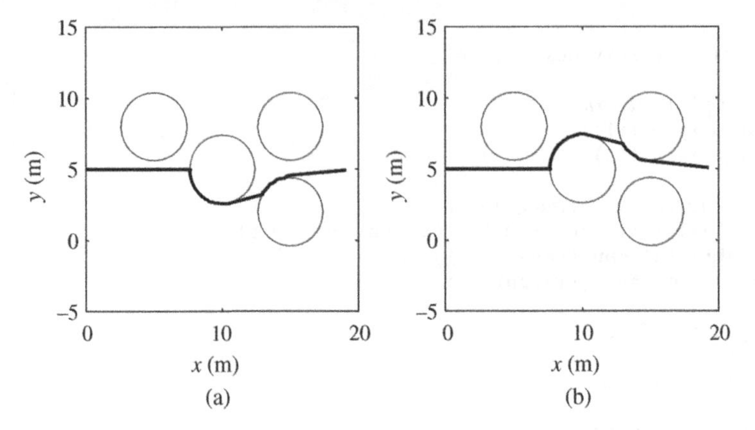

Figure 3.3 Two equally probably paths shown in (a) and (b), in which the obstacle configuration is *symmetric* along $y = 5$.

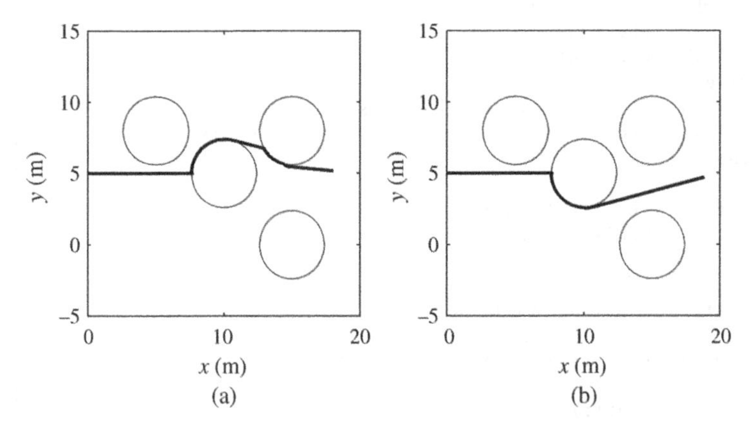

Figure 3.4 Two equally probably paths shown in (a) and (b), in which the obstacle configuration is *not symmetric*.

be better as it goes to the final destination with less turning manoeuvre compared to the upper path. This path planning algorithm cannot distinguish between these two paths at the point of solving the differential equation.

3.1.1.2 Python

The implementation in python is left as an exercise.

3.1.2 Graph Theory-Based Sampling Method

Graph theory is a field of mathematics to study graphs. A graph has two construction building blocks, called node and edge. Figure 3.5 shows an example of the

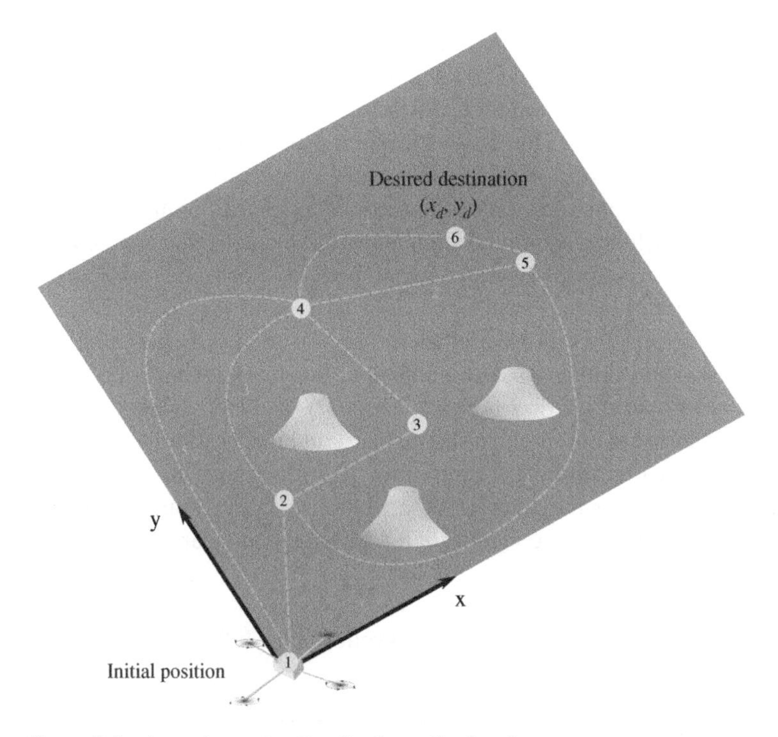

Figure 3.5 A graph construction for the path planning.

graph for path planning. In the figure, the circled numbers from 0 to 5 are nodes, and the nine edges are dashed lines connecting the nodes. The length of each edge is indicated by the number next to each dashed line. The distance in the path planning could be a generalized concept beyond the physical length. It might represent the physical distance, the risk, the energy consumption, the visibility of some area, and the combinations of these. As the path is bidirectional, the vehicle can fly from node 1 to node 4 or from node 4 to node 1, the matrix is symmetric. As there is no cost for staying in the same location, the diagonal terms are zero. If the quadcopter's electrical energy consumption is important, the cost penalty of staying at the same node should be introduced.

Construct the weighted adjacency matrix, A, as follows:

$$A = \begin{bmatrix} 0 & 2 & 0 & 5 & 0 & 0 \\ 2 & 0 & 1 & 3 & 4 & 0 \\ 0 & 1 & 0 & 4 & 0 & 0 \\ 5 & 3 & 4 & 0 & 2 & 4 \\ 0 & 4 & 0 & 2 & 0 & 1 \\ 0 & 0 & 0 & 4 & 1 & 0 \end{bmatrix} \tag{3.10}$$

where each row or column represents each node, i.e. the i-th row or column is for the i-th node, and the numbers in the matrix represent the distance between the nodes, e.g. the second row and the fourth column element, 3, is the distance between node 2 and node 4. One distinct characteristic of the adjacency matrix is sparsity. There are 18 zeros out of 36 elements in the matrix. Half of the elements are zero. The sparsity is one of the common properties of many existing network systems including communication networks, biomolecular interactions, social networks, and so forth. Typically, the proportion of zeros increases as the size of networks increases.

Both MATLAB and Python have efficient ways to process sparse matrices. There have been continuing research studies on the computational algorithm developments for sparse matrices (Press et al., 2007). Roughly saying, the zeros in the sparse matrix are not stored in the memory, and the locations of non-zero elements are stored instead. The computing algorithms exploit the location information and reduce the number of operations. The computational cost savings obtained in this way are significant for the large-scale sparse matrices.

3.1.2.1 MATLAB

Using the *sparse* command as in Program 3.3 converts the full matrix to a sparse matrix.

```
1 A_path_graph_full = [0 2 0 5 0 0;
2                      2 0 1 3 4 0;
3                      0 1 0 4 0 0;
4                      5 3 4 0 2 4;
5                      0 4 0 2 0 1;
6                      0 0 0 4 1 0];
7 A_path_graph_sparse = sparse(A_path_graph_full);
```

Program 3.3 (MATLAB) Construct sparse matrix

Type the sparse matrix in the command prompt window, and it prints

```
1 >> A_path_graph_sparse
2
3 A_path_graph_sparse =
4
5     (2,1)        2
6     (4,1)        5
7     (1,2)        2
8     (3,2)        1
9     ....
```

where row and column numbers and their element are printed.

3.1.2.2 Python

Using the *csr_matrix* command as in Program 3.4 converts the full matrix to the sparse matrix, where csr stands for compressed sparse row. Scipy has several functions constructing sparse matrices. Each method has its numerical advantages over the others. The details of the methods can be found in the Scipy reference manual (Virtanen et al., 2020).

```
1  import numpy as np
2  from scipy import sparse
3
4  A_path_graph_full = np.array([[0, 2, 0, 5, 0, 0],
5                                [2, 0, 1, 3, 4, 0],
6                                [0, 1, 0, 4, 0, 0],
7                                [5, 3, 4, 0, 2, 4],
8                                [0, 4, 0, 2, 0, 1],
9                                [0, 0, 0, 4, 1, 0]])
10
11 A_path_graph_sparse = sparse.csr_matrix(A_path_graph_full)
```

Program 3.4 (Python) Construct sparse matrix

Print the sparse matrix in the python command prompt using the *print* command:

```
1  In [108]: print(A_path_graph_sparse)
2    (0, 1)  2
3    (0, 3)  5
4    (1, 0)  2
5    (1, 2)  1
6    ...
```

Python prints the matrix for each column element in a fixed row, while MATLAB prints each row element for a fixed column. Construct the matrix using the *csc_matrix* in Scipy, where csc stands for the compressed sparse column and print the matrix. Then, the print sequence is the same as the one in the MATLAB sparse matrix.

In both the MATLAB and the Python programs, the sparse matrix is created from the full matrix. This method is not appropriate with large size matrices requiring large memory to build the full matrix first. The direct sparse matrix construction using the row and the column numbers for non-zero elements in the matrix is suitable for large matrices. The python script is shown here, where the upper triangular part is constructed and the lower triangular part is obtained by transposing the upper triangular part. A MATLAB implementation using row and column numbers is left as an exercise.

```
1  # sparse matrix from (row,column,values)
2  row_size = 6
3  col_size = 6
```

```
4
5 row = np.array([0, 0, 1, 1, 1, 2, 3, 3, 4])
6 col = np.array([1, 3, 2, 3, 4, 3, 4, 5, 5])
7 val = np.array([2, 5, 1, 3, 4, 4, 2, 4, 1])
8
9 A_path_graph_sparse = sparse.csc_matrix((val,(row,col)),shape=(
      row_size,col_size))
10 A_path_graph_sparse = A_path_graph_sparse+A_path_graph_sparse.
      transpose()
```

The computational advantage is not noticeable for the 6×6 matrix. The superiority of using sparse matrices is much evident for large size sparse matrices. In path planning using the graph, most computation time spends for calculating the distance values for edges. Given the graph constructed, several existing algorithms, known to be efficient for sparse matrix problems, solve the optimal path planning problem. Dijkstra's algorithm, Bellman–Ford Algorithm, and A* algorithms are the most frequently used algorithms for the shortest path planning.

3.1.2.3 Dijkstra's Shortest Path Algorithm

The algorithm solves the shortest path problem for the graphs with non-negative edges (Dijkstra, 1959). The algorithm is simple to implement with a few lines codes in MATLAB or Python. Understanding the algorithm and implementing it is not straightforward without a few or many debug iterations. Here, we take this as an opportunity for practising high-level design approaches. It is the same practice as off-the-shelf hardware component usages in hardware system implementations. Of course, for the correct integration, we must have proper high-level understandings of the off-the-shelf components.

What levels of details need to be understood varies depending on each case. In this case, acknowledging the following two would be enough: Dijkstra's algorithm is for the non-negative distance graph and returns the optimal paths from a single source node to all the other nodes. The latter is an unavoidable step as it is unknown whether the calculated path is optimal or not without checking all the remaining possible paths.

3.1.2.4 MATLAB

Although we covered earlier how to construct sparse matrices in MATLAB, MATLAB has the specialized data type for graphs. Creating a graph in MATLAB is similar to building a sparse matrix. The row and the column numbers and the values are given to the function, *graph*, as shown in Program 3.5. Then, *shortestpath* calculates the shortest path from the given start node and the end node. Several algorithms are available to choose from in *shortestpath*. The default algorithm is Dijkstra's one. The return variable, *opt_path*, gives the sequence of the nodes for the optimal path, which is ① → ② → ⑤ → ⑥, and

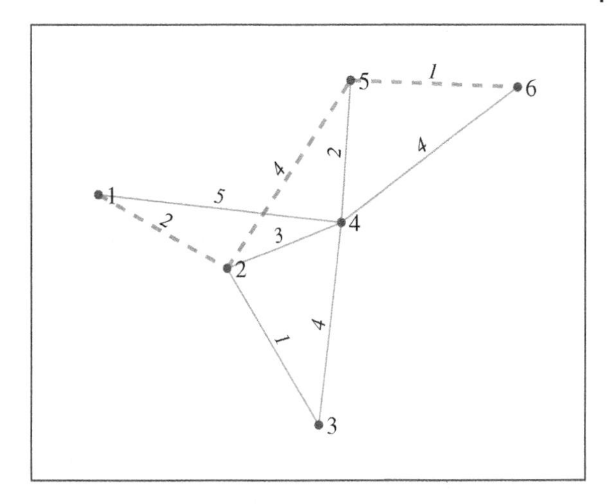

Figure 3.6 (MATLAB) Plot the graph and the shortest path.

opt_dist returns the corresponding distance, which is equal to 7. The last two lines of Program 3.5 shows how to draw the graph and the optimal path with little effort. The resulting plot is shown in Figure 3.6. The graph in the figure keeps only the topology the same, where the dashed line indicates the optimal path.

```
1  % Shortest path planning
2  st_node = 1;
3  end_node = 6;
4
5  row_size = 6;
6  col_size = 6;
7
8  row = [1 1 2 2 2 3 4 4 5];
9  col = [2 4 3 4 5 4 5 6 6];
10 val = [2 5 1 3 4 4 2 4 1];
11
12 G_path_graph = graph(row,col,val);
13 [opt_path,opt_dst] = shortestpath(G_path_graph,st_node,end_node);
14
15 % Plot the result
16 G_graph_plot = plot(G_path_graph, ...
17    'EdgeLabel',G_path_graph.Edges.Weight, ...
18          'NodeFontSize',14,'EdgeFontSize',12);
19 highlight(G_graph_plot,opt_path,'EdgeColor','r', ...
20          'LineWidth',2,'LineStyle','--');
```

Program 3.5 (MATLAB) Shortest path calculation using Dijkstra's algorithm

3.1.2.5 Python

Dijkstra's algorithm is available from Scipy. The *dijkstra* function imported from *scipy.sparse.csgraph* is called in line 6 in Program 3.6. The first two input arguments are the graph and the start node. The end node is not indicated as the algorithm calculates all optimal paths from the start node to every other node. The third input argument indicates that the path returns to the second output variable, 'pred'. The i-th element of 'pred' gives the previous index in the optimal path to arrive at the i-th element. For example, 'pred[5]' equal to 4 tells that the previous index is 4, i.e. from index 4 to index 5, which implies that the path is from node 5 to node 6 as the python index starts from zero. If we try to keep translating the i-th index in Python corresponding to the $i + 1$-th node in Figure 3.5, some confusion should occur when the translation is forgotten. *It would be convenient in python programming to assume that the numbering of the nodes always starts from zero.* Using the while-loop in the program, the optimal path from the end node is obtained in the backwards tracking. The list, *path*, is empty initially, and the node starting from the end node is added to the list by *path.append* until it reaches the start node. Applying *path.reverse*, the list starts from the start node. The first output variable, 'dist', is the list of the distance costs from the start node to all the other nodes in the graph.

> **Node index in Python**: To reduce confusion in node numbers, we always assume that the node starts from zero in Python.

```
1  # shortest path from node 1
2  start_node = 0
3  end_node = 5
4
5  from scipy.sparse.csgraph import dijkstra
6  dist, pred = dijkstra(A_path_graph_sparse, indices = start_node,
       return_predecessors=True)
7
8  # print out the distance from start_node to end_node
9  print(f"distance from {start_node} to {end_node}: {dist[end_node]}.
       ")
10
11 # construct the path
12 path = []
13 idx=end_node
14 while idx!=start_node:
15     path.append(idx)
16     idx = pred[idx]
17
18 path.append(start_node)
19 path.reverse()
20 print('path=',path)
```

Program 3.6 (Python) Shortest path calculation using Dijkstra's algorithm

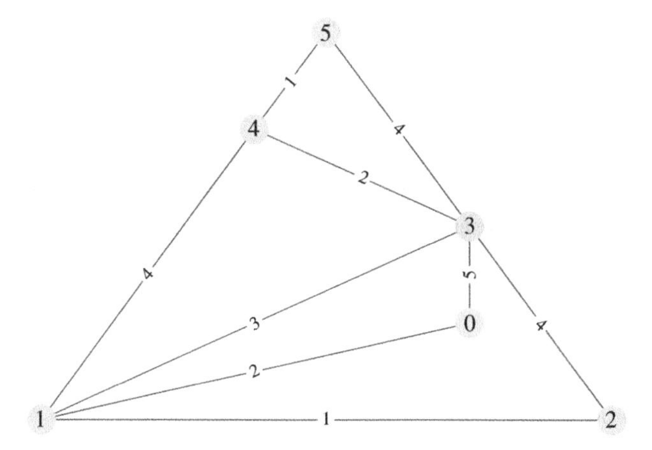

Figure 3.7 (Python) Plot the graph and the shortest path.

Figure 3.7 shows the graph and the optimal path drawn in Python. It is not as simple as two lines in MATLAB to draw the graph and the optimal path in Python. Program 3.7 shows how to use *networkx* with *matplotlib* to draw the graph. All the optimal edges belonging to the optimal path are constructed using the zip-command and assigned to *opt_path_edge*. The sparse matrix is converted to networkx data type using *from_scipy_sparce_matrix*. There are a few layout functions to choose from in *networkx*. *planar_layout* draws the graph avoiding edge overlaps. The edge labels and the colours are constructed, and they are passed to *draw* function in *networkx*.

```
1  opt_path_edge = [(i,j) for i,j in zip(path[0:-1],path[1::])]
2
3  # draw graph
4  import matplotlib.pyplot as plt
5  fig, ax = plt.subplots(nrows=1,ncols=1)
6
7  import networkx as nx
8  A_graph_nx = nx.from_scipy_sparse_matrix(A_path_graph_sparse)
9
10 #pos=nx.spring_layout(A_graph_nx)
11 pos=nx.planar_layout(A_graph_nx)
12
13 edge_labels=nx.get_edge_attributes(A_graph_nx,'weight')
14 edge_color = ['red' if key in opt_path_edge else 'green' for key in
       edge_labels.keys()]
15
16 nx.draw(A_graph_nx, pos, node_size=500, node_color='yellow',
      edge_color=edge_color, labels={node:node for node in A_graph_nx
      .nodes()})
```

```
17 nx.draw_networkx_edge_labels(A_graph_nx,pos=pos,edge_labels=
       edge_labels)
```

Program 3.7 (Python) Draw graph and optimal path

In line 15, construct *edge_color* list, whose element is red if the edge belongs to the optimal path or green if not. It is one of the distinct ways of programming in Python compared to MATLAB. Consider the following two lists:

```
F=[(0,1), (0,3), (1,2), (1,3)]
A=[(0,3), (1.3)]
```

and we want to find if each element of F is in A. The result must be [False, True, False, True], which indicates that the second and the last elements in F are in A. Some would try implementing a code similar to the following program using two for-loops:

```
C = [False, False, False, False]
for aa in A:
    idx=0;
    for ff in F:
        if aa==ff:
            C[idx] = True
        idx+=1
```

Instead, the following one line returns the same result:

```
C = [ True if ff in A else False for ff in F]
```

The code is easier to understand and possibly more efficient than the previous one with two for-loops. It takes one element from 'F', 'for ff in F', checks if it is in 'A', 'if ff in A', and returns True or False. These repeat for all elements in 'F'.

3.1.3 Complex Obstacles

Consider the map given in Figure 3.8, where two obstacles, one circular obstacle and another non-convex shape obstacle, are present. The initial position is at the origin, and the destination is $x = 9$ and $y = 4$. Because of the non-convex shape obstacle, it is not easy to apply the potential field method. Paths generated by the potential function method may never reach the destination. The graph-based path planning algorithm can solve these types of complex shape obstacle environments without too many difficulties. Voronoi diagram is a set of convex polygons dividing a two-dimensional space.

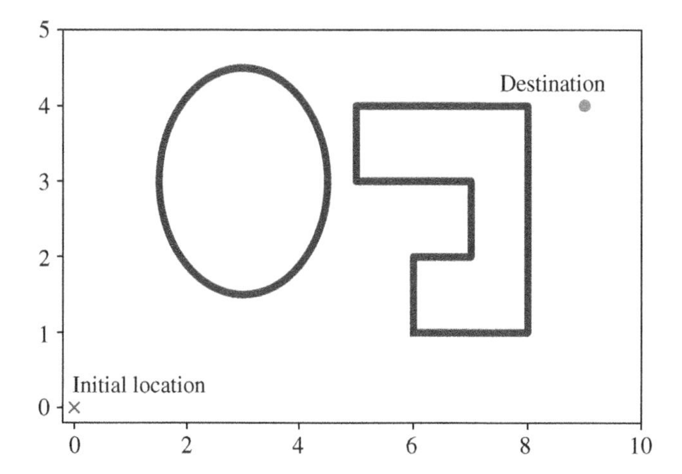

Figure 3.8 Three repulsive potential functions and one attractive potential function.

3.1.3.1 MATLAB

Program 3.8 distributes uniformly generated random numbers over the operation area defined by the map size. The function, *Voronoi*, constructs the Voronoi diagram based on the random numbers. Voronoi diagram divides the two-dimensional plane by a set of convex polygons (Klein, 2016). Each polygon includes only one point from the generated random points, which is the closest point to any location inside the polygon.

Figure 3.9 shows the random numbers and the corresponding Voronoi diagram. Each polygon includes only one point. The polygons at the boundary would

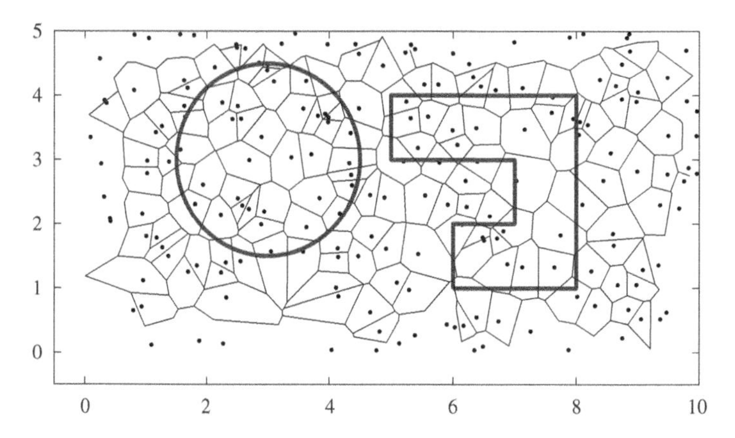

Figure 3.9 Random points and Voronoi diagram.

be open to infinity. To prevent the infinite size polygons, we remove the edges defining the polygons outside of the operational area. The random points near the map boundary do not belong to any polygons. Adding more random points would reduce these boundary effects caused by the removal.

```
1  % number of samples
2  num_sample = 200;
3
4  % map size
5  x_min = 0; x_max = 10;
6  y_min = 0; y_max = 5;
7
8  % starting point
9  xy_start = [0 0];
10 xy_dest  = [9 4];
11
12 % spread num_sample random points over the map area
13 xn=rand(1,num_sample)*(x_max-x_min) + x_min;
14 yn=rand(1,num_sample)*(y_max-y_min) + y_min;
15
16 % divide region using voronoi
17 [vx,vy] = voronoi(xn,yn);
18
19 % reject points outside the map region
20 idx = (vx(1,:) < x_min) | (vx(2,:) < x_min);
21 vx(:,idx) = [];
22 vy(:,idx) = [];
23 idx = (vx(1,:) > x_max) | (vx(2,:) > x_max);
24 vx(:,idx) = [];
25 vy(:,idx) = [];
26 idx = (vy(1,:) < y_min) | (vy(2,:) < y_min);
27 vx(:,idx) = [];
28 vy(:,idx) = [];
29 idx = (vy(1,:) > y_max) | (vy(2,:) > y_max);
30 vx(:,idx) = [];
31 vy(:,idx) = [];
32
33 % circular obstacle
34 th=0:0.01:2*pi;
35 c_cx = 3; c_cy = 3; c_r = 1.5;
36 xc=c_r*cos(th)+c_cx;
37 yc=c_r*sin(th)+c_cy;
38
39 % polygon obstacle
40 xv = [6; 8; 8; 5; 5; 7; 7; 6; 6];
41 yv = [1; 1; 4; 4; 3; 3; 2; 2; 1];
42
43 % draw sampling points & obstacles
44 figure(1); clf;
45 plot(xn,yn,'k.');
46 hold on;
47 plot(xc,yc,'r-','LineWidth',2);
```

```
48 plot(xv,yv,'r-','LineWidth',2);
49 axis equal;
50 axis([x_min-0.5 x_max y_min-0.5 y_max]);
51 plot(vx,vy,'b-');
```

Program 3.8 (MATLAB) Path planning for complex shape obstacle

Program 3.9 removes the edges inside the obstacles and adds the start and the destination points to the closest nodes in the graph, respectively. Use the following inequality for circular objects to find the edges to remove:

$$(x_i - x_c)^2 + (y_i - y_c)^2 < r_c^2$$

where (x_i, y_i) is the coordinate to define an edge, (x_c, y_c) is the centre of the circular obstacle, and r_c is the radius of the obstacle. Finding whether a point is inside an arbitrary shape polygon or not is a geometry problem. MATLAB has a function to solve the point-in-polygon problem, called *inpolygon*. Providing a polygon shape by a series of coordinates describing the boundary to *inpolygon*, it returns a Boolean value if the given point is inside (True) or outside (False) of the polygon.

The result with 200 random points is shown in Figure 3.10. If the number of random points is too low, the graph would be disjointed, and there might be no path between the start and the destination. The risk is negligible if a sufficient number of random points are spread across the mission area. The example shown in Figure 3.10 with 200 random points has the high possibility that the problem would occur. To reduce the risk, use more than 1000 points in this case.

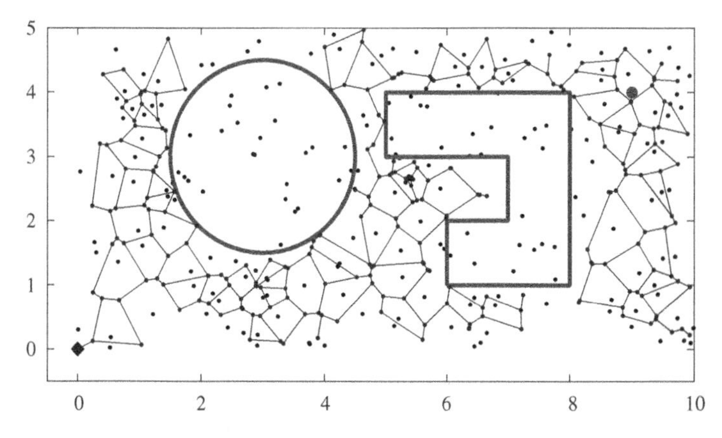

Figure 3.10 Edges inside the obstacles removed and add start and destination.

```
1 % remove vertices inside the circular object
2 vx1=vx(1,:);
3 vy1=vy(1,:);
```

```
4  r_sq = (vx1-c_cx).^2+(vy1-c_cy).^2;
5  idx1=(r_sq <) c_r^2);
6
7  vx2=vx(2,:);
8  vy2=vy(2,:);
9  r_sq = (vx2-c_cx).^2+(vy2-c_cy).^2;
10 idx2=(r_sq <) c_r^2);
11
12 idx= or(idx1,idx2);
13 vx(:,idx)=[];
14 vy(:,idx)=[];
15
16 % remove vertices inside the polygon
17 vx1=vx(1,:);
18 vy1=vy(1,:);
19 in = inpolygon(vx1,vy1,xv,yv);
20 vx(:,in)=[];
21 vy(:,in)=[];
22
23 vx2=vx(2,:);
24 vy2=vy(2,:);
25 in = inpolygon(vx2,vy2,xv,yv);
26 vx(:,in)=[];
27 vy(:,in)=[];
28
29 % add start &) destination points to the graph
30 vx1d = vx(:);
31 vy1d = vy(:);
32 dr = kron(ones(length(vx1d),1),xy_start) - [vx1d vy1d];
33 [~,min_id] = min(sum(dr.^2,2));
34 vx = [vx [xy_start(1); vx1d(min_id)]];
35 vy = [vy [xy_start(2); vy1d(min_id)]];
36
37 dr = kron(ones(length(vx1d),1),xy_dest) - [vx1d vy1d];
38 [~,min_id] = min(sum(dr.^2,2));
39 vx = [vx [xy_dest(1); vx1d(min_id)]];
40 vy = [vy [xy_dest(2); vy1d(min_id)]];
```

Program 3.9 (MATLAB) Remove the edges and add start and destination

Program 3.10 constructs the graph using the nodes and the vertices defined by the Voronoi diagram. To this end, firstly, the start nodes and the end nodes are stacked together in 'xy_12', whose size is (the number of edges) × 2. The first half of the nodes in the matrix are the start nodes, and the other half of the nodes in the matrix are the end nodes. Hence, the coordinate of one node would appear more than once in the matrix as one node would belong to more than one edge. To have a list of nodes where the same node does not appear multiple times, use the function, *unique*, as follows:

```
[node_coord,~,node_index]=unique(xy_12,'rows');
```

The option ’rows’ indicates that the row elements in the matrix are compared and returned the list of the unique row elements to the first return variable, ‘node_coord’. In addition, the second and the third return variables store the list of indices. These would be much easier to understand with an example. Assume that ‘xy_12’ is given by

```
xy_12 = [
   1.5  2.1;
   3.2  -4.2;
   1.5  2.1;
   4.2  4.3];
```

where the first and the third rows are the same. Call the unique function as follows:

```
>> [node_unique, unique_index, full_index] = unique(xy_12);
```

Each return variable has the following result:

```
node_unique =
   1.5   2.1
   3.2  -4.2
   4.2   4.3
unique_index =
   1
   2
   4
full_index =
   1
   2
   1
   3
```

The first matrix, ‘node_unique’, removes the third-row element in the original matrix, which is appeared twice. The second matrix, ‘unique_index’, includes the row index to construct the first matrix using the original matrix, i.e. ‘xy_12 (unique_index,:)’ returns the matrix the same as the first matrix, ‘node_unique’. Similarly, ‘node_unique(full_index,:)’ recovers the original matrix ‘xy_12’. In Program 3.10, the second output of *unique* is not needed, and the corresponding output variable name is replaced by ‘~’. For example,

```
>> [node_unique, ~, full_index] = unique(xy_12);
```

where the second output is not created.

These nodes list and indices create the row and the column numbers of the nodes defining bi-directional edges and their length, which are entered into the

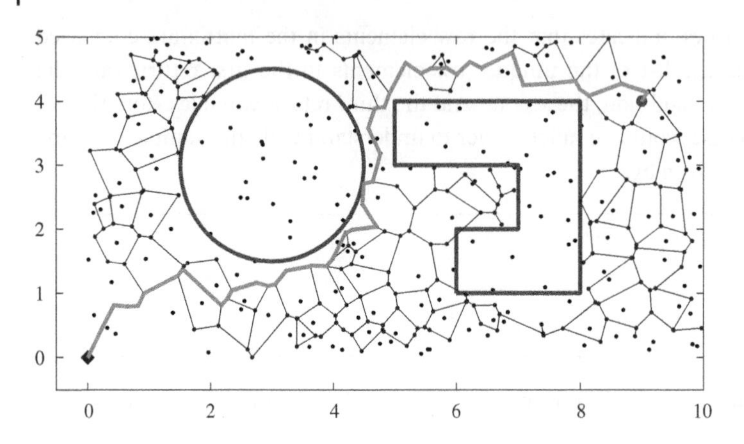

Figure 3.11 Shortest path in the graph constructed using Voronoi diagram.

graph function. Finally, the shortest path is obtained from the graph as shown in Figure 3.11.

```
1  %% construct graph
2  xy_1 = [vx(1,:); vy(1,:)];
3  xy_2 = [vx(2,:); vy(2,:)];
4  xy_12 = [xy_1 xy_2]';
5  [node_coord,~,node_index]=unique(xy_12,'rows');
6  st_node_index = node_index(1:length(vx));
7  ed_node_index = node_index(length(vx)+1:end);
8  dst_edges = sqrt(sum((xy_1-xy_2).^2));
9  st_node = node_index(length(vx)-1);
10 ed_node = node_index(length(vx));
11
12 row = [st_node_index(:); ed_node_index(:)];
13 col = [ed_node_index(:); st_node_index(:)];
14 val = kron([1;1],dst_edges(:));
15 G_path_graph = graph(row,col,val);
16
17 %% calculate optimal path and plot the path
18 [opt_path_idx,opt_dst] = shortestpath(G_path_graph,st_node,ed_node)
   ;
19 opt_path = node_coord(opt_path_idx,:);
20
21 % draw voronoi after removing points in the obstacles
22 figure(1); clf;
23 plot(xn,yn,'k.');
24 hold on;
25 plot(xc,yc,'r-','LineWidth',2);
26 plot(xv,yv,'r-','LineWidth',2);
27 axis equal;
28 plot(vx,vy,'b.-');
29 axis([x_min-0.5 x_max y_min-0.5 y_max]);
```

```
30 plot(xy_start(1),xy_start(2),'bx','MarkerSize',5,'LineWidth',5);
31 plot(xy_dest(1),xy_dest(2),'ro','MarkerSize',5,'MarkerFacecolor','
       red');
32 plot(opt_path(:,1),opt_path(:,2),'g-','LineWidth',2);
```

Program 3.10 (MATLAB) Shortest path in the graph

3.1.3.2 Python

Random points are spread uniformly over the designated map area. The initial location and the destination are included as the first and the second nodes, respectively. The boundary coordinates of the circular obstacle and the non-convex obstacle are defined. These two boundaries define two path objects using *Path* in the module *matplotlib.path*.

The Path object has a useful function called *contains_points* for checking whether a set of points are inside (return True) or outside (return False) of the objects. Negating the return values using '~' in front of the Boolean lists, we obtain the lists of True or False, indicating if the corresponding point is outside of the obstacles or inside of one of the obstacles. For the details, check the following Program 3.11.

```
1  # number of samples
2  num_sample = 2000
3
4  # map size
5  map_width = 10
6  map_height = 5
7
8  # x,y coordinates of start and destination of the path to be
       calculated
9  xy_start = np.array([0,0])
10 xy_dest = np.array([9,4])
11
12 # spread num_sample random points over the map area
13 xy_points = np.random.rand(num_sample,2)
14 xy_points[::,0] = xy_points[::,0]*map_width
15 xy_points[::,1] = xy_points[::,1]*map_height
16
17 # stacking them all together with start and destination
18 xy_points = np.vstack((xy_start,xy_dest,xy_points))
19 start_node = 0
20 end_node = 1
21
22 # circular obstacle at [3,3], radius 1.5 & define the boundary
23 obs_xy = [3,3]
24 obs_rad = 1.5
25 th = np.arange(0,2*np.pi+0.01,0.01)
26 x_obs_0 = obs_rad*np.cos(th)+obs_xy[0]
27 y_obs_0 = obs_rad*np.sin(th)+obs_xy[1]
28 xy_obs_0 = np.vstack((x_obs_0,y_obs_0)).T
```

```
29
30 # non-convex obstacle boundary
31 x_obs_1 = np.array([6,8,8,5,5,7,7,6,6])
32 y_obs_1 = np.array([1,1,4,4,3,3,2,2,1])
33 xy_obs_1 = np.vstack((x_obs_1,y_obs_1)).T
34
35 # define obstacle using Path in matplotlib.path
36 from matplotlib.path import Path
37 Obs_0 = Path(xy_obs_0)
38 Obs_1 = Path(xy_obs_1)
39
40 # found points are not inside the circular obstacle
41 mask_0 = ~Obs_0.contains_points(xy_points)
42 xy_points = xy_points[mask_0,::]
43
44 mask_1 = ~Obs_1.contains_points(xy_points)
45 xy_points = xy_points[mask_1,::]
```

Program 3.11 (Python) Path planning for complex shape obstacle

A graph is constructed in Program 3.12 using *delaunay*. The Delaunay triangulation is closely related to the Voronoi diagram (Klein, 2016). The Delaunay triangulation constructs triangles using a given set of points. Each of the points is the vertex of the triangles. The way of constructing the triangles is that each circle circumscribes only one triangle, and no other triangles are inside the circle. The Delaunay triangulation is closely related to the Voronoi diagram. Fortune (1995) provides further details.

The return value, *tri*, from the *Delaunay* function is the delaunay object. *tri.simplices* includes the list of vertices to define the triangles. Each row of *tri.simplices* is an array with three elements, which defines a triangle. The three elements define the three vertices of the triangle in each row of *tri.simplices*.

As we initially remove the points inside the obstacle, the vertices that cross the obstacle tend to be longer in length than the vertices outside the obstacle. The cut-off length is 1σ multiplied by the average length, and we remove the vertices longer than the cutoff length from the graph. This method has still the problem that parts of some vertices would pass through the obstacles. Some modifications of the algorithm could fix the problem. For example, sample additional points along each vertices and check if they are inside obstacles. Or, sample along the calculated optimal path, construct a new graph using the new sample points, and calculate the optimal path for the new graph.

```
1 # construct graph using delaunay
2 from scipy.spatial import Delaunay
3 tri = Delaunay(xy_points)
4
5 # found triangle definition index
6 temp_idx=tri.simplices[::,0]
```

```
 7 temp_jdx=tri.simplices[::,1]
 8 temp_kdx=tri.simplices[::,2]
 9 
10 # remove longer paths, which are likely passing through the
       obstacle
11 dist_ij = np.sqrt(np.sum((xy_points[temp_idx,::] - xy_points[temp_jdx
       ,::])**2,1))
12 dist_jk = np.sqrt(np.sum((xy_points[temp_jdx,::] - xy_points[temp_kdx
       ,::])**2,1))
13 dist_ki = np.sqrt(np.sum((xy_points[temp_kdx,::] - xy_points[temp_idx
       ,::])**2,1))
14 dd_all = np.hstack((dist_ij,dist_jk,dist_ki))
15 cut_dist = np.mean(dd_all)+np.std(dd_all)
16 
17 # distance threshold for removing longer paths
18 cut_mask_ij = dist_ij<cut_dist
19 cut_mask_jk = dist_jk<cut_dist
20 cut_mask_ki = dist_ki<cut_dist
21 temp_xy_ij = np.vstack((temp_idx[cut_mask_ij],temp_jdx[cut_mask_ij
       ]))
22 temp_xy_jk = np.vstack((temp_jdx[cut_mask_jk],temp_kdx[cut_mask_jk
       ]))
23 temp_xy_ki = np.vstack((temp_kdx[cut_mask_ki],temp_idx[cut_mask_ki
       ]))
24 
25 # corresponding distance to the paths
26 dist_ij = dist_ij[cut_mask_ij]
27 dist_jk = dist_jk[cut_mask_jk]
28 dist_ki = dist_ki[cut_mask_ki]
29 
30 # change format into row, column and the distance
31 xy_index = np.hstack((temp_xy_ij,temp_xy_jk,temp_xy_ki)).T
32 row_org = xy_index[::,0]
33 col_org = xy_index[::,1]
34 row = np.hstack((row_org,col_org))
35 col = np.hstack((col_org,row_org))
36 dist = np.hstack((dist_ij,dist_jk,dist_ki))
37 dist = np.hstack((dist,dist))
38 num_node = xy_points.shape[0]
39 
40 # construct the distance matrix
41 from scipy.sparse import csr_matrix
42 dist_sparse = csr_matrix((dist,(row,col)), shape=(num_node,num_node
       ))
43 
44 # calculate the shortest path
45 from scipy.sparse.csgraph import dijkstra
46 dist, pred = dijkstra(dist_sparse, indices = start_node,
       return_predecessors=True)
47 print(f'distance from node #{start_node:0d} to node #{end_node:0d}:
        {dist[end_node]:4.2f}')
48 
49 # obtain the shortest path
```

```
50 path = []
51 i=end_node
52 if np.isinf(dist[end_node]):
53     print('the path does not exist!')
54 else:
55     while i!=start_node:
56         path.append(i)
57         i = pred[i]
58     path.append(start_node)
59     print('path=',path[::-1])
60 
61 opt_path = np.asarray(path[::-1])
```

Program 3.12 (Python) Shortest path in the graph

Program 3.13 shows an example of re-sampling. Similar to the previous method, we remove the edges with longer lengths to prevent the edges from overlapping with the obstacles. Figure 3.12a compares the original path (the dashed–dotted line) with the updated path (the solid line) by the re-samplings. The re-sampling points are shown in Figure 3.12b. The updated path avoids the top-left corner, unlike the original path that goes into the corner of the obstacle. Implementing the program to generate Figure 3.12 is left as an exercise.

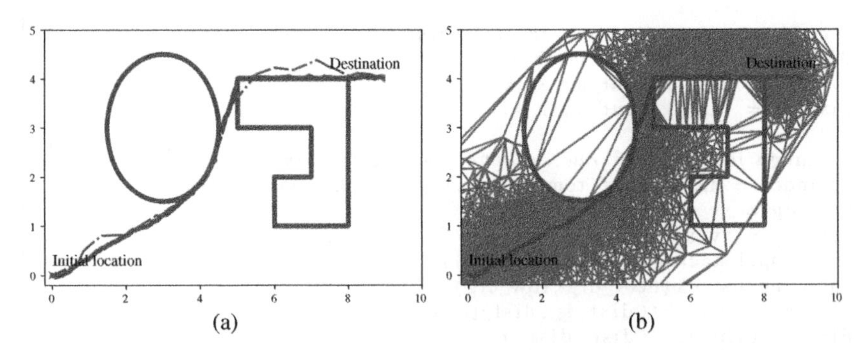

Figure 3.12 (a) Original optimal path in the dash-dotted line and the updated optimal path in the solid line and (b) updated optimal path shown with the re-sampled graph.

```
1 xy_opt_points = xy_points[opt_path,:]
2 dxy_opt_dist = np.sqrt(np.sum((xy_opt_points[0:-1]-xy_opt_points
      [1::])**2,1))
3 N_new_samp = 1000
4 
5 xy_samp = np.empty((0,2))
6 
7 for crd, dst in zip(xy_opt_points,dxy_opt_dist):
8     xy_samp = np.append(xy_samp,crd + np.random.randn(N_new_samp,2)
          *dst,axis=0)
```

Program 3.13 (Python) Re-sampling around the optimal path

The existence of diverse path planning scenarios makes it challenging to construct one unique path planning algorithm outperforming over all the other algorithms for any situation. Many heuristic approaches and variations are possible in path planning problems. Understanding what types of operation scenarios are considered and what assumptions are introduced is vital in path planning algorithm design.

3.2 Moving Target Tracking

Consider a fixed-wing unmanned aerial vehicle (UAV) equipped with a vision sensor to keep a moving ground target in the camera's field of view, as shown in Figure 3.13. Keeping the altitude in the z-axis as high as possible and the plane distance from the UAV to the target in the x–y axes as close as possible maximizes the camera field of view. Each UAV has the maximum altitude it could reach, and the vision sensor performance, e.g. camera resolution, and the weather condition, e.g. clouds, also limit the possible maximum altitude. Given these limitations, the best choice for the UAV in the z-direction is simply the maximum allowable altitude, and the target tracking problem becomes one in the two-dimensional space, i.e. x–y plane.

3.2.1 UAV and Moving Target Model

A simplified aircraft dynamics with a constant altitude shown in Figure 3.14 is given by

$$\dot{x}_a = v_x, \quad \dot{y}_a = v_y, \quad \dot{v}_x = u_x, \quad \dot{v}_y = u_y$$

where x_a and y_a are the x and y coordinates of aircraft location in metres, v_x and v_y are the x and y directional velocities of aircraft in m/s, and u_x and u_y are the control

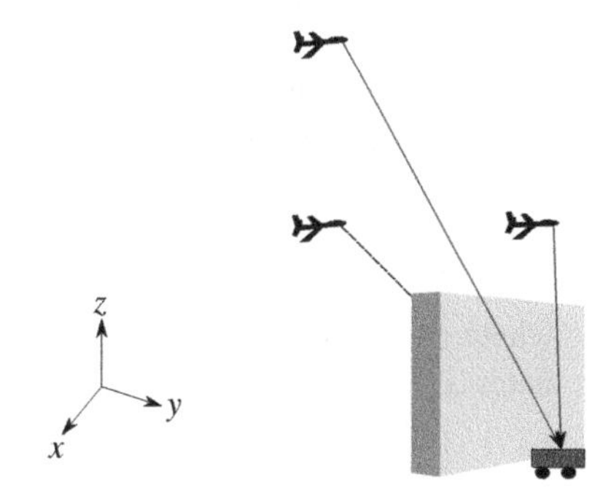

Figure 3.13 The higher altitude and the closer distance relative to the target are the optimal positions of UAVs in maximizing the field of view.

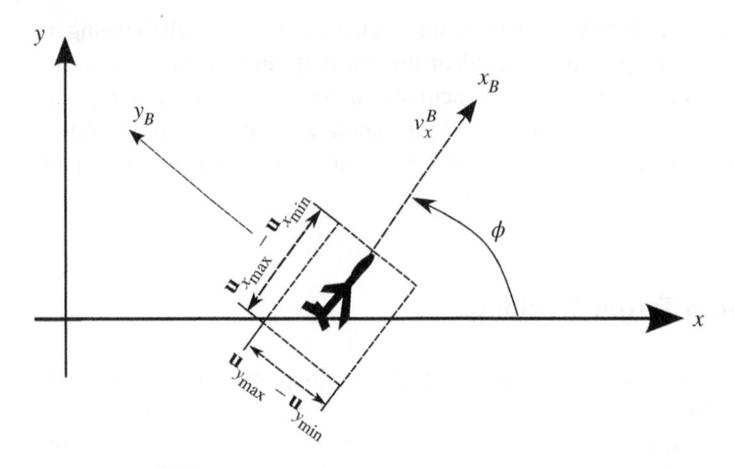

Figure 3.14 Global (x–y) and UAV local (x_B–y_B) coordinates with the control input magnitude constraint indicated by the dotted box.

input in N, respectively. *All quantities are expressed in the global coordinates,* x and y, as shown in Figure 3.14. In the state-space form,

$$\dot{\mathbf{x}}_a = \begin{bmatrix} 0_2 & I_2 \\ 0_2 & 0_2 \end{bmatrix} \mathbf{x}_a + \begin{bmatrix} 0_2 \\ I_2 \end{bmatrix} \mathbf{u} = A_a \mathbf{x}_a + B_a \mathbf{u} \tag{3.11a}$$

$$\mathbf{y} = \begin{bmatrix} I_2 & 0_2 \end{bmatrix} \mathbf{x}_a = C_A \mathbf{x}_a \tag{3.11b}$$

where $\mathbf{x}_a = [x_a,\ y_a,\ ,v_x,\ v_y]^T$ and $\mathbf{u} = [u_x,\ u_y]^T$.

To design an optimal guidance algorithm, consider the following important constraints on the aircraft:

- The velocity for x-direction in the body coordinates must satisfy the following inequality:

$$0 < v_{\min} \leq v_x^B \leq v_{\max} \tag{3.12}$$

where v_x^B is the aircraft velocity in the body coordinates, and the aircraft attitude is assumed to coincide with the velocity vector, and v_y^B is always, hence, equal to zero. Unlike quadcopters, as the fixed-wing UAV cannot hover, the minimum velocity magnitude, $v_{\min}$, is always greater than zero. As the velocity in the global coordinates is given by

$$v_x = v_x^B \cos\phi, \quad v_y = v_x^B \sin\phi$$

where ϕ is equal to $\tan^{-1}(v_y/v_x)$, the inequality, (3.12), in the global coordinates is

$$v_{\min}^2 \leq v_x^2 + v_y^2 \leq v_{\max}^2 \tag{3.13}$$

- The control input magnitudes are constrained as follows:

$$u_{x_{\min}} \leq u_x^B \leq u_{x_{\max}} \tag{3.14a}$$

$$u_{y_{\min}} \leq u_y^B \leq u_{y_{\max}} \tag{3.14b}$$

where u_x^B and u_y^B are the control input expressed in the aircraft body coordinates. The inequalities given in (3.14) becomes

$$u_{x_{\min}} \leq u_x \cos\phi + u_y \sin\phi \leq u_{x_{\max}} \tag{3.15a}$$

$$u_{y_{\min}} \leq -u_x \sin\phi + u_y \cos\phi \leq u_{x_{\max}} \tag{3.15b}$$

in the global coordinates.
- The aircraft turn radius must be larger than the minimum radius turn of the aircraft. The radius of curvature of the flight path must be smaller than the inverse of the minimum radius as follows:

$$\frac{\left|v_x u_y - v_y u_x\right|}{\left(v_x^2 + v_y^2\right)^{3/2}} \leq \frac{1}{r_{\min}} \tag{3.16}$$

where $r_{\min}$ is the radius of the circle corresponding to the minimum radius turn, and the left-hand side of the inequality is the curvature equation for curves in the two-dimensional space.

The target tracking algorithm to provide the command acceleration input, u_x and u_y, must satisfy the three constraints, i.e. (3.13), (3.15), and (3.16).

Let the target dynamics be equal to the following equations:

$$\dot{x}_t = w_x \tag{3.17a}$$

$$\dot{y}_t = w_y \tag{3.17b}$$

where x_t and y_t are the x and y coordinates of the target in metres and w_x and w_y are the x and y directional velocities in m/s, respectively. From the target tracking design aspect, how the ground target would behave is unknown. The target velocity in x and y coordinates, i.e. w_x and w_y, cannot be perfectly known, in general. Using sensors in UAVs, a Kalman filter to estimate the target position and the velocity could be designed. This is out of the scope of this chapter, and the reader is referred to Julier and Uhlmann (2004) and Zhan and Wan (2007).

In the state-space form,

$$\dot{\mathbf{x}}_t = I_2\mathbf{w} = B_t\mathbf{w} \tag{3.18a}$$

$$\mathbf{z} = I_2\mathbf{x}_t = C_t\mathbf{x}_t \tag{3.18b}$$

where $\mathbf{x}_t = [x_t, y_t]^T$ and $\mathbf{w} = [w_x, w_y]^T$. The target modelled as the first-order system can sharply change its velocity. As ground moving targets change their velocity much faster than aircraft, it would be a reasonable assumption.

The velocity range is bounded by

$$0 \leq w_x^2 + w_y^2 \leq w_{\max}^2 \tag{3.19}$$

where $w_{\max}$ greater than 0 is the maximum target speed.

3.2.2 Optimal Target Tracking Problem

The cost function to be minimized is the distance between the target and the aircraft as follows:

$$\underset{\mathbf{w}(t)\in\mathbb{W}}{\text{Maximize}}\ \underset{\mathbf{u}(t)\in\mathbb{U}}{\text{Minimize}} J = \int_{t=t_0}^{t=t_f} [\mathbf{y}(t) - \mathbf{z}(t)]^T [\mathbf{y}(t) - \mathbf{z}(t)]\, dt \tag{3.20}$$

subject to

$$\dot{\mathbf{x}}_a = A_a \mathbf{x}_a + B_a \mathbf{u} \tag{3.21a}$$

$$\dot{\mathbf{x}}_t = B_t \mathbf{w} \tag{3.21b}$$

$$\mathbf{y} = C_a \mathbf{x}_a \tag{3.21c}$$

$$\mathbf{z} = C_t \mathbf{x}_t \tag{3.21d}$$

and

$$v_{\min}^2 \leq v_x^2 + v_y^2 \leq v_{\max}^2 \tag{3.22a}$$

$$u_{x_{\min}} \leq u_x \cos\phi + u_y \sin\phi \leq u_{x_{\max}} \tag{3.22b}$$

$$u_{y_{\min}} \leq -u_x \sin\phi + u_y \cos\phi \leq u_{y_{\max}} \tag{3.22c}$$

$$-\frac{1}{r_{\min}}\left(v_x^2 + v_y^2\right)^{3/2} \leq v_x u_y - v_y u_x \leq \frac{1}{r_{\min}}\left(v_x^2 + v_y^2\right)^{3/2} \tag{3.22d}$$

$$0 \leq w_x^2 + w_y^2 \leq w_{\max}^2 \tag{3.22e}$$

where $\mathbb{W}$ and $\mathbb{U}$ are the feasible control input sets for the target and the aircraft, respectively, which are not empty, t_0 and t_f are the initial time and the final time, respectively, and the initial condition, $\mathbf{x}_a(t_0)$ and $\mathbf{x}_t(t_0)$, are given. The min–max problem represents the tracking problem, where the UAV maximizes the target tracking by the camera on the UAV while the target tries to evade it.

The above optimization problem is not easy to solve because of the input constraints, u_x and u_y, and the state constraints, v_x and v_y, simultaneously. To simplify the problem, the governing differential equation is discretized as follows:

$$\mathbf{x}_a(k+1) = F_a \mathbf{x}_a(k) + G_a \mathbf{u}(k) \tag{3.23a}$$

$$\mathbf{x}_t(k+1) = F_t \mathbf{x}_t(k) + G_t \mathbf{w}(k) \tag{3.23b}$$

$$\mathbf{y}(k) = C_a \mathbf{x}_a(k) \tag{3.23c}$$

$$\mathbf{z}(k) = C_t \mathbf{x}_t(k) \tag{3.23d}$$

where F_a, G_a, F_t, G_t, H_a, and H_t are the matrices corresponding to the discretized system of the continuous system using the zero-order holder. For example,

$$F_a = \begin{bmatrix} 1 & 0 & \Delta t & 0 \\ 0 & 1 & 0 & \Delta t \\ 0 & 0 & 1 & 0 \\ 0 & 0 & 0 & 1 \end{bmatrix}, \quad G_a = \begin{bmatrix} 0 & 0 \\ 0 & 0 \\ \Delta t & 0 \\ 0 & \Delta t \end{bmatrix}, \quad F_t = I_2, \quad G_t = \Delta t I_2 \tag{3.24}$$

The integral of the cost function is approximated as a finite sum as follows:

$$\underset{\mathbf{w}(0),\mathbf{w}(1)\in\mathbb{W}}{\text{Maximize}} \; \underset{\mathbf{u}(0)\in\mathbb{U}}{\text{Minimize}} \frac{J}{\Delta t/2} \approx \sum_{k=1}^{2} [\ell(k)]^2$$

where Δt is equal to $t_f - t_0$, and

$$\ell(k) = \mathbf{y}(k) - \mathbf{z}(k)$$

We would approximate the above cost with a much smaller Δt, e.g. $\Delta t = (t_f - t_0)/n$, where n is the number of sub-intervals, then the decision variables include all $\mathbf{u}(k)$ for $k = 0,1,\dots,n-2$. For simplicity, the two-step approximation is used, and the extension for $n > 2$ is straightforward. $n = 2$ is the minimum interval number to make the control input appear in the approximated cost function, and such n is called *the relative degree* of the system.

3.2.2.1 MATLAB

Program 3.14 performs symbolic operations to obtain the approximated cost function. Figure 3.15 shows the output of *pretty()*. It displays symbolic equations in a better format to read.

```
  4    2         3            3             3             2            2              4    2
Dt  ux0  + (4 Dt  vxa0  - 2 Dt  wx0 - 2 Dt  wx1 + 2 Dt  xa0  - 2 Dt  xt0) ux0 + Dt  uy0

            3             3             3             2           2                  2     2        2
   + (4 Dt  vya0 - 2 Dt  wy0 - 2 Dt  wy1 + 2 Dt  ya0  - 2 Dt  yt0) uy0 + 5 Dt  vxa0  - 6 Dt  vxa0 wx0

         2                  2     2        2                   2                 2    2        2                2    2
   - 4 Dt  vxa0 wx1 + 5 Dt  vya0  - 6 Dt  vya0 wy0 - 4 Dt  vya0 wy1 + 2 Dt  wx0  + 2 Dt  wx0 wx1 + Dt  wx1

         2    2        2               2    2
   + 2 Dt  wy0  + 2 Dt  wy0 wy1 + Dt  wy1  + 6 Dt vxa0 xa0 - 6 Dt vxa0 xt0 + 6 Dt vya0 ya0 - 6 Dt vya0 yt0

   - 4 Dt wx0 xa0 + 4 Dt wx0 xt0 - 2 Dt wx1 xa0 + 2 Dt wx1 xt0 - 4 Dt wy0 ya0 + 4 Dt wy0 yt0 - 2 Dt wy1 ya0

                          2                  2          2                   2
   + 2 Dt wy1 yt0 + 2 xa0  - 4 xa0 xt0 + 2 xt0  + 2 ya0  - 4 ya0 yt0 + 2 yt0
```

Figure 3.15 MATLAB *pretty()* function output.

```
1 clear;
2
3 % define time interval
4 syms Dt real;
```

```
5
6  % aircraft & target dynamics
7  Fa = eye(4) + [zeros(2) Dt*eye(2); zeros(2,4)];
8  Ga = [zeros(2); Dt*eye(2)];
9  Ca = eye(2,4);
10
11 Ft = eye(2);
12 Gt = Dt*eye(2);
13 Ct = eye(2);
14
15 % define symbols for aircraft's and target's control inputs
16 syms ux0 uy0 ux1 uy1 real;
17 syms wx0 wy0 wx1 wy1 real;
18
19 u_vec_0 = [ux0 uy0]';
20 w_vec_0 = [wx0 wy0]';
21
22 u_vec_1 = [ux1 uy1]';
23 w_vec_1 = [wx1 wy1]';
24
25 % define symbols for the initial conditions
26 syms xa0 ya0 vxa0 vya0 real;
27
28 syms xt0 yt0 real;
29
30 xa_vec_0 = [xa0 ya0 vxa0 vya0]';
31 xt_vec_0 = [xt0 yt0]';
32
33 xa_k_plus_1 = Fa*xa_vec_0     + Ga*u_vec_0;
34 xa_k_plus_2 = Fa*xa_k_plus_1 + Ga*u_vec_1;
35 y_k_plus_1 = Ca*xa_k_plus_1;
36 y_k_plus_2 = Ca*xa_k_plus_2;
37
38 xt_k_plus_1 = Ft*xt_vec_0     + Gt*w_vec_0;
39 xt_k_plus_2 = Ft*xt_k_plus_1 + Gt*w_vec_1;
40 z_k_plus_1 = Ct*xt_k_plus_1;
41 z_k_plus_2 = Ct*xt_k_plus_2;
42
43 dyz_1=(y_k_plus_1-z_k_plus_1);
44 dyz_2=(y_k_plus_2-z_k_plus_2);
45 J_over_dt_2 = simplify(expand(dyz_1'*dyz_1+dyz_2'*dyz_2));
46
47 pretty(collect(J_over_dt_2,[ux0 uy0 ux1 uy1]))
```

Program 3.14 (MATLAB) Obtain the approximated cost function using symbolic manipulations

The cost function in Figure 3.15 is written in a compact form as follows:

$$\frac{J}{\Delta t/2} = \Delta t^4 \left[u_x^2(0) + \alpha u_x(0) + u_y^2(0) + \beta u_y(0)\right] + \gamma \tag{3.25}$$

where α, β, and γ are the functions of the initial conditions, $\mathbf{w}(0)$ and $\mathbf{w}(1)$. Program 3.15 obtains the functions. In the last line, the program constructs the cost function using α, β, and γ, compares it with the original cost function to confirm the obtained α, β, and γ are correct. The min–max problem given in (3.25) is impossible to solve as it requires the full knowledge of the target positions, which is not available in general. The cost function could evaluate tracking algorithms after the simulations are completed, where the target positions are all known. In addition, we can check how close the achieved cost is to the true optimal.

```
1  ux_poly = coeffs(J_over_dt_2, ux0);
2  alpha = ux_poly(2)/ux_poly(3);
3
4  uy_poly = coeffs(ux_poly(1), uy0);
5  beta = uy_poly(2)/uy_poly(3);
6
7  gama = uy_poly(1);
8
9  poly_recover=(alpha*(Dt^4)*ux0+Dt^4*ux0^2+beta*(Dt^4)*uy0+Dt^4*uy0
       ^2)+gama;
10
11 % check alpha, beta, gama are correct: the following must return
       zero
12 zero_check = eval(expand(poly_recover-J_over_dt_2));
13 fprintf('Is this zero? %4.2f \n',zero_check);
```

Program 3.15 (MATLAB) Obtain α, β, and γ in (3.25)

3.2.2.2 Python

Program 3.16 is the python script to calculate the cost function symbolically. Remind that the input argument to *np.zeros()* is the tuple. To make the 3 ×4 zero matrix, it must be *np.zeros((2,3))*. *np.zeros(2,3)* produces an error. Also, the matrix multiplication is '@'. In the Python sympy, the output of symbolic expression is printed automatically in a figure format, and a function similar to *pretty()* in MATLAB is not required.

```
1  import numpy as np
2  from sympy import symbols, simplify, expand
3
4  Dt, ux0, uy0, ux1, uy1, wx0, wy0, wx1, wy1 = symbols('Dt ux0 uy0
       ux1 uy1 wx0 wy0 wx1 wy1 ')
5  xa0, ya0, vxa0, vya0, xt0, yt0, th, w_max = symbols('xa0 ya0 vxa0
       vya0 xt0 yt0 th w_max')
6
7  # Dynamics
8  Fa = np.eye(4)+np.vstack((np.hstack((np.zeros((2,2)),Dt*np.eye(2)))
       ,np.zeros((2,4))))
9  Ga = np.vstack((np.zeros((2,2)),Dt*np.eye(2)))
```

```
10 Ca = np.eye(2,4)
11
12 Ft = np.eye(2)
13 Gt = Dt*np.eye(2)
14 Ct = np.eye(2)
15
16 # control inputs
17 u_vec_0 = np.array([[ux0], [uy0]])
18 w_vec_0 = np.array([[wx0], [wy0]])
19 u_vec_1 = np.array([[ux1], [uy1]])
20 w_vec_1 = np.array([[wx1], [wy1]])
21
22 # initial conditions
23 xa_vec_0 = np.array([[xa0], [ya0], [vxa0], [vya0]])
24 xt_vec_0 = np.array([[xt0], [yt0]])
25
26 # state propagation
27 xa_k_plus_1 = Fa@xa_vec_0     + Ga@u_vec_0
28 xa_k_plus_2 = Fa@xa_k_plus_1 + Ga@u_vec_1;
29 y_k_plus_1 = Ca@xa_k_plus_1;
30 y_k_plus_2 = Ca@xa_k_plus_2;
31
32 xt_k_plus_1 = Ft@xt_vec_0     + Gt@w_vec_0
33 xt_k_plus_2 = Ft@xt_k_plus_1 + Gt@w_vec_1
34 z_k_plus_1 = Ct@xt_k_plus_1
35 z_k_plus_2 = Ct@xt_k_plus_2
36
37 #-----------------------------------------------------------
38 # calculate the cost function in the original form
39 #-----------------------------------------------------------
40 dyz_1 = y_k_plus_1-z_k_plus_1
41 dyz_2 = y_k_plus_2-z_k_plus_2
42 J_over_dt_2 = dyz_1.T@dyz_1+dyz_2.T@dyz_2
43 J_over_dt_2 = simplify(expand(J_over_dt_2[0][0]))
44
45 alpha = J_over_dt_2.coeff(ux0,1)/(Dt**4)
46
47 temp = J_over_dt_2.coeff(ux0,0)
48 beta = temp.coeff(uy0,1)/(Dt**4)
49 gama = temp.coeff(uy0,0)
50
51 poly_recover = alpha*(Dt**4)*ux0 + (Dt**4)*(ux0**2) + beta*(Dt**4)*
       uy0 + (Dt**4)*(uy0**2) + gama
52
53 # check alpha, beta, gama are correct: the following must return
       zero
54 zero_check = float(expand(poly_recover-J_over_dt_2))
55 print(f"Is this zero? {zero_check:4.2f}\n")
```

Program 3.16 (Python) Obtain the approximated cost function using symbolic manipulations

Given that current and future positions of the target are known, (3.25) is simply a minimization problem of the quadratic function with the constraints in (3.22). The optimal solution is

$$\frac{\partial \bar{J}}{\partial u_x(0)} = 0 \rightarrow u_x(0) = -\frac{\alpha}{2}$$
$$\frac{\partial \bar{J}}{\partial u_y(0)} = 0 \rightarrow u_y(0) = -\frac{\beta}{2}$$

for $u_x(0)$ and $u_y(0)$ satisfying the constraints or inspecting the cost functions at the boundary of the constraints.

3.2.2.3 Worst-Case Scenario

Consider the min–max problem from a different aspect, i.e. the target point of view. Assume that the UAV has the optimal tracking algorithm. What would be the best policy for the target to maximize the sum of the relative distances? Construct the worst-case target movement to maximize the cost function. The aircraft moves from the position at $k = 0$ to the position at $k = 1$ as shown in Figure 3.16. When the target is to decide which direction it has to move, the best choice would be the same direction as the following vector with the maximum velocity:

$$\Delta \mathbf{r}_{T_0A_1} = [x_t(0) - x_a(1)]\mathbf{i} + [y_t(0) - y_a(1)]\mathbf{j}$$

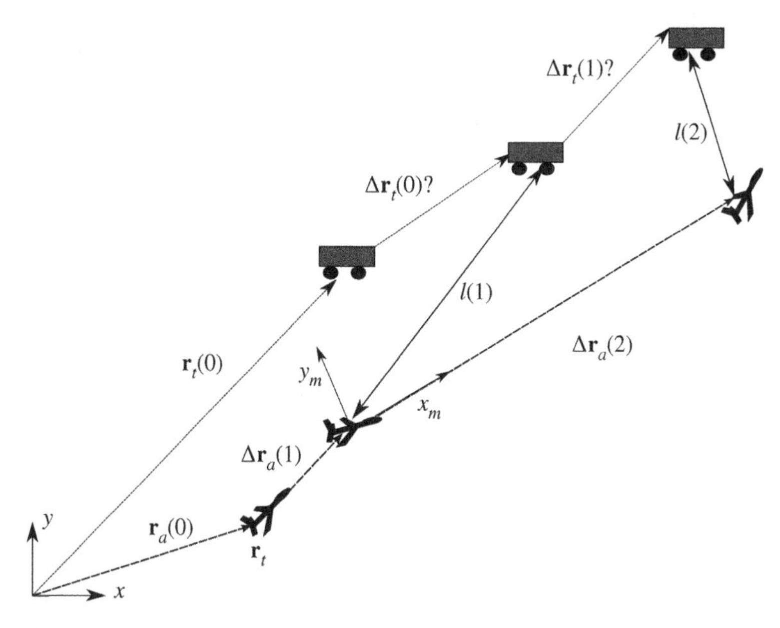

Figure 3.16 UAV target tracking in two-step prediction.

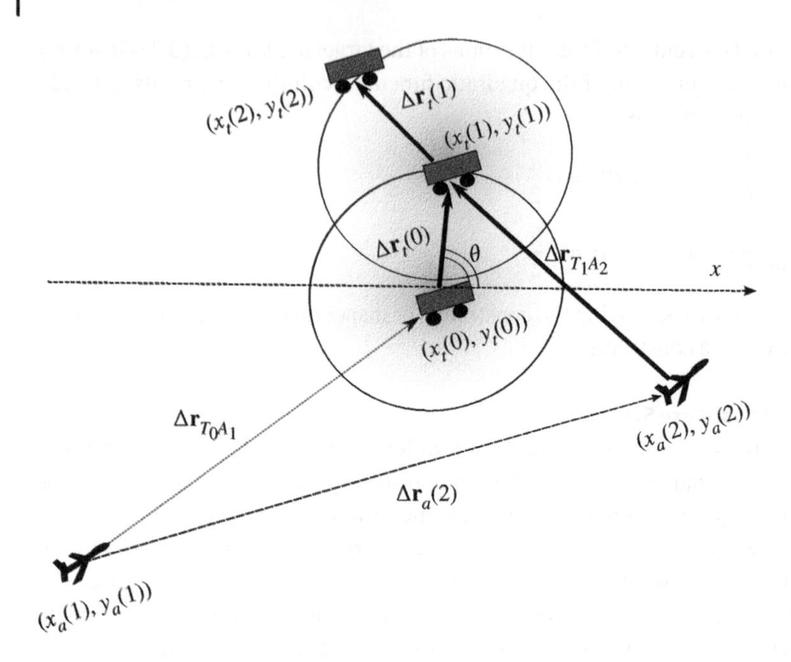

Figure 3.17 UAV target tracking in two-step prediction in the mission coordinates.

This direction, however, would not be optimal for the target to maximize the cost function as it affects the distance to be achieved in the next step at $k = 2$. Introduce the angle, θ, as the optimization parameter to maximize the cost function. The distance at $k = 1$, i.e. $\ell(1)$, indicated in Figure 3.17 is obtained as follows:

$$\ell(1) = \|\Delta\mathbf{r}_{T_0A_1} + \Delta\mathbf{r}_t(0)\| = \|\Delta\mathbf{r}_{T_0A_1} + w_{\max}\Delta t(\cos\theta\mathbf{i} + \sin\theta\mathbf{j})\|$$

For $k = 2$, the best choice for the target is to move away from the aircraft position at $k = 2$ at full speed as indicated in Figure 3.17. Hence, the distance at $k = 2$, i.e. $\ell(2)$, is obtained as follows:

$$\ell(2) = \|\Delta\mathbf{r}_{T_1A_2} + \Delta\mathbf{r}_t(1)\| = \|\Delta\mathbf{r}_{T_1A_1}\| + \|\Delta\mathbf{r}_t(1)\| = \|\Delta\mathbf{r}_{T_1A_1}\| + w_{\max}\Delta t$$

where the fact that the two vectors are parallel makes it possible to calculate the distance separately and add them. From minimizing the length, minimizing $\ell(2)$ is the same as minimizing the following function:

$$\overline{\ell}(2) = \|\Delta\mathbf{r}_{T_1A_1}\| = \|\Delta\mathbf{r}_{T_0A_1} - \Delta\mathbf{r}_a(2) + \Delta\mathbf{r}_t(0)\|$$

Minimizing the following cost function is equivalent to the original minimization problem:

$$\overline{J} = [\ell(1)]^2 + [\overline{\ell}(2)]^2 \tag{3.26}$$

The symbolic expression of $[\overline{\ell}(2)]^2$ is much simpler than that of $[\ell(2)]^2$. $[\ell(2)]^2$ is given by

$$[\ell(2)]^2 = \|\Delta \mathbf{r}_{T_1A_1}\|^2 + 2w_{\max}\Delta t\|\Delta \mathbf{r}_{T_1A_1}\| + w_{\max}^2 \Delta t^2$$

and the magnitude of $\Delta \mathbf{r}_{T_1A_1}$ in the second term takes the square root of $(\Delta \mathbf{r}_{T_1A_1} \cdot \Delta \mathbf{r}_{T_1A_1})$, which results in a complex expression. Noticing the equivalency of $\ell(2)$ and $\overline{\ell}(2)$ provides significant computational benefits in the following steps.

```
1  clear;
2  
3  %----------------------------------------------
4  % define symbols
5  %----------------------------------------------
6  % define time interval
7  syms Dt real;
8  
9  % define symbols for aircraft's and target's control inputs
10 syms ux0 uy0 ux1 uy1 real;
11 syms wx0 wy0 wx1 wy1 real;
12 
13 % define symbols for the initial conditions
14 syms xa0 ya0 vxa0 vya0 real;
15 syms xt0 yt0 real;
16 
17 % target characteristics
18 syms th w_max real;
19 
20 %----------------------------------------------
21 % Dynamics
22 %----------------------------------------------
23 % aircraft & target dynamics
24 Fa = eye(4) + [zeros(2) Dt*eye(2); zeros(2,4)];
25 Ga = [zeros(2); Dt*eye(2)];
26 Ca = eye(2,4);
27 
28 Ft = eye(2);
29 Gt = Dt*eye(2);
30 Ct = eye(2);
31 
32 % control inputs
33 u_vec_0 = [ux0 uy0]';
34 w_vec_0 = [wx0 wy0]';
35 u_vec_1 = [ux1 uy1]';
36 w_vec_1 = [wx1 wy1]';
37 
38 % initial conditions
39 xa_vec_0 = [xa0 ya0 vxa0 vya0]';
40 xt_vec_0 = [xt0 yt0]';
41 
42 % state propagation
43 xa_k_plus_1 = Fa*xa_vec_0      + Ga*u_vec_0;
```

```
44 xa_k_plus_2 = Fa*xa_k_plus_1 + Ga*u_vec_1;
45 y_k_plus_1 = Ca*xa_k_plus_1;
46 y_k_plus_2 = Ca*xa_k_plus_2;
47
48 xt_k_plus_1 = Ft*xt_vec_0   + Gt*w_vec_0;
49 xt_k_plus_2 = Ft*xt_k_plus_1 + Gt*w_vec_1;
50 z_k_plus_1 = Ct*xt_k_plus_1;
51 z_k_plus_2 = Ct*xt_k_plus_2;
52
53 %------------------------------------------------
54 % calculate the cost function with the worst target manoeuvre
55 %------------------------------------------------
56 xa1 = y_k_plus_1(1);
57 ya1 = y_k_plus_1(2);
58 xa2 = y_k_plus_2(1);
59 ya2 = y_k_plus_2(2);
60
61 r_T0A1 = [xt0 - xa1; yt0 - ya1];
62 Delta_rt_0 = [Dt*w_max*cos(th); Dt*w_max*sin(th)];
63 r_A2A1 = [xa2-xa1; ya2-ya1];
64
65 ell_1 = r_T0A1 + Delta_rt_0;
66 ell_1_squared = ell_1(:)'*ell_1(:);
67
68 r_T1A2 = ell_1 - r_A2A1;
69 ell_2_squared = r_T1A2(:)'*r_T1A2(:);% + (Dt*w_max)^2;
70
71 J_cost_worst = (ell_1_squared + ell_2_squared);
72 dJdth_worst = simplify(diff(J_cost_worst,th));
73 coeff_cos_sin = coeffs(simplify(expand(dJdth_worst)),[cos(th) sin(
       th)]);
74
75 % calculate the worst cost function
76 a_triangle = coeffs(dJdth_worst,cos(th));
77 a_triangle = -a_triangle(2);  % do not forget the minus sign
78 b_triangle = coeffs(dJdth_worst,sin(th));
79 b_triangle = b_triangle(2);
80 check_a_b = expand(-a_triangle*cos(th)+b_triangle*sin(th)-
       dJdth_worst);
81 fprintf('Check [a*cos(th)+b*sin(th)]-dJdth_worst equal to zero?
       %4.2f\n', check_a_b);
82
83 c_triangle = sqrt(a_triangle^2 + b_triangle^2);
84 J_cost_worst = eval(J_cost_worst);
85 J_cost_worst = subs(J_cost_worst,sin(th),a_triangle/c_triangle);
86 J_cost_worst = subs(J_cost_worst,cos(th),b_triangle/c_triangle);
87 J_cost_worst = expand(J_cost_worst);
88
89 dJdux0 = simplify(diff(J_cost_worst,ux0));
90 dJduy0 = simplify(diff(J_cost_worst,uy0));
```

Program 3.17 (MATLAB) The cost function, $\bar{J}$, with the worst target manoeuvre

3.2.2.4 MATLAB

We can obtain the expression of $\bar{J}$ in terms of the initial conditions using the symbolic computations shown in Program 3.17. Taking the derivative with respect to θ gives the following equation:

$$\frac{d\bar{J}}{d\theta} = 0 \Rightarrow -a_{\triangle} \cos\theta^* + b_{\triangle} \sin\theta^* = 0$$

$$\Rightarrow \tan\theta^* = \frac{u_y(0)(\Delta t)^2 + 3v_y(0)\Delta t + 2y_a(0) - 2y_t(0)}{u_x(0)(\Delta t)^2 + 3v_x(0)\Delta t + 2x_a(0) - 2x_t(0)} = \frac{a_{\triangle}}{b_{\triangle}} \tag{3.27}$$

where θ^* is the worst direction of the target to make for the given configuration. Note that there is the minus sign in front of $a_{\triangle}$. Take caution to use the function *coeffs()*. The coefficients returning from the function is ordered from the lowest to the highest. For example, the function returns the following array for the function, $2\cos\theta - 3\sin\theta + 5$:

```
1 >> syms theta real;
2 >> f=2*cos(theta)-3*sin(theta) + 5;
3 >> coeffs(f)
4 [-3*sin(theta)+5, 2]
```

where the coefficient of the cosine function, 2, is the second element of the return array.

The sine and the cosine functions in $\bar{J}$ are replaced by

$$\sin\theta^* = \frac{a_{\triangle}}{\sqrt{a_{\triangle}^2 + b_{\triangle}^2}}, \quad \cos\theta^* = \frac{b_{\triangle}}{\sqrt{a_{\triangle}^2 + b_{\triangle}^2}} \tag{3.28}$$

Program 3.18 constructs $\bar{J}$ and its derivatives using the symbolic manipulation functions in MATLAB. The worst possible cost for the tracking aircraft, $\bar{J}$, is now a function of the aircraft control input at $k = 0$, i.e. $u_x(0)$ and $u_y(0)$, only.

In line 47 of Program 3.18, the *eval()* function evaluates the symbolic expression of the cost function, obtained in Program 3.17. Replacing the symbols by the given values makes *J_cost_uxuy0* a function of $u_x(0)$ and $u_y(0)$. Calling the *eval()* function for a fixed set of the control input values shows a cost function contour plot. It evaluates the symbolic expression many times and is slower than usual numerical evaluations. To speed up the symbolic evaluations, the symbolic expressions are converted to a function using the *matlabFunction()*. Once it converts into a function, it allows evaluating the function with vector or matrix inputs. Hence, a set of all combinations of the control inputs is evaluated in one line, line 51.

```
1 %--------------------------------------------------------
2 % evaluate the cost function for test scenario values
3 %--------------------------------------------------------
4 
5 % initial target position
6 xt0 = (2*rand(1)-1)*200;  %[m]
```

```
7  yt0 = (2*rand(1)-1)*200;  %[m]
8
9  % initial uav position
10 xa0 = (2*rand(1)-1)*100; %[m]
11 ya0 = (2*rand(1)-1)*100; %[m]
12
13 % initial uav velocity
14 tha0 = rand(1)*2*pi; %[radian]
15 current_speed = 25; %[m/s]
16 vxa0 = current_speed*cos(tha0);
17 vya0 = current_speed*sin(tha0);
18
19 % uav minimum & maximum speed
20 v_min = 20; v_max = 40;
21
22 % time interval for the cost approximation
23 Dt = 2; % [seconds]
24
25 % target maximum speed
26 w_max = 60*1e3/3600; %[m/s]
27
28 % uav flying path curvature constraint
29 r_min = 400; %[m]
30
31 % control acceleration input magnitude constraints
32 ux_max = 10; % [m/s^2]
33 ux_min = -1; % [m/s^2]
34 uy_max = 2;  % [m/s^2]
35 uy_min = -2; % [m/s^2]
36
37 ux_max_org = ux_max;
38 ux_min_org = ux_min;
39
40 % evaluate the cost function over the ux0-uy0 control input
41 num_idx = 20;
42 num_jdx = 19;
43 min_max_u_plot = 20;
44 ux_all = linspace(-min_max_u_plot,min_max_u_plot,num_idx);
45 uy_all = linspace(-min_max_u_plot,min_max_u_plot,num_jdx);
46
47 J_cost_uxuy0 = eval(J_cost_worst);
48 J_cost_uxuy0_function = matlabFunction(J_cost_uxuy0);
49
50 [UX0,UY0]=meshgrid(ux_all,uy_all);
51 J_cost_worst_val=J_cost_uxuy0_function(UX0,UY0);
```

Program 3.18 (MATLAB) Evaluate the cost function, $\bar{J}$, in the control input space

Run the following program and compare the execution time. The part with *matlabFunction()* is around 100 times faster than the nested for-loop part.

```
1  clear;
2
3  syms x1 x2 real;
4  f_x1x2 = x1 + x2^2;
5
6  row = 100; col = 99;
7
8  x1_list = linspace(-10,10,row);
9  x2_list = linspace(-10,10,col);
10
11 f_ij_1 = zeros(row,col);
12 tic
13 for idx=1:row
14     x1 = x1_list(idx);
15     for jdx=1:col
16         x2 = x2_list(jdx);
17         f_ij_1(idx,jdx) = eval(f_x1x2);
18     end
19 end
20 toc
21
22 tic
23 f_x1x2_fun = matlabFunction(f_x1x2);
24 [X1,X2]=meshgrid(x1_list,x2_list);
25 f_ij_2 = f_x1x2_fun(X1',X2');
26 toc
```

3.2.2.5 Python

Program 3.19 calculates $\bar{J}$ and $d\bar{J}/d\theta$ symbolically. The derivative obtained in line 68 can be collected by the trigonometric terms as follows:

```
1 In [19]: from sympy import collect
2
3 In [20]: print(collect(dJdth_worst,[sin(th),cos(th)]))
4 (2.0*Dt**3*ux0*w_max + 6.0*Dt**2*vxa0*w_max + 4.0*Dt*w_max*xa0 - 4*
      Dt*w_max*xt0)*sin(th) + (-2.0*Dt**3*uy0*w_max - 6.0*Dt**2*vya0*
      w_max - 4.0*Dt*w_max*ya0 + 4*Dt*w_max*yt0)*cos(th)
```

which confirms that the worst θ occurs at (3.27). Program 3.19 declares the symbols as the real type in lines 12 and 13. As Δt and $w_{\max}$ are always positive real values, the real and the positive flags are set to True. This helps to simplify the square root of square variables, for example, $\sqrt{x^2}$, into x instead of $|x|$.

Declaring the symbol as real simplifies the rest of the symbolic calculation compared to the computation time when the symbol type is complex. Unlike MATLAB, $\partial J/\partial u_x(0)$ and $\partial J/\partial u_y(0)$ at the end of Program 3.19 do not call *simplify()* in the

Sympy. The MATLAB *simplify()* cannot make the derivative results simpler and gives up on the tasks quickly. On the other hand, *simplify()* in the Sympy takes a long computation without any meaningful simplification.

```
1  import numpy as np
2  import matplotlib.pyplot as plt
3  from matplotlib import path
4
5  from sympy import symbols, simplify, expand
6  from sympy import cos, sin, sqrt, diff
7  from sympy.utilities.lambdify import lambdify
8
9  import time
10 from scipy.optimize import minimize, fsolve
11
12 ux0, uy0, ux1, uy1, wx0, wy0, wx1, wy1 = symbols('ux0 uy0 ux1 uy1
       wx0 wy0 wx1 wy1', real=True)
13 xa0, ya0, vxa0, vya0, xt0, yt0, th = symbols('xa0 ya0 vxa0 vya0 xt0
        yt0 th', real=True)
14
15 Dt, w_max = symbols('Dt w_max', real=True, positive=True)
16
17 #------------------------------------------------
18 # Dynamics
19 #------------------------------------------------
20 Fa = np.eye(4)+np.vstack((np.hstack((np.zeros((2,2)),Dt*np.eye(2)))
        ,np.zeros((2,4))))
21 Ga = np.vstack((np.zeros((2,2)),Dt*np.eye(2)))
22 Ca = np.eye(2,4)
23
24 Ft = np.eye(2)
25 Gt = Dt*np.eye(2)
26 Ct = np.eye(2)
27
28 # control inputs
29 u_vec_0 = np.array([[ux0], [uy0]])
30 w_vec_0 = np.array([[wx0], [wy0]])
31 u_vec_1 = np.array([[ux1], [uy1]])
32 w_vec_1 = np.array([[wx1], [wy1]])
33
34 # initial conditions
35 xa_vec_0 = np.array([[xa0], [ya0], [vxa0], [vya0]])
36 xt_vec_0 = np.array([[xt0], [yt0]])
37
38 # state propagation
39 xa_k_plus_1 = Fa@xa_vec_0     + Ga@u_vec_0
40 xa_k_plus_2 = Fa@xa_k_plus_1 + Ga@u_vec_1;
41 y_k_plus_1 = Ca@xa_k_plus_1;
42 y_k_plus_2 = Ca@xa_k_plus_2;
43
44 xt_k_plus_1 = Ft@xt_vec_0     + Gt@w_vec_0
45 xt_k_plus_2 = Ft@xt_k_plus_1 + Gt@w_vec_1
```

```
46 z_k_plus_1 = Ct@xt_k_plus_1
47 z_k_plus_2 = Ct@xt_k_plus_2
48 
49 #------------------------------------------------------------
50 # calculate the cost function with the worst target manoeuvre
51 #------------------------------------------------------------
52 xa1 = y_k_plus_1[0][0]
53 ya1 = y_k_plus_1[1][0]
54 xa2 = y_k_plus_2[0][0]
55 ya2 = y_k_plus_2[1][0]
56 
57 r_T0A1 = np.array([[xt0 - xa1], [yt0 - ya1]])
58 Delta_rt_0 = np.array([[Dt*w_max*cos(th)], [Dt*w_max*sin(th)]])
59 r_A2A1 = np.array([[xa2-xa1], [ya2-ya1]])
60 
61 ell_1 = r_T0A1 + Delta_rt_0
62 ell_1_squared = (ell_1.T@ell_1)[0][0]
63 
64 r_T1A2 = ell_1 - r_A2A1
65 ell_2_squared = (r_T1A2.T@r_T1A2)[0][0]
66 
67 J_cost_worst = expand(ell_1_squared + ell_2_squared)
68 dJdth_worst = expand(simplify(diff(J_cost_worst,th)))
69 coeff_cos = dJdth_worst.coeff(cos(th))
70 coeff_sin = dJdth_worst.coeff(sin(th))
71 
72 # calculate the worst coast function
73 a_triangle = -coeff_cos # do not forget the minus sign
74 b_triangle = coeff_sin
75 c_triangle = simplify(sqrt(a_triangle**2 + b_triangle**2))
76 check_a_b = float(expand(-a_triangle*cos(th)+b_triangle*sin(th)-
       dJdth_worst))
77 print(f'Check [-a*cos(th)+b*sin(th)]-dJdth_worst equal to zero? {
       check_a_b:4.2f}')
78 
79 J_cost_worst = J_cost_worst.subs(sin(th),a_triangle/c_triangle)
80 J_cost_worst = J_cost_worst.subs(cos(th),b_triangle/c_triangle)
81 
82 dJdux0 = diff(J_cost_worst,ux0)
83 dJduy0 = diff(J_cost_worst,uy0)
84 
85 #------------------------------------------------------------
86 # evaluate the cost function for test scenario values
87 #------------------------------------------------------------
88 
89 # initial target position
90 xt0_v = (2*np.random.rand(1)-1)*200*0+150  #[m]
91 yt0_v = (2*np.random.rand(1)-1)*200*0  #[m]
92 
93 # initial uav position
94 xa0_v = (2*np.random.rand(1)-1)*100*0 #[m]
95 ya0_v = (2*np.random.rand(1)-1)*100*0 #[m]
96 
```

```
 97 # initial uav velocity
 98 tha0 = np.random.rand(1)*2*np.pi*0 #[radian]
 99 current_speed = 25 #[m/s]
100 vxa0_v = current_speed*np.cos(tha0)
101 vya0_v = current_speed*np.sin(tha0)
102 
103 # uav minimum & maximum speed
104 v_min = 20  #[m/s]
105 v_max = 40  #[m/s]
106 
107 # time interval for the cost approximation
108 Dt_v = 2 # [seconds]
109 
110 # target maximum speed
111 w_max_v = 60*1e3/3600 #[m/s]
112 
113 # uav flying path curvature constraint
114 r_min = 400 #[m]
115 
116 # control acceleration input magnitude constraints
117 ux_max = 10 # [m/s^2]
118 ux_min = -1 # [m/s^2]
119 uy_max = 2  # [m/s^2]
120 uy_min = -2 # [m/s^2]
121 
122 ux_max_org = ux_max
123 ux_min_org = ux_min
124 uy_max_org = uy_max
125 uy_min_org = uy_min
126 
127 # evaluate the cost function over the ux0-uy0 control input
128 num_idx = 20
129 num_jdx = 19
130 min_max_u_plot = 20
131 ux_all = np.linspace(-min_max_u_plot,min_max_u_plot,num_idx)
132 uy_all = np.linspace(-min_max_u_plot,min_max_u_plot,num_jdx)
133 
134 values = [(Dt,Dt_v), (xa0,xa0_v[0]), (ya0,ya0_v[0]), (vxa0,vxa0_v
        [0]), (vya0,vya0_v[0]),
135           (xt0,xt0_v[0]), (yt0,yt0_v[0]), (w_max,w_max_v)]
136 
137 J_cost_uxuy0 = J_cost_worst.subs(values)
138 J_cost_uxuy0_function = lambdify([ux0,uy0],J_cost_uxuy0)
139 
140 UX0,UY0=np.meshgrid(ux_all,uy_all)
141 J_cost_worst_val=J_cost_uxuy0_function(UX0,UY0)
142 
143 # replace symbols by values
144 Dt = Dt_v
145 xa0 = xa0_v[0]
146 ya0 = ya0_v[0]
147 vxa0 = vxa0_v[0]
148 vya0 = vya0_v[0]
```

```
149 xt0 = xt0_v[0]
150 yt0 = yt0_v[0]
151 w_max = w_max_v
```

Program 3.19 (Python) The cost function, $\bar{J}$, with the worst target manoeuvre

Program 3.19 uses *subs()* to substitute values into the symbols in sympy. After the substitutions, the result is still a symbol, unlike floating-point values in MATLAB. Consider the following Python commands:

```
1 In [1]: from sympy import symbols
2 In [2]: x=symbols('x')
3 In [3]: f=(x+4)**2
```

To evaluate 'f' at 'x=3.0'

```
1  In [4]: z=f.subs([(x,3.0)])
2  In [5]: z2=49.0
3  In [6]: whos
4  Variable    Type        Data/Info
5  ------------------------------
6  f           Pow         (x + 4)**2
7  symbols     function    <function symbols at 0x7f12f6c32160>
8  x           Symbol      x
9  z           Float      49.0000000000000000
10 z2          float      49.0
```

The result, *z*, is not a usual floating-point value but a symbolic float type, while *z2* is a floating-point value. To convert the symbolic value to the floating-point value, use *float()* as follows:

```
1 In [7]: z=float(z)
```

Or, make a function using *lambdify()* as follows:

```
1  In [8]: from sympy.utilities.lambdify import lambdify
2  In [9]: g=lambdify([x],f)
3  In [10]: z=g(3.0)
4  In [11]: whos
5  Variable    Type        Data/Info
6  ------------------------------
7  f           Pow         (x + 4)**2
8  g           function    <function _lambdifygenerated at 0
       x7f12daa26ee0>
9  lambdify    function    <function lambdify at 0x7f12f6dc1820>
10 symbols     function    <function symbols at 0x7f12f6c32160>
```

```
11 x              Symbol     x
12 z              float      49.0
13 z2             float      49.0
```

This is how the function is constructed and called on lines 138 and 141 of Program 3.19.

3.2.2.6 Optimal Control Input

The next step is to calculate the control input to minimize the worst cost, and it completes the min–max optimization problem. Assume that v_x and v_y at $k = 0$ satisfy the magnitude constraint. We check the velocity at $k = 1$ as follows:

$$\frac{v_x^2(1) + v_y^2(1)}{(\Delta t)^2} = \left[\frac{v_x(0)}{\Delta t} + u_x(0)\right]^2 + \left[\frac{v_y(0)}{\Delta t} + u_y(0)\right]^2 = r_v^2 \tag{3.29}$$

It is the equation of a circle centred at $(-v_x(0)/\Delta t, -v_y(0)/\Delta t)$ in the $u_x(0)$–$u_y(0)$ domain. The radius, r_v, is between $v_{\min}/\Delta t$ and $v_{\max}/\Delta t$.

Figure 3.18 shows the two circles indicating the circular constraints. In the figure, the control input magnitude given by (3.14) is the dashed line box. The curvature limitation given by (3.16) adds two line restrictions parallel to $u_x^B(0)$ indicated in the figure.

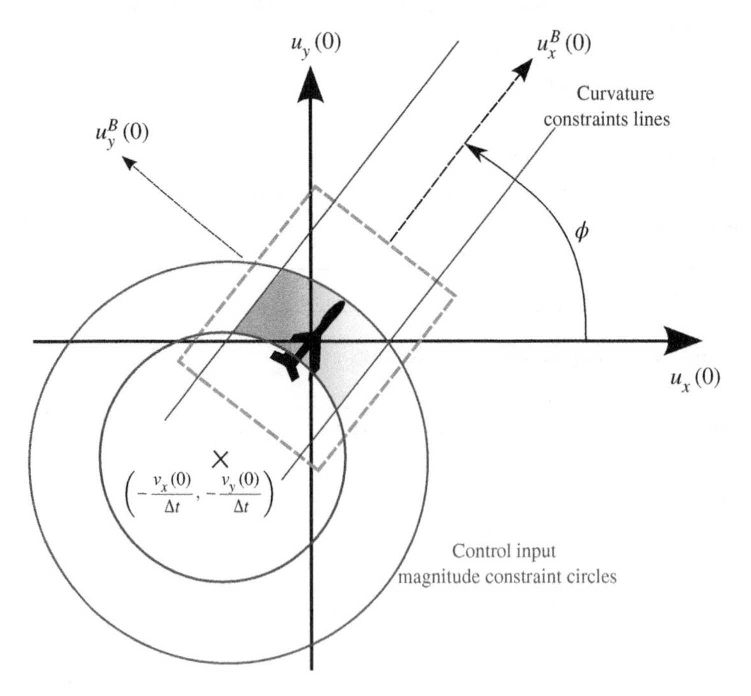

Figure 3.18 Feasible control input space indicated by the painted region.

As $v_x(0)$ is strictly greater than zero, i.e. safe for division, the following line equation is obtained:

$$u_y(0) = m_{\text{cvt}} u_x(0) \pm c_{\text{cvt}} \tag{3.30}$$

where

$$m_{\text{cvt}} = \tan\phi = \frac{v_y(0)}{v_x(0)}, \quad c_{\text{cvt}} = \frac{1}{r_{\min} v_x(0)} \left[v_x(0)^2 + v_y(0)^2 \right]^{3/2}$$

For $v_x(0)$ closer to zero implying $v_y(0)$ away from zero, the lines are

$$u_x(0) = n_{\text{cvt}} u_y(0) \mp d_{\text{cvt}}$$

where

$$n_{\text{cvt}} = \tan\psi = \frac{v_x(0)}{v_y(0)} = \tan\left(\frac{\pi}{2} - \phi\right), \quad d_{\text{cvt}} = \frac{1}{r_{\min} v_y(0)} \left[v_x(0)^2 + v_y(0)^2 \right]^{3/2}$$

Including all restrictions, the feasible input space is indicated by the darker (or gray) region in Figure 3.18.

For the following example values: $v_{\min} = 20\,\text{m/s}$, $v_{\max} = 40\,\text{m/s}$, $u_{x_{\min}} = -1\,\text{m/s}^2$, $u_{x_{\max}} = 10\,\text{m/s}^2$, $u_{y_{\min}} = -2\,\text{m/s}^2$, $u_{y_{\max}} = 2\,\text{m/s}^2$, $r_{\min} = 400\,\text{m}$, $w_{\max} = 60\,\text{km/h}$, $\Delta t = 2\,\text{s}$, $x_a(0) = -84.86\,\text{m}$, $y_a(0) = 66.72\,\text{m}$, $v_x(0) = 19.12\,\text{m/s}$, $v_y(0) = -16.10\,\text{m/s}$, $x_t(0) = 50.15\,\text{m}$, and $y_t(0) = 125.02\,\text{m}$, the control input constraints over the cost function are shown in Figure 3.19. For this example set, the control input magnitude constraints, the curvature constraints, and the maximum velocity constraint are all active, and the minimum occurs at the boundary of the intersection area of the constraints.

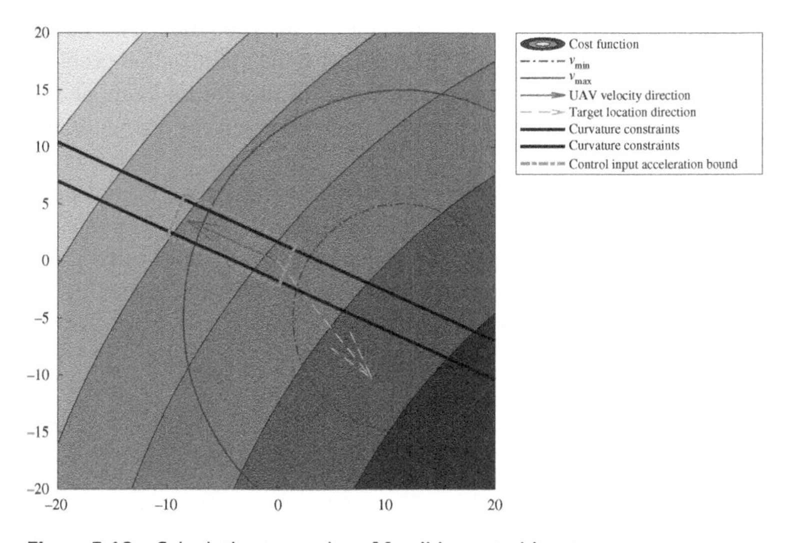

Figure 3.19 Calculation examples of feasible control input space.

In Program 3.17, we obtain the cost function derivative for the control input. In addition, we would try to solve

$$\frac{d\bar{J}}{du_x(0)} = 0, \quad \frac{d\bar{J}}{du_y(0)} = 0$$

by the following MATLAB command:

```
>> solve([dJdux0==0,dJduy0==0],[ux0 uy0])
```

The equations are complex and less likely would have analytical solutions except for trivial cases. After all variables in the first derivatives are set to the values given in Program 3.18 using *eval()* as follows:

```
>> solve([eval(dJdux0)==0,eval(dJduy0)==0],[ux0 uy0])
```

It returns the solution solved numerically. Then, we check the solution if it is inside the constraints. For the given configurations of the aircraft and the target, the optimal control would be inside the constraints for only a few limited cases. It is a waste of computing resources as the results are mostly rejected.

Instead, we assume that the solution would be outside of the constraints. If the solution is outside the bounds of the constraint, the minimum occurs at the boundary. Sampling points along the boundary of the constraints, calculating the cost values for the points, and selecting the control input corresponding to the minimum cost value among the samples. The next step is checking if the assumption is correct. Sampling points inside the constraints, calculating the cost values, finding the minimum cost, and checking if it is smaller than the minimum cost found at the boundary. If it is smaller than the one at the boundary, the global minimum occurs inside the boundary. Then, we solve the optimization problem with the initial guess that we just found. Otherwise, the minimum found on the boundary is optimal. For example, the optimal input occurs at the boundary in a scenario shown in Figure 3.20.

For solving the optimization problem inside the boundary, we do not need to consider the constraint as long as the minimization algorithm converges to the minimum inside. We guarantee convergence based on the convexity of the cost function. The convexity of the cost function is checked numerically by constructing the Hessian matrix as follows:

$$H[u_x(0), u_y(0)] = \left. \begin{bmatrix} \dfrac{\partial^2 \bar{J}}{\partial u_x^2} & \dfrac{\partial^2 \bar{J}}{\partial u_x \partial u_y} \\ \dfrac{\partial^2 \bar{J}}{\partial u_x \partial u_y} & \dfrac{\partial^2 \bar{J}}{\partial u_y^2} \end{bmatrix} \right|_{\substack{u_x = u_x(0) \\ u_y = u_y(0)}} \tag{3.31}$$

calculating the eigenvalues and inspecting if the minimum eigenvalue is positive.

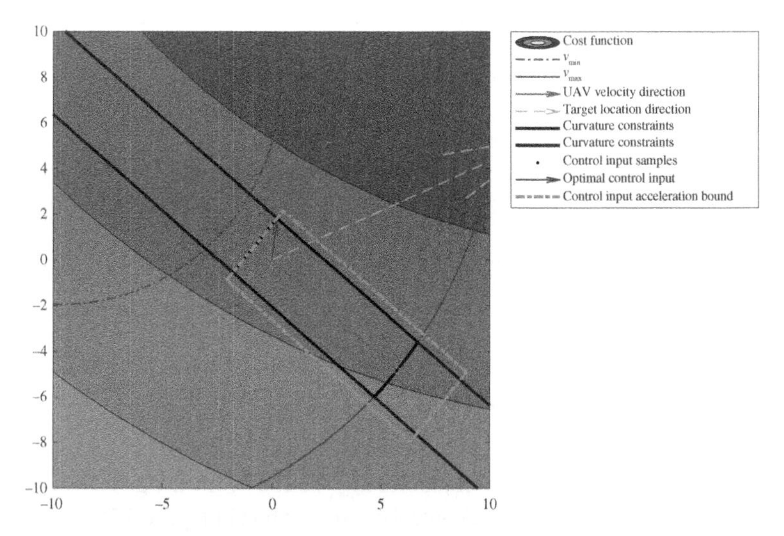

Figure 3.20 Optimal control input for the target tracking.

3.3 Tracking Algorithm Implementation

Algorithm 3.1 summarizes the tracking algorithm. The gradient descent method with Armijo's rule in Algorithm 3.2 solves the minimization problem when the solution is inside the boundary. The gradient vector, $\mathbf{g}_k$, determines the search direction in the gradient descent, and Armijo's rule by adjusting α_{amj} determines how far it moves in the search direction. We discuss step-by-step implementations of the tracking algorithm in MATLAB and Python.

3.3.1 Constraints

3.3.1.1 Minimum Turn Radius Constraints

The curvature constraint in (3.30) given in the global frame becomes a lot simpler if it is in the body frame. As shown in Figure 3.18, the UAV velocity is aligned with the x-body axis and $v_y^B(0) = 0$. The curvature lines in the body frame become

$$u_y^B = \pm c_{\text{cvt}}^B$$

as the slope is zero in the body frame, where

$$c_{\text{cvt}}^B = \frac{1}{r_{\min} v_x^B(0)} \left\{ \left[v_x^B(0)\right]^2 + 0 \right\}^{3/2} = \frac{\left[v_x^B(0)\right]^2}{r_{\min}}$$

If c_{cvt}^B is smaller than $u_{y_{\max}}$ or $u_{y_{\min}}$, the corresponding bound is replaced by c_{cvt}^B. In the program, we assume that the UAV is symmetric along the x-body axis, hence, $u_{y_{\min}} = -u_{y_{\max}}$.

Algorithm 3.1 Optimal target tracking control input

1: Initialize the prediction interval and the UAV/target constraints

$$\Delta t,\ v_{\max},\ v_{\min},\ u_{x_{\max}},\ u_{x_{\min}},\ u_{y_{\max}},\ u_{y_{\min}},\ r_{\min},\ w_{\max}$$

2: Set the initial position/velocity of UAV and the target position

$$x_a(0),\ y_a(0),\ v_x(0), v_y(0),\ x_t(0),\ y_t(0)$$

3: Calculate $v_s(0) = \sqrt{v_x(0)^2 + v_y(0)^2}$

4: Calculate $u_{\text{cvt}} = v_s(0)/r_{\min}$

5: **if** $u_{\text{cvt}} < u_{y_{\max}}$ or $u_{\text{cvt}} < |u_{y_{\min}}|$ **then** replace

$$u_{y_{\max}} = u_{\text{cvt}},\quad u_{y_{\min}} = -u_{\text{cvt}}$$

6: **end if**

7: **if** $v_s(0)/\Delta t + u_{x_{\max}} > v_{\max}/\Delta t$ **then** replace $u_{\max}$ bound by the arc given by the larger circle intersecting with the control constraint box in Figure 3.18.

8: **end if**

9: **if** $v_s(0)/\Delta t + u_{x_{\min}} < v_{\min}/\Delta t$ **then** replace $u_{\min}$ bound by the arc given by the smaller circle intersecting with the control constraint box in Figure 3.18.

10: **end if**

11: Sample points *along the boundary of the constraints*, and calculate the cost for the samples; Find the optimal control corresponding to the minimum cost, J^*_{bd}, among the samples

12: Sample points *inside the boundary of the constraints* and calculate the cost for the samples; Find the optimal control corresponding to the minimum cost, J^*_{in}, among the samples

13: **if** $J^*_{\text{bd}} \leq J^*_{\text{in}}$ **then**

14: $u_x(0)$ and $u_y(0)$ corresponding to J^*_{bd} are optimal

15: **else**

16: Let $u_x(0)$ and $u_y(0)$ corresponding to J^*_{in} be the initial guess of the minimization of (3.26), where the sine and the cosine functions are substituted by (3.28). The optimization can be solved using the gradient descent method with Armijo's rule (Armijo, 1966) summarized in Algorithm 3.2.

17: **end if**

Programs 3.20 and 3.21 are part of the tracking algorithm implementation of the algorithm in MATLAB and Python, respectively. Considering the input constraints in the body frame simplifies the calculations, and later, they can transform into ones in the global frame when it needs them to be expressed in the global coordinates.

Algorithm 3.2 The gradient descent with Armijo's rule

1: Initialize $\mathbf{u}_k = [u_x(0), u_y(0)]^T, s_{\text{amj}} = 0.01, \beta_{\text{amj}} = 0.5, \sigma_{\text{amj}} = 10^{-5}, \Delta u = 10$ and $\epsilon = 10^{-6}$. Be careful as the solution may diverge if s_{amj} is set to a large value.
2: **while** $\Delta u > \epsilon$ **do**
3: Calculate $\bar{J}(\mathbf{u}_k)$ given in (3.26)
4: $\mathbf{g}_k \leftarrow \partial \bar{J}/\partial \mathbf{u}_k$
5: $\alpha_{\text{amj}} \leftarrow s_{\text{amj}}$
6: $\mathbf{u}_{k+1} \leftarrow \mathbf{u}_k + \alpha_{\text{amj}} \mathbf{g}_k$
7: **while** $\bar{J}(\mathbf{u}_{k+1}) > \bar{J}(\mathbf{u}_k) + \sigma_{\text{amj}} \alpha_{\text{amj}} \left(\mathbf{g}_k^T \mathbf{g}_k\right)$ **do**
8: $\alpha_{\text{amj}} \leftarrow \beta_{\text{amj}} \alpha_{\text{amj}}$
9: $\mathbf{u}_{k+1} \leftarrow \mathbf{u}_k + \alpha_{\text{amj}} \mathbf{g}_k$
10: **end while**
11: $\Delta u \leftarrow \|\mathbf{u}_k - \mathbf{u}_{k+1}\|$
12: $\mathbf{u}_k \leftarrow \mathbf{u}_{k+1}$
13: $\alpha_{\text{amj}} \leftarrow s_{\text{amj}}$
14: **end while**

3.3.1.2 Velocity Constraints

Two circles in Figure 3.21 represent the maximum and the minimum velocity constraints given in (3.29) equal to the following:

$$\left[\frac{v_x^B(0)}{\Delta t} + u_x^B(0)\right]^2 + [u_y^B(0)]^2 = r_v^2$$

in the body frame. For the larger circle, i.e. the maximum velocity constraint at $u_x^B(0) = u_{x_{\text{max}}}$ and $u_y^B(0) = 0$ provides

$$\left[\frac{v_x^B(0)}{\Delta t} + u_{x_{\text{max}}}\right]^2 > \left(\frac{v_{\text{max}}}{\Delta t}\right)^2 \Rightarrow \frac{\|v\|}{\Delta t} + u_{x_{\text{max}}} > \frac{v_{\text{max}}}{\Delta t}$$

where $\|v\|$ is the current speed of the UAV. If the inequality is true, the maximum velocity circle passes through the control magnitude constraint as illustrated in Figure 3.21. The x_B coordinate of the square dot is smaller than the one of the star dot. Therefore, the arc of the larger circle in the figure becomes the x_B-direction maximum control input bound.

To find the x_B-coordinate of the square dot, establish the following equation:

$$\left[\frac{\|v\|}{\Delta t} + u_{x_{\text{max}}}\right]^2 + u_{y_{\text{max}}}^2 = \left(\frac{v_{\text{max}}}{\Delta t}\right)^2$$

where $u_{x_{\text{max}}}$ is *not* the original maximum control input in the x_B-direction but the x_B-direction control input corresponding to $u_{y_{\text{max}}}$ illustrated in Figure 3.22. The purpose of finding this value is to sample the dots along the arc as shown

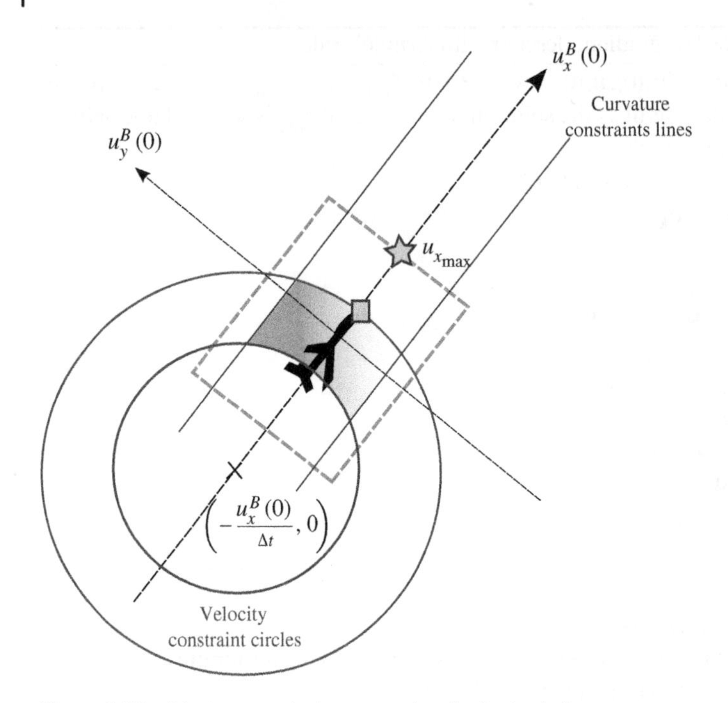

Figure 3.21 Maximum velocity constraints in the body frame.

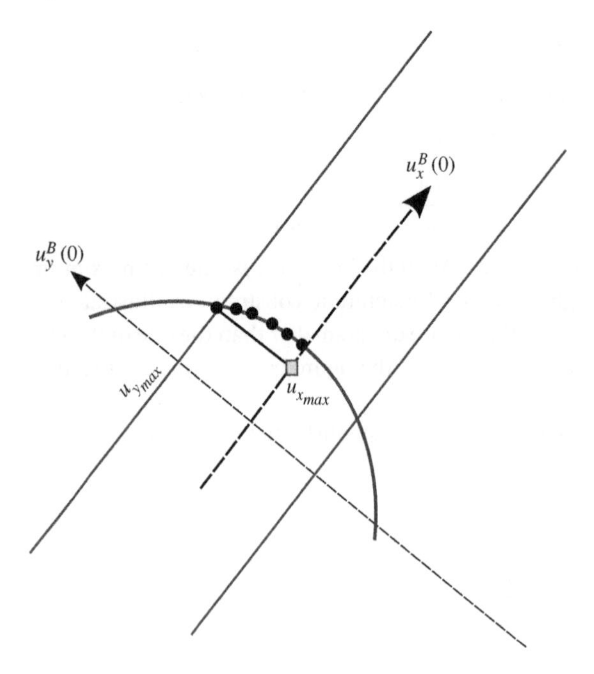

Figure 3.22 Maximum velocity arc sampling.

in the figure. Expand the equation

$$u_{x_{\max}}^2 + 2\frac{\|v\|}{\Delta t}u_{x_{\max}} + \left[\left(\frac{\|v\|}{\Delta t}\right)^2 + u_{y_{\max}}^2 - \left(\frac{v_{\max}}{\Delta t}\right)^2\right] = 0$$

and solve for $u_{x_{\max}}$

$$u_{x_{\max}} = -\frac{\|v\|}{\Delta t} \pm \sqrt{\left(\frac{v_{\max}}{\Delta t}\right)^2 - u_{y_{\max}}^2} \tag{3.32}$$

where we take the larger $u_{x_{\max}}$. A similar derivation is performed for a smaller circle, i.e. the minimum velocity constraint. Programs 3.20 and 3.21 implement these, where the direction cosine matrix to transform between the global and the body frames are calculated as follows:

$$\begin{bmatrix} u_x(k) \\ u_y(k) \end{bmatrix} = \begin{bmatrix} \cos\phi & -\sin\phi \\ \sin\phi & \cos\phi \end{bmatrix} \begin{bmatrix} u_x^B(k) \\ u_y^B(k) \end{bmatrix}$$

```
1  %% Optimal control input
2  %-------------------------------------------------------
3  % find optimal control
4  %-------------------------------------------------------
5  
6  % check the curvature constraint in the body frame
7  u_curvature = current_speed^2/r_min;
8  if u_curvature < uy_max
9      % active constraint & replace the uy bound
10     uy_max = u_curvature;
11     uy_min = -u_curvature;
12 end
13 
14 % active the maximum velocity constraint
15 vmax_active = false;
16 if current_speed/Dt+ux_max > v_max/Dt
17     ux_max = -current_speed/Dt+sqrt((v_max/Dt)^2-uy_max^2);
18     vmax_active = true;
19 end
20 
21 % active the minimum velocity constraint
22 vmin_active = false;
23 if current_speed/Dt+ux_min < v_min/Dt
24     ux_min = -current_speed/Dt+sqrt((v_min/Dt)^2-uy_max^2);
25     vmin_active = true;
26 end
27 
28 th_flight = atan2(vya0,vxa0);
29 dcm_from_body_to_global = [cos(th_flight) -sin(th_flight); sin(
       th_flight) cos(th_flight)];
```

Program 3.20 (MATLAB) Turn-radius and velocity constraints in the body frame

```
## Optimal control input
#------------------------------------------------
# find optimal control
#------------------------------------------------

# check the curvature constraint in the body frame
u_curvature = current_speed**2/r_min
if u_curvature < uy_max:
    # active constraint & replace the uy bound
    uy_max = u_curvature
    uy_min = -u_curvature

# active the maximum velocity constraint
vmax_active = False
if current_speed/Dt+ux_max > v_max/Dt:
    ux_max = -current_speed/Dt+np.sqrt((v_max/Dt)**2-uy_max**2)
    vmax_active = True

# active the minimum velocity constraint
vmin_active = False
if current_speed/Dt+ux_min < v_min/Dt:
    ux_min = -current_speed/Dt+np.sqrt((v_min/Dt)**2-uy_max**2)
    vmin_active = True

th_flight = np.arctan2(vya0,vxa0)
dcm_from_body_to_global = np.array([
  [np.cos(th_flight), -np.sin(th_flight)],
        [np.sin(th_flight), np.cos(th_flight)]])
```

Program 3.21 (Python) Turn-radius and velocity constraints in the body frame

3.3.2 Optimal Solution

3.3.2.1 Control Input Sampling

Depending on which constraints are active, the feasible control input region has four possible shapes shown in Figure 3.23. Each case has four sides, and they are called the upper, the lower, the left, and the right sides, respectively, in Programs 3.22 and 3.23. Sampling along the straight lines are trivial. Figure 3.22 indicates the samples along the arc. Whenever the maximum velocity constraint is active, $u_{x_{\max}}$ is set to the u_x^B coordinates corresponding to $u_{y_{\min}}$ in Figure 3.22. The u_x^B-coordinates of the samples on the arc are between $u_{x_{\max}}$ and the point where the arc and u_x^B axis meet. The u_x^B coordinate of the point, where the arc meets u_x^B axis, satisfies the equation for the circle with $u_y^B = 0$, i.e.

$$u_x^B = \frac{v_{\max} - \|v\|}{\Delta t} \tag{3.33}$$

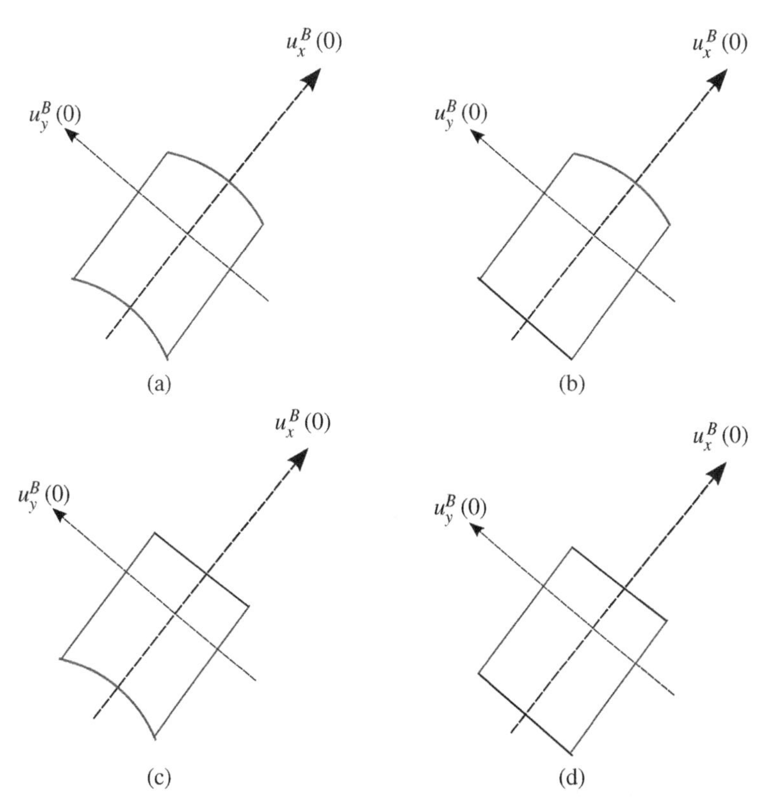

Figure 3.23 Four possible shapes of the feasible control input set: (a) the maximum and the minimum velocity constraints active; (b) the maximum velocity constraint active; (c) the minimum velocity constraint active; (d) no velocity constraints active.

The following equation from the circle equation obtains the u_y^B-coordinates of the samples:

$$u_{y_{\text{sample}}} = \pm\sqrt{\left(\frac{v_{\max}}{\Delta t}\right)^2 - \left[\frac{\|v\|}{\Delta t} + u^B_{x_{\text{sample}}}\right]^2}$$

where $u^B_{x_{\text{sample}}}$ is the sample from the interval in $[u_{x_{\max}}$ in (3.32), u_x^B in (3.33)]. The samples for the minimum velocity arc are obtained similarly.

We evaluate the cost function corresponding to samples along the boundary and find its minimum. The control input is transformed to the coordinates in the global frame using the direction cosine matrix. These are implemented in Programs 3.22 and 3.23.

```
1  % (continue)
2  % find the optimal solution along the boundary
3  n_sample = 50;
4  ux_sample = linspace(ux_min, ux_max, n_sample);
5  upper_line = [ux_sample; ones(1,n_sample)*uy_max];
6  lower_line = [ux_sample; ones(1,n_sample)*uy_min];
7
8  if vmax_active
9      ux_sample = linspace(ux_max,(v_max-current_speed)/Dt,n_sample);
10     uy_sample = sqrt((v_max/Dt)^2-(ux_sample+current_speed/Dt).^2);
11     right_line = [ux_sample ux_sample(end-1:-1:1); uy_sample -
           uy_sample(end-1:-1:1)];
12 else
13     uy_sample = linspace(uy_min,uy_max,n_sample);
14     right_line = [ones(1,n_sample)*ux_max; uy_sample];
15 end
16
17 if vmin_active
18     ux_sample = linspace(ux_min,(v_min-current_speed)/Dt,n_sample);
19     uy_sample = sqrt((v_min/Dt)^2-(ux_sample+current_speed/Dt).^2);
20     left_line = [ux_sample ux_sample(end-1:-1:1); uy_sample -
           uy_sample(end-1:-1:1)];
21 else
22     uy_sample = linspace(uy_min,uy_max,n_sample);
23     left_line = [ones(1,n_sample)*ux_min; uy_sample];
24 end
25
26 all_samples_in_body_frame = [upper_line lower_line right_line
       left_line];
27 all_samples_in_global_frame = dcm_from_body_to_global*
       all_samples_in_body_frame;
28 ux0_sample = all_samples_in_global_frame(1,:);
29 uy0_sample = all_samples_in_global_frame(2,:);
30
31 J_cost_uxuy0 = eval(J_cost_worst);
32 J_cost_uxuy0_function = matlabFunction(J_cost_uxuy0);
33 J_val = J_cost_uxuy0_function(all_samples_in_global_frame(1,:),
       all_samples_in_global_frame(2,:));
34
35 [J_val_opt,opt_idx]=min(J_val);
36 uxy_opt_body = all_samples_in_body_frame(:,opt_idx);
37 uxy_opt_global = all_samples_in_global_frame(:,opt_idx);
```

Program 3.22 (MATLAB) Sampling along the control boundary

```
1  # continue
2  # find the optimal solution along the boundary
3  n_sample = 50
4  ux_sample = np.linspace(ux_min, ux_max, n_sample)
5  upper_line = np.vstack((ux_sample,np.ones(n_sample)*uy_max))
6  lower_line = np.vstack((ux_sample,np.ones(n_sample)*uy_min))
7
```

```
 8 if vmax_active:
 9     ux_sample = np.linspace(ux_max,(v_max-current_speed)/Dt,
           n_sample)
10     uy_sample = np.sqrt((v_max/Dt)**2-(ux_sample+current_speed/Dt)
           **2)
11     right_line = np.vstack((np.hstack((ux_sample,np.flip(ux_sample)
           )), np.hstack((uy_sample,-np.flip(uy_sample)))))
12 else:
13     uy_sample = np.linspace(uy_min,uy_max,n_sample)
14     right_line = np.vstack((np.ones(n_sample)*ux_max, uy_sample))
15
16 if vmin_active:
17     ux_sample = np.linspace(ux_min,(v_min-current_speed)/Dt,
           n_sample)
18     uy_sample = np.sqrt((v_min/Dt)**2-(ux_sample+current_speed/Dt)
           **2)
19     left_line = np.vstack((np.hstack((ux_sample,np.flip(ux_sample))
           ), np.hstack((uy_sample,-1*np.flip(uy_sample)))))
20 else:
21     uy_sample = np.linspace(uy_min,uy_max,n_sample)
22     left_line = np.vstack((np.ones(n_sample)*ux_min, uy_sample))
23
24 all_samples_in_body_frame = np.hstack((upper_line,lower_line,
       right_line,left_line))
25 all_samples_in_global_frame =
       dcm_from_body_to_global@all_samples_in_body_frame
26 ux0_sample = all_samples_in_global_frame[0,:]
27 uy0_sample = all_samples_in_global_frame[1,:]
28 J_val = J_cost_uxuy0_function(ux0_sample,uy0_sample)
29
30 J_val_opt = J_val.min()
31 opt_idx = J_val.argmin()
32 uxy_opt_body = all_samples_in_body_frame[:,opt_idx]
33 uxy_opt_global = all_samples_in_global_frame[:,opt_idx]
```

Program 3.23 (Python) Sampling along the control boundary

3.3.2.2 Inside the Constraints

If the minimum of the cost function occurs inside the constraints, there are values of the cost function smaller than the minimum value found on the boundary. To check if the minimum is inside the boundary, we generate samples inside the constraints, calculate the cost function values for the samples, and compare the values with the minimum found on the boundary.

A polygon to describe the feasible control set must be defined. Both MATLAB and Python have polygon functions to describe polygons and check the points if they are inside or outside of polygons. The following MATLAB commands create the polygon corresponding to arbitrary generated 10 points. From the 10 points, the centre of the points is calculated. Construct a vector from the centre to each of 10 points. The angles of the vectors from the horizontal axis are calculated. Then, the points are ordered by the magnitude of the angles. Finally, the first point is added at the end of the array to define the polygon as the polygon must be closed to return to the first point.

```
1  % matlab
2  polygon_points = randn(2,10);
3  polygon_center = mean(polygon_points,2);
4  pc_vector = polygon_points-polygon_center ;
5  th_pc = atan2(pc_vector(2,:),pc_vector(1,:));
6  [~, idx_pc] = sort(th_pc);
7  polygon_points = polygon_points(:,idx_pc);
8  polygon_points = [polygon_points polygon_points(:,1)];
9  figure;
10 plot(polygon_points(1,:),polygon_points(2,:),'o')
11 hold on
12 plot(polygon_points(1,:),polygon_points(2,:),'r')
```

In addition, the following is the python script to construct the polygon, where the *path* module in *matplotlib* is used to construct the polygon, which has *contains_points()* function to check if a point is inside the polygon or not. Unlike MATLAB, the polygon points do not have to include the same points at the beginning and the end of the array to close the path.

```
1  # python
2  from matplotlib import path
3  polygon_points = np.random.randn(2,10)
4  polygon_center = polygon_points.mean(axis=1)
5  pc_vector = polygon_points-polygon_center[:,np.newaxis]
6  th_pc = np.arctan2(pc_vector[1,:],pc_vector[0,:])
7  idx_pc = th_pc.argsort()
8  polygon_points = polygon_points[:,idx_pc]
9  polygon = path.Path(polygon_points.transpose())
10 plt.plot(polygon_points[0,:],polygon_points[1,:],'o')
11 plt.plot(polygon_points[0,:],polygon_points[1,:],'r-')
```

Once random points are generated around the feasible control input sets, they are checked using *inpolygon()* in MATLAB or *contains_points()* in Python as in Program 3.24 or 3.25.

```
1  % continue
2  % check the cost function inside the constraint
3  polygon_points = [ux0_sample(:)'; uy0_sample(:)'];
4  polygon_center = mean(polygon_points,2);
5  pc_vector = polygon_points-polygon_center ;
6  th_pc = atan2(pc_vector(2,:),pc_vector(1,:));
7  [~, idx_pc] = sort(th_pc);
8  polygon_points = polygon_points(:,idx_pc);
9  polygon_points = [polygon_points polygon_points(:,1)];
```

```
10
11 n_inside_sample = 1000;
12 x_sample = min(polygon_points(1,:)) + ...
13     (max(polygon_points(1,:))-min(polygon_points(1,:)))*rand(1,
         n_inside_sample);
14 y_sample = min(polygon_points(2,:)) + ...
15     (max(polygon_points(2,:))-min(polygon_points(2,:)))*rand(1,
         n_inside_sample);
16
17 [in,on] = inpolygon(x_sample,y_sample,polygon_points(1,:),
     polygon_points(2,:));
18 x_sample = x_sample(in);
19 y_sample = y_sample(in);
20 J_val_inside = J_cost_uxuy0_function(x_sample,y_sample);
21 J_val_inside = J_val_inside(J_val_inside<J_val_opt);
```

Program 3.24 (MATLAB) Samples inside the control boundary

```
1 # continue
2 # check the cost function inside the constraint
3 polygon_points = np.vstack((ux0_sample,uy0_sample))
4 polygon_center = polygon_points.mean(axis=1)
5 pc_vector = polygon_points-polygon_center[:,np.newaxis]
6 th_pc = np.arctan2(pc_vector[1,:],pc_vector[0,:])
7 idx_pc = th_pc.argsort()
8 polygon_points = polygon_points[:,idx_pc]
9
10 from matplotlib import path
11 polygon = path.Path(polygon_points.transpose())
12
13 n_inside_sample = 1000;
14 x_sample = polygon_points[0,:].min() \
15     + (polygon_points[0,:].max()-polygon_points[0,:].min())*np.
         random.rand(n_inside_sample)
16 y_sample = polygon_points[1,:].min() \
17     + (polygon_points[1,:].max()-polygon_points[1,:].min())*np.
         random.rand(n_inside_sample)
18 xy_sample=np.vstack((x_sample,y_sample)).transpose()
19
20 in_out = polygon.contains_points(xy_sample)
21 x_sample = x_sample[in_out]
22 y_sample = y_sample[in_out]
23 J_val_inside = J_cost_uxuy0_function(x_sample,y_sample)
24 J_val_inside = J_val_inside[J_val_inside<J_val_opt]
```

Program 3.25 (Python) Sample inside the control boundary

3.3.2.3 Optimal Input

Programs 3.26 and 3.27 solves the optimization problem using three different methods for the case that the minimum occurs inside the boundary. The following three methods are implemented to obtain the optimal control inside the boundary:

- Minimize the cost function, $\bar{J}$, in (3.26) using *fminunc* in MATLAB and *minimize()* in Python, which solves the unconstrained minimization problems
- Solve the first-optimality condition, i.e. $\partial\bar{J}/\partial u_x(0) = 0$ and $\partial\bar{J}/\partial u_y(0) = 0$, using *fsolve()* in MATLAB or Python, which seeks the roots of a set of algebraic equations
- Minimize the cost function, $\bar{J}$, in (3.26) using Armijo's rule given in Algorithm 3.2

Considering that ultimately we implement the algorithm in the on-board computer of an autonomous vehicle, we test further the third implementation as it is less likely that the minimization or the root-finding functions are available in the on-board computer.

```
1  % continue
2  if ~isempty(J_val_inside)
3      [J_val_opt,min_idx] = min(J_val_inside);
4  
5      tic
6      J_cost_minimize=@(x)J_cost_uxuy0_function(x(1),x(2));
7      uxy_opt_global_1 = fminunc(J_cost_minimize,[ux0_sample(min_idx)
           uy0_sample(min_idx)]);
8      toc
9  
10     tic
11     dJdux0_fun=matlabFunction(eval(dJdux0));
12     dJduy0_fun=matlabFunction(eval(dJduy0));
13     dJduxy=@(x)[dJdux0_fun(x(1),x(2)); dJduy0_fun(x(1),x(2))];
14     uxy_opt_global_2 = fsolve(dJduxy,[ux0_sample(min_idx)
         uy0_sample(min_idx)]);
15     toc
16 
17     tic
18     s_amj = 0.5;
19     alpha_amj = s_amj; beta_amj = 0.5; sigma_amj = 1e-5;
20     u_xy_current = [ux0_sample(min_idx) uy0_sample(min_idx)];
21     J_current = J_cost_minimize(u_xy_current);
22     dJdu = dJduxy(u_xy_current);
23     while true
24         u_xy_update = u_xy_current - alpha_amj*dJdu(:)';
25         J_update = J_cost_minimize(u_xy_update);
26         if J_update < (J_current + sigma_amj*alpha_amj*sum(dJdu.^2)
             )
27             if norm(u_xy_current-u_xy_update)<1e-6
28                 break
29             end
30             alpha_amj = s_amj;
31             J_current = J_cost_minimize(u_xy_update);
32             dJdu = dJduxy(u_xy_update);
33             u_xy_current = u_xy_update;
34         else
```

```
35             alpha_amj = beta_amj*alpha_amj;
36         end
37
38     end
39     toc
40
41     uxy_opt_global = u_xy_current(:);
42     uxy_opt_body = dcm_from_body_to_global'*uxy_opt_global;
43
44     [   uxy_opt_global_1(:)';
45         uxy_opt_global_2(:)';
46         uxy_opt_global(:)']
47
48 end
```

Program 3.26 (MATLAB) Optimal tracking command solution

```
1  # continue
2  if J_val_inside.shape[0]!=0:
3      J_val_opt = J_val_inside.min()
4      min_idx = J_val_inside.argmin()
5
6      t0 = time.time()
7      J_cost_minimize=lambda x: J_cost_uxuy0_function(x[0],x[1])
8      sol_opt=minimize(J_cost_minimize,[ux0_sample[min_idx],
           uy0_sample[min_idx]],method='BFGS')
9      uxy_opt_global_1 = sol_opt.x
10     tf = time.time() - t0
11     print(f'minimization: {tf:10.8f} [s]\n')
12
13     t0 = time.time()
14     dJdux0_fun = lambdify([ux0,uy0],dJdux0.subs(values))
15     dJduy0_fun = lambdify([ux0,uy0],dJduy0.subs(values))
16     dJduxy=lambda x: np.array([dJdux0_fun(x[0],x[1]), dJduy0_fun(x
           [0],x[1])])
17     uxy_opt_global_2 = fsolve(dJduxy,[ux0_sample[min_idx],
           uy0_sample[min_idx]])
18     tf = time.time() - t0
19     print(f'fsolve: {tf:10.8f} [s]\n')
20
21     t0 = time.time()
22     s_amj = 0.01
23     alpha_amj = s_amj; beta_amj = 0.5; sigma_amj = 1e-5
24     u_xy_current = np.array([ux0_sample[min_idx], uy0_sample[
           min_idx]])
25     J_current = J_cost_minimize(u_xy_current)
26     dJdu = dJduxy(u_xy_current)
27     while True:
28         u_xy_update = u_xy_current - alpha_amj*dJdu
29         J_update = J_cost_minimize(u_xy_update)
30         if J_update < (J_current + sigma_amj*alpha_amj*np.sum(dJdu
               **2)):
31             if np.linalg.norm(u_xy_current-u_xy_update)<1e-6:
```

```
32                  break
33              alpha_amj = s_amj
34              J_current = J_cost_minimize(u_xy_update)
35              dJdu = dJduxy(u_xy_update)
36              u_xy_current = u_xy_update
37          else:
38              alpha_amj = beta_amj*alpha_amj
39
40      tf = time.time() - t0
41      print(f'Gradient Descent with Armijo\'s Rule: {tf:10.8f} [s]\n'
            )
42
43      uxy_opt_global = u_xy_current
44      uxy_opt_body = dcm_from_body_to_global.T@uxy_opt_global
45
46      print(uxy_opt_global_1)
47      print(uxy_opt_global_2)
48      print(uxy_opt_global)
```

Program 3.27 (Python) Optimal tracking command solution

3.3.3 Verification Simulation

The next step in algorithm development is to test the algorithm to verify the design. There are different levels of verification, which could be flight tests, hardware-in-the-loop simulations, computer simulations, etc. Zhu et al. (2017) and Chaves et al. (2018). One of the vital aspects of control algorithm verification is to test the algorithms against various cases, which are not perfectly satisfying all assumptions made for the algorithm developments. For example, the target tracking algorithm runs with inaccurate target position knowledge. Or, the maximum target acceleration assumption in the algorithm is overestimated or underestimated of the true target acceleration bounds.

The following verification is one of the preliminary levels. The only difference from the assumptions made for the algorithm development is how the target moves. Instead of the worst movement, the target velocity direction is set to a random direction with the maximum velocity as follows:

$$\dot{x}_t = w_{\max} \cos \theta_t \tag{3.34}$$

$$\dot{y}_t = w_{\max} \sin \theta_t \tag{3.35}$$

where θ_t is a random number between 0 and 2π with the uniform distribution, and the velocity is updated at every Δt.

In computer simulations, we have the advantage of knowing the full true states. As the algorithm assumes the worst movement of the target by the random velocity direction, we would be interested in how the performance varies when the target has different strategies. The optimal control corresponding to the actual movement of the target is obtained. We compare the control input from the worst-case assumption-based algorithm with the true optimal control using the following two measures:

$$\cos\theta^* = \frac{\mathbf{u}\cdot\mathbf{u}^*}{\|\mathbf{u}\|\|\mathbf{u}^*\|}$$
$$\Delta u^* = \frac{\|\mathbf{u}\|}{\|\mathbf{u}^*\|}$$

where $\mathbf{u} = [u_x(0),\ u_y(0)]^T$ is from the optimal tracking algorithm, and $\mathbf{u}^* = [u_x^*(0),\ u_y^*(0)]^T$ is the true optimal based on the actual target movement. The closer the values of $\cos\theta^*$ or Δu^* to one, the closer the algorithm input to the true optimal.

Figure 3.24 shows the UAV and the target trajectories for three 20 minute intervals. The control input in terms of the relative direction and magnitude with respect to the truth remains close to the true optimal more than 60% of one hour time interval.

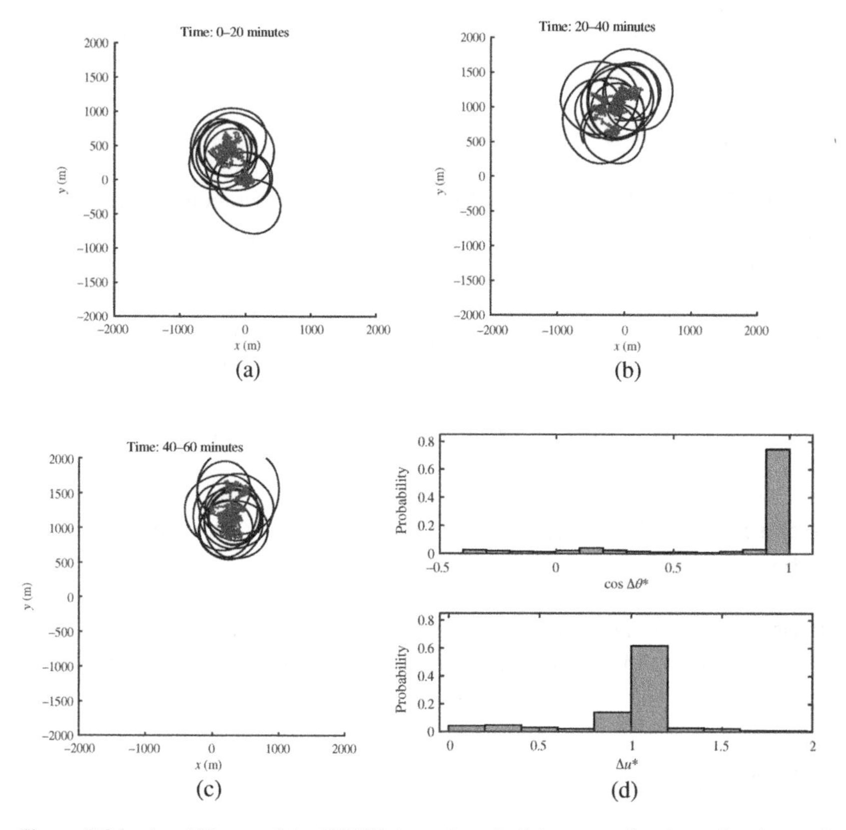

Figure 3.24 (a–c) Traces of the UAV flight path and of the target for three time intervals, respectively, and (d) the direction and the magnitude comparison to the true optimal.

Exercises

Exercise 3.1 Derive the attractive and the repulsive forces by the potential functions, (3.2), in the **y** direction.

Exercise 3.2 (MATLAB) Construct the sparse matrix in Program 3.3 using the row and the column numbers for the non-zero elements.

Exercise 3.3 (MATLAB) Convert Program 3.10 to obtain the shortest path by constructing the graph using the *delaunay* function.

Exercise 3.4 (Python) Convert Program 3.12 to obtain the shortest path by constructing the graph using the *voronoi* function.

Exercise 3.5 (MATLAB/Python) Implement a resampling method to improve the optimal path obtained from the graph constructed by the Voronoi or the Delaunay function and plot the figures shown in Figure 3.12.

Exercise 3.6 Derive (3.15) from (3.14).

Exercise 3.7 (MATLAB/Python) Obtain (3.27) using symbolic computations and discuss when $a_{\triangle}$ and $b_{\triangle}$ are equal to zero at the same time and how to prevent it from happening.

Exercise 3.8 (MATLAB/Python) Using the following values: $v_{\text{min}} = 20\,\text{m/s}$, $v_{\text{max}} = 40\,\text{m/s}$, $u_{x_{\text{min}}} = -1\,\text{m/s}^2$, $u_{x_{\text{max}}} = 10\,\text{m/s}^2$, $u_{y_{\text{min}}} = -2\,\text{m/s}^2$, $u_{y_{\text{max}}} = 2\,\text{m/s}^2$, $r_{\text{min}} = 400\,\text{m}$, $w_{\text{max}} = 60\,\text{km/h}$, $\Delta t = 2\,\text{s}$, $x_a(0) = -74.60\,\text{m}$, $y_a(0) = 82.68\,\text{m}$, $v_x(0) = -16.84\,\text{m/s}$, $v_y(0) = -18.48\,\text{m/s}$, $x_t(0) = 125.89\,\text{m}$, and $y_t(0) = 162.32\,\text{m}$, draw Figure 3.19.

Exercise 3.9 (MATLAB/Python) Using (3.31), check the convexity of the cost function for the values given in Exercise 3.8.

Exercise 3.10 (MATLAB/Python) Simulate the random target movement using (3.34), test the optimal tracking algorithm, and generate a figure similar to Figure 3.24.

Bibliography

Larry Armijo. Minimization of functions having Lipschitz continuous first partial derivatives. *Pacific Journal of Mathematics*, 16(1):1–3, 1966. https://doi.org/pjm/1102995080.

Lennon Chaves, Iury V. Bessa, Hussama Ismail, Adriano Bruno dos Santos Frutuoso, Lucas Cordeiro, and Eddie Batista de Lima Filho. DSVerifier-aided verification applied to attitude control software in unmanned aerial vehicles. *IEEE Transactions on Reliability*, 67(4):1420–1441, 2018. https://doi.org/10.1109/TR.2018.2873260.

H. Chou, P. Kuo, and J. Liu. Numerical streamline path planning based on log-space harmonic potential function: a simulation study. In *2017 IEEE International Conference on Real-time Computing and Robotics (RCAR)*, pages 535–542, 2017. https://doi.org/10.1109/RCAR.2017.8311918.

Edsger W. Dijkstra. A note on two problems in connexion with graphs. *Numerische Mathematik*, 1(1):269–271, 1959.

Steven Fortune. Voronoi diagrams and delaunay triangulations. *Lecture Notes Series on Computing: Volume 1 Computing in Euclidean Geometry*, pages 193–233, 1992. https://doi.org/10.1142/9789814355858 0006

S. J. Julier and J. K. Uhlmann. Unscented filtering and nonlinear estimation. *Proceedings of the IEEE*, 92(3):401–422, 2004. https://doi.org/10.1109/JPROC.2003.823141.

Rolf Klein. *Voronoi Diagrams and Delaunay Triangulations*, pages 2340–2344. Springer, New York, NY, 2016. ISBN 978-1-4939-2864-4. https://doi.org/10.1007/978-1-4939-2864-4_507.

W. H. Press, S. A. Teukolsky, W. T. Vetterling, and B. P. Flannery. *Numerical Recipes 3rd Edition: The Art of Scientific Computing*. Cambridge University Press, 2007. ISBN 9780521880688.

Pauli Virtanen, Ralf Gommers, Travis E. Oliphant, Matt Haberland, Tyler Reddy, David Cournapeau, Evgeni Burovski, Pearu Peterson, Warren Weckesser, Jonathan Bright, Stéfan J. van der Walt, Matthew Brett, Joshua Wilson, K. Jarrod Millman, Nikolay Mayorov, Andrew R. J. Nelson, Eric Jones, Robert Kern, Eric Larson, C. J. Carey, İlhan Polat, Yu Feng, Eric W. Moore, Jake VanderPlas, Denis Laxalde, Josef Perktold, Robert Cimrman, Ian Henriksen, E. A. Quintero, Charles R. Harris, Anne M. Archibald, Antônio H. Ribeiro, Fabian Pedregosa, Paul van Mulbregt, and SciPy 1.0 Contributors. SciPy 1.0: Fundamental algorithms for scientific computing in python. *Nature Methods*, 17:261–272, 2020. https://doi.org/10.1038/s41592-019-0686-2.

S. Waydo and R. M. Murray. Vehicle motion planning using stream functions. In *2003 IEEE International Conference on Robotics and Automation (Cat. No.03CH37422)*, volume 2, pages 2484–2491, 2003. https://doi.org/10.1109/ROBOT.2003.1241966.

Ronghui Zhan and Jianwei Wan. Iterated unscented Kalman filter for passive target tracking. *IEEE Transactions on Aerospace and Electronic Systems*, 43(3):1155–1163, 2007. https://doi.org/10.1109/TAES.2007.4383605.

Chuangchuang Zhu, Xiaolong Liang, Lvlong He, and Liu Liu. Demonstration and verification system for UAV formation control. In *2017 3rd IEEE International Conference on Control Science and Systems Engineering (ICCSSE)*, pages 56–60, 2017. https://doi.org/10.1109/CCSSE.2017.8087894.

4

Biological System Modelling

4.1 Biomolecular Interactions

The genetic information of living organisms resides in deoxyribonucleic acid (DNA), which is a double-stranded molecule. DNA is composed of a series of nucleotides, which are sugar-phosphate molecules with nitrogen bases. Part of DNA includes gene information, and it is transcribed into ribonucleic acid (RNA) when it is activated by internal and/or external stimuli. RNA is translated into a protein, which interacts with other proteins or evokes further interactions. This process illustrated in Figure 4.1 is ubiquitous for all living organisms, and it is called the central dogma of molecular biology. The interactions are stochastic spatial-temporal and include non-linear complex feedback loops. Modelling biomolecular networks becomes a quickly daunting task in dealing with the interactions between hundreds or thousands of different molecular species. It needs some level of approximation and simplification in the modelling and analysis depending on the purposes of each study for a particular part of biomolecular network interactions. Each modelling method has its advantages and limitations.

4.2 Deterministic Modelling

Biomolecular interactions can be modelled as a set of ordinary differential equations (ODEs) as shown in the ligand–receptor example in Section 1.2. The interactions are derived from chemical theory or experiments. Some of the reaction rates in the interactions are directly measured or found by numerical optimization to fit the model to experimental data. Usually, these experiments are performed about culture cells, i.e. an isogenic cell population.

Dynamic System Modelling and Analysis with MATLAB and Python: For Control Engineers, First Edition. Jongrae Kim.

Companion Website: www.wiley.com/go/kim/dynamicmodeling

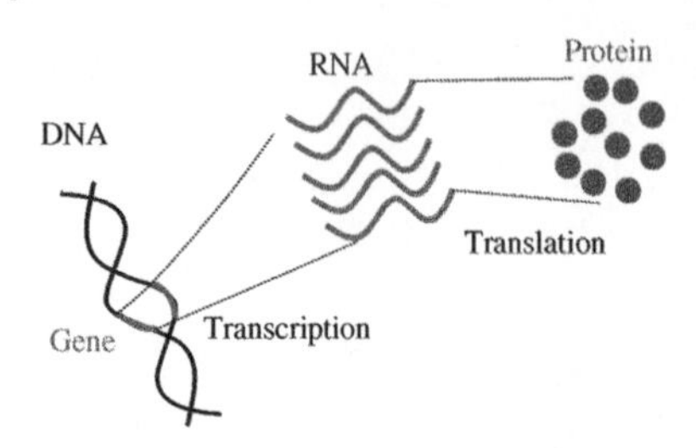

Figure 4.1 Transcription from DNA to RNA and translation from RNA to protein.

4.2.1 Group of Cells and Multiple Experiments

Cell-to-cell variation is greatly reduced in culture measurements, but the response is also dependent on environmental conditions. Morohashi et al. (2002) has presented two categories of the parameters of cell responses: parameters robust to frequent variations and parameters hypersensitive to uncommon perturbations (Carlson and Doyle, 2000).

Consider the following ODE:

$$\frac{d\mathbf{x}(t)}{dt} = \mathbf{f}[\mathbf{x}(t), \mathbf{p}_N + \mathbf{v}_N, \mathbf{p}_E + \mathbf{c}_E + \mathbf{v}_E] + \mathbf{w}(t)$$

where $\mathbf{x}(t)$ is an n-dimensional non-negative real vector, each element of $\mathbf{x}(t)$ is the concentration of a molecular species, t is time, $\mathbf{p}_N$ is the nominal values of the kinetic parameters in the model affected by $\mathbf{v}_N$, which is the stochastic fluctuation given by the zero mean random constant with unknown distribution, varies from cell to cell, $\mathbf{p}_E$ is the nominal values of the adaptive kinetic parameters affected by $\mathbf{c}_E$, which varies with environmental changes such as different nutrition concentrations and temperature, and $\mathbf{v}_E$ is the fluctuation in the adaptive parameters, which is the zero-mean random constant and its distribution is unknown.

$\mathbf{v}_N$ and $\mathbf{v}_E$ are independent of each other and exist always. $\mathbf{w}(t)$ is the process noise, i.e. unmodelled dynamics, which is the zero-mean Gaussian white noise. In addition, $\mathbf{f}(\cdot, \cdot, \cdot)$ is a non-linear function in general. After dividing the parameters into two types, we test whether the kinetic model is robust for a range of perturbations or not and whether the adaptive parameters change appropriately with the environmental changes (Morohashi et al., 2002).

Consider that the k-number of isogenic cells are observed in the same condition and the kinetics for each cell is given by

$$\frac{d\mathbf{x}_1(t)}{dt} = \mathbf{f}[\mathbf{x}_1(t), \mathbf{p}_N + \mathbf{v}_{N_1}, \mathbf{p}_E + \mathbf{c}_E + \mathbf{v}_{E_1}] + \mathbf{w}_1(t) \tag{4.1a}$$

$$\frac{d\mathbf{x}_2(t)}{dt} = \mathbf{f}[\mathbf{x}_2(t), \mathbf{p}_N + \mathbf{v}_{N_2}, \mathbf{p}_E + \mathbf{c}_E + \mathbf{v}_{E_2}] + \mathbf{w}_2(t) \tag{4.1b}$$

$$\vdots$$

$$\frac{d\mathbf{x}_k(t)}{dt} = \mathbf{f}[\mathbf{x}_k(t), \mathbf{p}_N + \mathbf{v}_{N_k}, \mathbf{p}_E + \mathbf{c}_E + \mathbf{v}_{E_k}] + \mathbf{w}_k(t) \tag{4.1c}$$

The resulting time histories, $\mathbf{x}_1(t), \mathbf{x}_2(t), \ldots, \mathbf{x}_k(t)$, would not be very different from each other as all the kinetics describe the same type of cells in the same environment. Hence, the time history of $\mathbf{x}_i(t)$ for $i = 1, 3, \ldots, k$ is written as

$$\mathbf{x}_i(t) = \bar{\mathbf{x}}(t) + \delta\mathbf{x}_i(t) \tag{4.2}$$

where $\bar{\mathbf{x}}(t)$ is the average equal to $[\sum_{i=1}^{k} \mathbf{x}_i(t)]/k$ and $\delta\mathbf{x}(t)$ is small compared to $\bar{\mathbf{x}}(t)$. Take the Taylor series expansion of (4.1) up to the first-order terms,

$$\begin{aligned}\frac{d\mathbf{x}_i(t)}{dt} \approx\; & \mathbf{f}[\bar{\mathbf{x}}(t), \mathbf{p}_N, \mathbf{p}_E + \mathbf{c}_E] + \left.\frac{\partial \mathbf{f}[\mathbf{x}(t), \mathbf{p}_N, \mathbf{p}_E + \mathbf{c}_E]}{\partial \mathbf{x}}\right|_{\mathbf{x}=\bar{\mathbf{x}}(t)} \delta\mathbf{x}_i(t) \\ & + \left.\frac{\partial \mathbf{f}[\bar{\mathbf{x}}(t), \mathbf{p}, \mathbf{p}_E + \mathbf{c}_E]}{\partial \mathbf{p}}\right|_{\mathbf{p}=\mathbf{p}_N} \mathbf{v}_{N_i} + \left.\frac{\partial \mathbf{f}[\bar{\mathbf{x}}(t), \mathbf{p}_N, \mathbf{p}]}{\partial \mathbf{p}}\right|_{\mathbf{p}=\mathbf{p}_E+\mathbf{c}_E} \mathbf{v}_{E_i} + \mathbf{w}_i(t)\end{aligned}$$

for $i = 1, 2, \ldots, k$. The average of the time derivatives is

$$\frac{d[\sum_{i=1}^{k} \mathbf{x}_i(t)]/k}{dt} = \frac{d\bar{\mathbf{x}}(t)}{dt} \approx \mathbf{f}[\bar{\mathbf{x}}(t), \mathbf{p}_N, \mathbf{p}_E + \mathbf{c}_E] \tag{4.3}$$

where

$$\frac{1}{k}\sum_{i=1}^{k} \delta\mathbf{x}_i(t) \to 0, \quad \frac{1}{k}\sum_{i=1}^{k} \mathbf{v}_{N_i} \to 0, \quad \frac{1}{k}\sum_{i=1}^{k} \mathbf{v}_{E_i} \to 0, \quad \frac{1}{k}\sum_{i=1}^{k} \mathbf{w}_i(t) \to 0$$

as k increases. The number of cells in one cell culture, k, is typically around several million (Papadimitriou and Lelkes, 1993).

The main assumptions for the model, (4.3), to be valid are as follows:

- the same type of cells are in a culture, i.e. $\mathbf{f}(\cdot, \cdot, \cdot)$ and $\mathbf{p}_N$ are the same for all the cells in the culture
- the cells are in the same environment, $\mathbf{p}_E$ and $\mathbf{c}_E$ are the same for all the cells in the culture
- the unmodelled dynamics, $\mathbf{w}_i(t)$ for $i = 1, 2, \ldots, k$, is the same zero-mean Gaussian process for all the cells in the culture

As long as $\|\delta\mathbf{x}_i(t)\|$ remains small, the error of the average model converges to zero, i.e. the average model represents the behaviour of every single cell well. However, if the effect of noise is significant to make the trajectories of the differential equations different from each other, then the estimated parameters for the kinetics from culture measurements could be very different from the actual values in each cell. In that case, the correct values in a cell cannot be extrapolated from the parameters identified from the population or culture measurements and to identify the correct values separate measurements for each cell has to be obtained (Elowitz et al., 2002, Colman-Lerner et al., 2005). A single cell modelling is considered later in this chapter.

With the assumption that the states of most of the cells in an isogenic population are close to each other, the environmental effect, $\mathbf{c}_E$, can be identified by exposing the same culture to different environmental conditions. When L-different environmental conditions are applied to subgroups of the same culture, the kinetics for the subgroups are given by

$$\frac{d\bar{\mathbf{x}}_1(t)}{dt} = \mathbf{f}[\bar{\mathbf{x}}_1(t), \mathbf{p}_N, \mathbf{p}_E + \mathbf{c}_{E_1}] \tag{4.4a}$$

$$\frac{d\bar{\mathbf{x}}_2(t)}{dt} = \mathbf{f}[\bar{\mathbf{x}}_2(t), \mathbf{p}_N, \mathbf{p}_E + \mathbf{c}_{E_2}] \tag{4.4b}$$

$$\vdots \tag{4.4c}$$

$$\frac{d\bar{\mathbf{x}}_L(t)}{dt} = \mathbf{f}[\bar{\mathbf{x}}_L(t), \mathbf{p}_N, \mathbf{p}_E + \mathbf{c}_{E_L}] \tag{4.4d}$$

where the initial conditions for each $\bar{\mathbf{x}}_1(t_0), \bar{\mathbf{x}}_2(t_0), \ldots,$ and $\bar{\mathbf{x}}_L(t_0)$ are different from each other in general.

4.2.1.1 Model Fitting and the Measurements

For each experiment, $i = 1, 2, \ldots, L$, finding the unknown model parameters is the optimization problem given by

$$\underset{\mathbf{p}_N, \mathbf{p}_E + \mathbf{c}_{E_i}}{\text{Minimize}}\, J = \sum_{j=1}^{r} \left\| \bar{\mathbf{y}}_i(t_j) - \tilde{\mathbf{y}}_i(t_j) \right\|^2 \tag{4.5}$$

subject to (4.4), where $\bar{\mathbf{y}}_i(t_j)$ and $\tilde{\mathbf{y}}_i(t_j)$ are the values from the model and the post-processed measurement at the time t_j, respectively, and r is the number of measurement for each experiment. The measurement before the post-processing, $\tilde{\mathbf{y}}_i^*(t_j)$, is given by

$$\tilde{\mathbf{y}}_i^*(t_j) = \mathbf{h}[\bar{\mathbf{x}}_i(t_j), t_j] \approx \alpha \bar{\mathbf{x}}_i(t_j)$$

where $\mathbf{h}(\cdot, \cdot)$ is an unknown non-linear function and α, which is frequently unknown, is constant. To obtain the linear approximation from the non-linear function requires a significant amount of time and effort. Through experiment preparation steps, cell sampling methods, post data processing, it could be considered that the measurements, $\tilde{\mathbf{y}}_i^*(t_j)$, are linearly proportional, α, to the quantity that we want to measure, $\bar{\mathbf{x}}_i(t_j)$. For the measurement, $\tilde{\mathbf{y}}_i^*(t_j)$, to have a linear relationship with the molecular concentration, $\bar{\mathbf{x}}_i(t_j)$, careful experiment design, preparation, and completion are required.

We rarely have molecular concentrations measured directly by counting the number of molecules. In a Western blot, the size and colour intensity indicate molecular concentration (Pediredla and Seelamantula, 2011). In the fluorescence resonance energy transfer (FRET), the colour represents the interactions of proteins (Sekar and Periasamy, 2003). Frequently, the size and the colour intensity would be normalized by the maximum size or the maximum intensity. In these cases, the post-processing measurement is given by

$$\bar{\mathbf{y}}_i(t_j) = \frac{\bar{\mathbf{y}}_i^*(t_j)}{\max_{s=1}^{r} \bar{\mathbf{y}}_i^*(t_s)}$$

where the measurement is normalized by the maximum measurement value. Or, given that the measurements reach a steady state, they could be normalized by the last measurement as follows:

$$\bar{\mathbf{y}}_i(t_j) = \frac{\bar{\mathbf{y}}_i^*(t_j)}{\bar{\mathbf{y}}_i^*(t_r)}$$

Table 4.1 shows the data extracted from the experiments in Yanofsky and Horn (1994). The experiments have shown the adaptive responses of the bacteria, *Escherichia coli*, to the three different environmental conditions. *E. coli*, in short, is one of the model organisms widely used in molecular biology to study biomolecular interactions. *E. coli* is a bacteria found in diverse environments, including the intestine of humans. The cylindrical cells grow from 1 to 7 μm in length and have a doubling time of about 20 minutes (Osella et al., 2014).

The procedures and the nature of biological measurements are mostly very different from the ones in engineering systems. The enzyme measurements for each experiment in Table 4.1, for example, are not from the same cells. A sample is taken from the same culture every 30 minutes or so and frozen. Each frozen sample is thawed and suspended before the enzyme activity is measured fluorometrically. During the procedures, the cell would grow and divide. Some

Table 4.1 *E. coli* enzyme responses for three different environmental Conditions.

Experiment A		Experiment B		Experiment C	
Time (min)	Active Enzyme (a.u.)	Time (min)	Active Enzyme (a.u.)	Time (min)	Active Enzyme (a.u.)
0	25	0	0	0	0
20	657	29	1370	29	754
38	617	60	1362	58	888
59	618	89	1291	88	763
89	577	179	913	118	704
119	577			178	683
149	567				

a) (a.u.), (arbitrary units).
Source: The data are extracted from the figures in Yanofsky and Horn (1994).

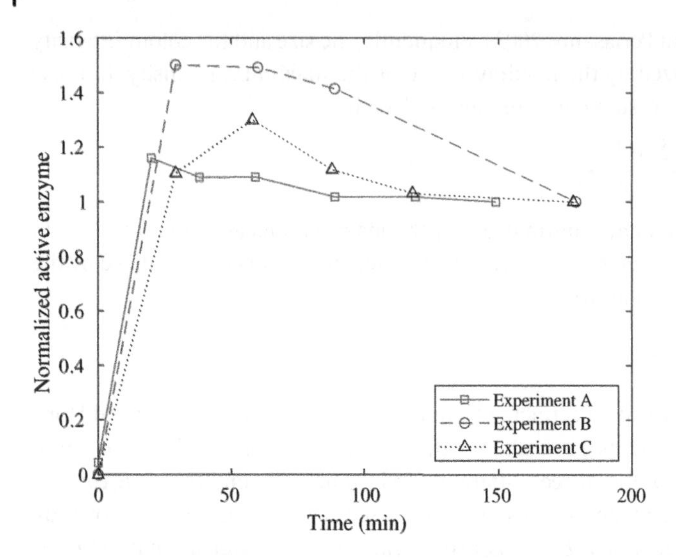

Figure 4.2 *E. coli* tryptophan measurements normalized by the steady state.

proper considerations are applied to normalize these effects on the concentration measurements. See the detailed descriptions of the a priori data processing in Yanofsky and Horn (1994).

As the unit for the active enzyme is arbitrary, which implies that α is unknown, the values in Table 4.1 do not have immediate physical meanings, and we cannot compare them directly. For example, we cannot conclude that the enzyme activity in Experiment B is higher than the one in Experiment A by comparing the values in the table. They would be normalized by the last value assuming that all three reach the steady states. Figure 4.2 shows the normalized time histories. In the figure, their dynamic responses show the differences from each other in the rising time, the overshoot, and the converging time.

Solving the optimization problem identifies the model parameters in the model fitting. The parameters describing the dynamical interactions, $\mathbf{p}_N$ and $\mathbf{p}_E + \mathbf{c}_{E_j}$, are not distinguishable, however, from single experiment or multiple experiments in the same environment. We do not know which parameters are for adapting to environmental changes. Also, $\mathbf{p}_E$ and $\mathbf{c}_{E_j}$ cannot be separable from a single or multiple experiments in the same environment. Separate identification of the three-parameter types would be possible from multiple experiments for different environmental conditions as the data given in Table 4.1.

4.2.1.2 Finding Adaptive Parameters

After solving the optimization problem for each experimental set, the optimal parameter combination fitting each experimental data is obtained. Denote the optimal parameter set as $\mathscr{P}_i = \{p_1, p_2, \dots, p_k\}$ for $i = 1, 2, \dots, L$, where p_j for $j = 1, 2, \dots, k$ includes all elements in the vector parameters, $\mathbf{p}_N$ and $\mathbf{p}_E + \mathbf{c}_{E_i}$. However, we do not know whether p_j belongs to $\mathbf{p}_N$ or $\mathbf{p}_E + \mathbf{w}_{E_i}$. The average, m_j, and the variance, σ_j^2, for each parameter, p_j, are calculated from the L-set of measurements obtained from the L different environment conditions for $j = 1, 2, \dots, k$. Noise, η_j, and noise strength, φ_j, for each parameter are defined by Kærn et al. (2005):

$$\eta_j = \frac{\sigma_j}{|m_j|} \tag{4.6a}$$

$$\varphi_j = \frac{\sigma_j^2}{|m_j|} \tag{4.6b}$$

for $j = 1, 2, \dots, k$. As p_j in $\mathbf{p}_N$ fluctuates only by the stochastic cell to cell variations, the noises are not stronger than p_j in $\mathbf{p}_E + \mathbf{c}_{E_j}$, which adjusts the responses to adapt to environmental changes. Hence, the parameters in $\mathbf{p}_E + \mathbf{c}_{E_j}$ would change significantly and the variance would be large. By inspecting the magnitudes of the noise and/or the noise strength, we could distinguish if p_j belongs to the non-adaptive parameters, $\mathbf{p}_N$, or the adaptive parameters, $\mathbf{p}_E + \mathbf{c}_{E_j}$.

In Section 4.2.2, we apply the procedures described above to biomolecular networks in bacteria.

4.2.2 *E. coli* Tryptophan Regulation Model

Tryptophan is an essential amino acid for *E. coli*. The tryptophan operon is a cluster of the genes in the DNA molecules and is transcribed by an mRNAP (messenger RNA Polymerase) to produce tryptophan (Alberts et al., 2015). The tryptophan operon regulation mechanisms for *E. coli* have been studied intensively, and several important feedback mechanisms are revealed. Yanofsky and Horn (1994) have presented the experimental results for *E. coli* tryptophan regulation mechanism to external changes. The tryptophan responses are measured indirectly via the concentration changes of the active enzyme related to the tryptophan changes as shown in Table 4.1.

Santillán and Mackey (2001) have presented a mathematical model for the tryptophan regulation mechanisms, which include the repression, feedback enzyme inhibition, and transcription attenuation. The model is given in four non-linear ODEs. The first equation is for the kinetics of the free operon, $O_F(t)$, given by

$$\boxed{\frac{dO_F(t)}{dt} = \frac{K_r}{K_r + R_A[T(t)]} h(O, O_F, P) - \mu O_F(t)} \tag{4.7}$$

where

$$h(O, O_F, P) = \mu O - k_p P O_F(t) + k_p P O_F(t - \tau_p) e^{-\mu \tau_p}$$

$$R_A[T(t)] = \frac{R\,T(t)}{K_t + T(t)}$$

O is the total operon concentration, $O_F(t)$ is the free operon concentration, P is the mRNAP concentration to bind and transcribe the free operon, $T(t)$ is the tryptophan concentration, K_r is the repression equilibrium constant, μ is the growth rate of the cell, k_p is the DNA-mRNAP isomerization rate, τ_p is the time taking that mRNAP binding the DNA moves away and frees the operon, $R_A[T(t)]$ is the active repressor, R is the total repressor concentration, and K_t is the rate equilibrium constant between the total repressor and the active repressor.

$R_A[T(t)]$ has a common structure in many biomolecular models. Consider enzyme (E), substrate (S), enzyme–substrate (ES) complex, and protein (P) interactions given by Alberts et al. (2015)

$$\mathrm{E} + \mathrm{S} \underset{k_{\mathrm{off}}}{\overset{k_{\mathrm{on}}}{\rightleftharpoons}} \mathrm{ES} \xrightarrow{k_{\mathrm{cat}}} \mathrm{E} + \mathrm{P} \tag{4.8}$$

The speed of the protein production, v, is

$$v = \frac{d[\mathrm{P}]}{dt} = k_{\mathrm{cat}}[\mathrm{ES}]$$

The speed of the ES complex production is

$$\frac{d[\mathrm{ES}]}{dt} = k_{\mathrm{on}}[\mathrm{E}][\mathrm{S}] - k_{\mathrm{off}}[\mathrm{ES}] - -k_{\mathrm{cat}}[\mathrm{ES}]$$

and $\mathrm{E} = \mathrm{E}_0 - \mathrm{ES}$, where E_0 is the total enzyme. Assume that [ES] reaches the steady state faster than the other reactions and

$$\frac{d[\mathrm{ES}]}{dt} \approx 0 \Rightarrow [\mathrm{ES}]_{\mathrm{ss}} = \frac{k_{\mathrm{on}}}{k_{\mathrm{off}} + k_{\mathrm{cat}}}[\mathrm{E}][\mathrm{S}] = \frac{1}{K_m}\left([\mathrm{E}_0] - [\mathrm{ES}]_{\mathrm{ss}}\right)[\mathrm{S}]$$

Solve for the steady-state ES complex concentration, $[\mathrm{ES}]_{\mathrm{ss}}$,

$$[\mathrm{ES}]_{\mathrm{ss}} = \frac{[\mathrm{E}_0][\mathrm{S}]}{K_m + [\mathrm{S}]}$$

Hence, the protein production speed is given by

$$v = \frac{k_{\mathrm{cat}}[\mathrm{E}_0][\mathrm{S}]}{K_m + [\mathrm{S}]} = \frac{v_{\mathrm{max}}[\mathrm{S}]}{K_m + [\mathrm{S}]}$$

where $v_{\mathrm{max}} = k_{\mathrm{cat}}[\mathrm{E}_0]$, and it is called *the Michaelis–Menten equation*. Several observations on this equation are as follows:

- when $[\mathrm{S}] = 0$, $v = 0$
- when $[\mathrm{S}] \ll K_m$, v is linearly proportional to [S], i.e. $v \approx v_{\mathrm{max}}[S]/K_m$
- when $[\mathrm{S}] = k_m$, v is the half of v_{max}
- when $[\mathrm{S}] \gg K_m$, v is equal to v_{max}

The same interpretation can be made for $R_A[T(t)]$.

The second equation of the model is for the kinetics of free mRNA concentration. The free mRNA, $M_F(t)$, is produced by the mRNAP, P, binding to the genes and transcribing the free tryptophan operon, $O_F(t)$.

$$\boxed{\begin{aligned}\frac{dM_F(t)}{dt} &= k_p P O_F(t-\tau_m)e^{-\mu\tau_m}\left[1-b\left(1-e^{-T(t)/c}\right)\right] \\ &\quad - k_\rho \rho\left[M_F(t)-M_F(t-\tau_\rho)e^{-\mu\tau_\rho}\right]-\left(k_d D+\mu\right)M_F(t)\end{aligned}} \tag{4.9}$$

where $M_F(t)$ is free mRNA concentration, τ_m is the time taking that mRNA produces after mRNAP bound to the DNA, b and c are the constants for the transcriptional attenuation, k_ρ is the mRNA-ribosome isomerization rate, ρ is the ribosomal concentration, τ_ρ is the time taking that ribosome binds to mRNA and initiates translation, k_d is the mRNA destroying rate, and D is the mRNA destroying enzyme concentration.

The third equation is for the kinetics of the enzyme produced by the mRNA, $M_F(t)$, given by

$$\boxed{\frac{dE(t)}{dt} = \frac{1}{2}k_\rho \rho M_F(t-\tau_e)e^{-\mu\tau_e}-(\gamma+\mu)E(t)} \tag{4.10}$$

where $E(t)$ is the total enzyme concentration, τ_e is a ribosome binding rate delay for the enzyme, and γ is the enzymatic degradation rate constant.

The last equation is for the kinetics of the tryptophan production, $T(t)$, given by

$$\boxed{\frac{dT(t)}{dt} = KE_A(E,T)-\frac{g\,T(t)}{K_g+T(t)}+d\frac{T_{\text{ext}}}{e+T_{\text{ext}}\left[1+T(t)/f\right]}-\mu T(t)} \tag{4.11}$$

where K is the tryptophan production rate, which is proportional to the active enzyme concentration, $E_A(E, T)$, g is the maximum tryptophan consumption rate, d, e, and f are the parameters for modelling the external tryptophan uptake rate, T_{ext} is the external tryptophan uptake, and the internal tryptophan consumption modelled by the Michaelis–Menten type equation with the constant K_g.

The active enzyme concentration, $E_A(E, T)$, is given by

$$E_A(E,T) = \frac{K_i^{n_H}}{K_i^{n_H}+T^{n_H}(t)}E(t) \tag{4.12}$$

where K_i is the equilibrium constant for the *Trp* feedback inhibition of anthranilate synthase reaction, which is modelled by the Hill equation with the coefficient, n_H (Alberts et al., 2015). The Hill equation is another common model structure in biomolecular system modelling. The tryptophan reduces the production rate of the active enzyme, which is from the fact that two tryptophans attach to and inactivate the enzyme. n_H is called the Hill coefficient to represent cooperative bindings. For example, when two molecules bind to another molecule at the same time,

n_H is equal to 2. As the cooperative level increases, the rate reduces or increases, depending on the sign in front of the Hill equation, faster than the non-cooperative case, i.e. $n_H = 1$, or the lower cooperative cases. For enzyme inhibition with two tryptophans, the bindings do not occur simultaneously when n_H is between 1 and 2. Detailed model explanations are found in Santillán and Mackey (2001).

4.2.2.1 Steady-State and Dependant Parameters

The steady-state concentration is for all derivatives equal to zero. However, the algebraic equation from (4.11) for the steady state of $T(t)$ is as follows:

$$0 = KE_A(\bar{E}, \bar{T}) - \frac{g\,\bar{T}}{K_g + \bar{T}} - \mu\bar{T} \tag{4.13}$$

has the T^{n_H} terms in $\bar{E}_A = E_A(\bar{E}, \bar{T})$, where the external tryptophan is equal to zero, $\bar{E}$ is the steady-state concentration of $E(t)$, and $\bar{T}$ is the steady-state concentration of $T(t)$. No analytic solution exists for this case in general. In the supplementary material of Santillán and Mackey (2001), they have argued that the active enzyme, E_A, at the steady state, i.e. $\bar{E}_A$, is half of the amount of the steady-state enzyme, $\bar{E}$. From the Hill equation in (4.12), the steady state of T, i.e. $\bar{T}$, must be

$$\bar{T} = K_i$$

so that $\bar{E}_A$ is equal to $\bar{E}/2$. Substitute $\bar{T} = K_i$ into (4.13) and solve for K

$$K = \frac{2(\bar{G} + \mu K_i)}{\bar{E}}$$

where

$$\bar{G} = \frac{gK_i}{K_g + K_i}$$

The maximum internal tryptophan consumption rate, g, is given by

$$g = \frac{T_{\text{cr}}(K_i + K_g)}{K_i}$$

where T_{cr} is the steady-state internal tryptophan consumption rate to be estimated.

The algebraic equation from (4.10) for the steady state of $E(t)$ is given by

$$0 = \frac{1}{2}k_\rho \rho \bar{M}_F e^{-\mu\tau_e} - (\gamma + \mu)\bar{E}$$

and the steady state, $\bar{E}$, is equal to

$$\bar{E} = \frac{k_\rho \rho \bar{M}_F e^{-\mu\tau_e}}{2(\gamma + \mu)}$$

After the steady state, no time delay effect occurs, i.e. $\bar{M}(t)_F$ equal to $\bar{M}_F(t - \tau_e)$. The steady-state condition for $M_f(t)$ is

$$0 = k_p P\bar{O}_F e^{-\mu\tau_m}\left[1 - b\left(1 - e^{-K_i/c}\right)\right] - k_\rho \rho(1 - e^{-\mu\tau_\rho})\bar{M}_F - \left(k_d D + \mu\right)\bar{M}_F$$

and $\bar{M}_F$ is given by

$$\bar{M}_F = \frac{k_p P \bar{O}_F e^{-\mu\tau_m} \left[1 - b\left(1 - e^{-K_i/c}\right)\right]}{k_\rho \rho (1 - e^{-\mu\tau_\rho}) + k_d D + \mu}$$

The steady state of O_F is

$$0 = \frac{K_r}{K_r + RK_i/(K_t + K_i)} \left(\mu O - k_p P \bar{O}_F + k_p P \bar{O}_F e^{-\mu\tau_p}\right) - \mu \bar{O}_F$$

and

$$\bar{O}_F = \frac{K_r \mu O}{K_r k_p P(1 - e^{-\mu\tau_p}) + \mu\left[K_r + RK_i/(K_t + K_i)\right]}$$

The other dependent parameters in the model are as follows:

$$k_\rho = \frac{1}{\rho\tau_\rho}, \quad k_p = \frac{1}{\tau_p P}, \quad k_d D = \frac{\rho\, k_\rho}{30}, \quad K_g = \frac{\bar{T}}{20}$$

and

$$K_r = \frac{k_{-r}}{k_{+r}}, \quad K_i = \frac{k_{-i}}{k_{+i}}, \quad K_t = \frac{k_{-t}}{k_{+t}}$$

4.2.2.2 Padé Approximation of Time-Delay

There are four time delay terms in the differential equations. Solving delay differential equations is expensive compared to the ones without time delay. As delayed states affect the derivatives of states, solvers must deal with possible discontinuities in the derivatives and need to access the past states from the memory.

We are to search the best parameter set for the delay differential equations and the experimental data by solving the differential equations many times through an optimization process to be explained later. Short computation time for solving the delay differential equation for each parameter set is preferable over the accuracy of the solutions as long as they are in the tolerable range.

Consider the following delay equation

$$y(t) = x(t - \tau)$$

for $t \in [0, \infty)$, where $y(t)$ is the delay of $x(t)$ by τ, which is strictly positive, and $x(t - \tau) = 0$ for $t < \tau$. Take the Laplace transform,

$$Y(s) = e^{-\tau s} X(s) \tag{4.14}$$

where $X(s)$ and $Y(s)$ are the Laplace transform of $x(t)$ and $y(t)$, respectively. The time delay becomes the exponential function in the Laplace domain, s. Substitute $s = j\omega$, where $j = \sqrt{-1}$ and $\omega \in [0, \infty)$, into the exponential function and apply Euler's formula

$$e^{-j\tau\omega} = \cos(\tau\omega) - j\sin(\tau\omega)$$

The time delay does not affect the magnitude of the signal as the magnitude of time delay is always 1 for all frequencies, $\omega \in [0, \infty)$, but only affects the phase of the signal.

For the relatively small τ and $X(s)$ in the low-frequency ranges, the following Padé approximation replaces the exponential function (Franklin et al., 2015):

$$e^{-\tau s} \approx \frac{1 - (\tau s)/2 + (\tau s)^2/12}{1 + (\tau s)/2 + (\tau s)^2/12} \tag{4.15}$$

which is $(p, q) = (2, 2)$ Padé approximation, where p is the order of the polynomial in the numerator and q is the order of the polynomial in the denominator. For $\tau = 1.32$ minutes, which would be twice the longest time delay in Table 4.2, the phase angle comparison between (2, 2)-Padé approximation and the exponential function is shown in Figure 4.3. The phase angles are matched well to each other up to 2 rad/min for $\tau = 1.32$ minutes.

4.2.2.3 State-Space Realization

There is no direct way to implement the transfer functions on the computer using elementary operations such as summation and multiplication. The state-space form is the *realization* of the transfer function as it is the first-order differential equation to be solved by the elementary operations.

For the realization, we use one of the properties of the Laplace transform. The differentiation of functions in the time domain and the corresponding Laplace transform satisfy the following equation:

$$sY(s) - y(0) = \int_{t=0}^{t=\infty} e^{-st}\dot{y}(t)dt$$

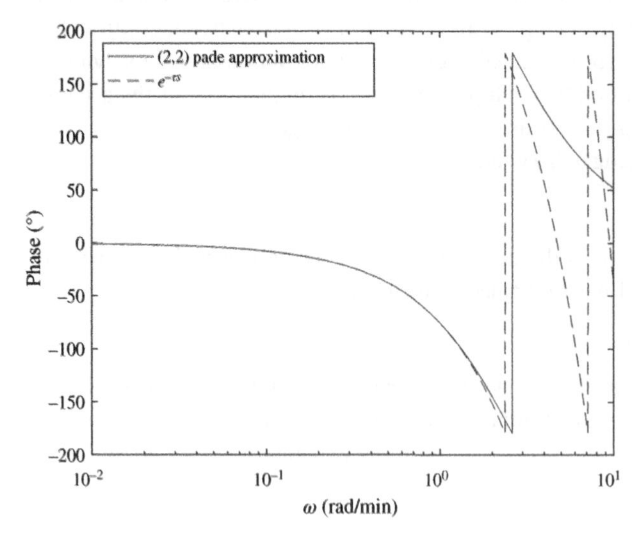

Figure 4.3 Comparison phase angles of (2,2)-Padé approximation and $e^{-\tau s}$.

We can show the following relation using the equation recursively:

$$s^n Y(s) - s^{n-1} y(0) - \cdots - s\frac{d^{n-2}y(0)}{dt^{n-2}} - \frac{d^{n-1}y(0)}{dt^{n-1}} = \int_{t=0}^{t=\infty} e^{-st}\frac{d^n y(t)}{dt^n}dt$$

Setting all the initial conditions to zero simply results in

$$s^n Y(s) = \int_{t=0}^{t=\infty} e^{-st}\frac{d^n y(t)}{dt^n}dt$$

Substitute the Padé approximation, (4.15), into (4.14)

$$Y(s) = \frac{1 - (\tau s)/2 + (\tau s)^2/12}{1 + (\tau s)/2 + (\tau s)^2/12}X(s)$$

Rearrange it

$$Y(s) = \left[1 - (\tau s)/2 + (\tau s)^2/12\right] Z(s)$$

where

$$Z(s) = \frac{1}{1 + (\tau s)/2 + (\tau s)^2/12}X(s)$$

Multiply both sides of the equation between $Z(s)$ and $X(s)$ by the denominator.

$$\left[1 + (\tau s)/2 + (\tau s)^2/12\right] Z(s) = X(s)$$

Assuming all initial conditions are zero, the corresponding differential equation is given by

$$z + \frac{\tau}{2}\dot{z} + \frac{\tau^2}{12}\ddot{z} = x$$

Define the state vector, $\mathbf{z}$,

$$\mathbf{z} = \begin{bmatrix} z \\ \dot{z} \end{bmatrix}$$

Take the time derivative

$$\dot{\mathbf{z}} = \begin{bmatrix} 0 & 1 \\ -\dfrac{12}{\tau^2} & -\dfrac{6}{\tau} \end{bmatrix}\mathbf{z} + \begin{bmatrix} 0 \\ \dfrac{12}{\tau^2} \end{bmatrix} x = A\mathbf{z} + Bx$$

The output equation from the differential equation for $y(t)$ and $z(t)$ is

$$y = z - \frac{\tau}{2}\dot{z} + \frac{\tau^2}{12}\ddot{z} = z - \frac{\tau}{2}\dot{z} + \left(x - z - \frac{\tau}{2}\dot{z}\right) = -\tau\dot{z} + x$$
$$= \begin{bmatrix} 0 & -\tau \end{bmatrix}\mathbf{z} + x = C\mathbf{z} + Dx$$

In MATLAB and scipy in Python, *tf2ss()* returns the state-space form for the given transfer function. Running the following in MATLAB

```
1 >> tau = 1.32;
2 >> num = [tau^2/12 -tau/2 1];
3 >> den = [tau^2/12 tau/2 1];
4 >> [A,B,C,D] = tf2ss(num,den)
```

or the following in Python

```
1 In [1]: from scipy.signal import tf2ss
2 In [2]: tau=1.32
3 In [3]: num=[tau**2/12, -tau/2, 1]
4 In [4]: den=[tau**2/12, tau/2,1]
5 In [5]: A,B,C,D=tf2ss(num,den)
```

returns

$$A = \begin{bmatrix} -4.54 & -6.89 \\ 1 & 0 \end{bmatrix}, \quad B = \begin{bmatrix} 1 \\ 0 \end{bmatrix}, \quad C = \begin{bmatrix} -9.09 & 0 \end{bmatrix}, \quad D = 1$$

The results are different from what we expect from the state-space form derived as follows:

$$A = \begin{bmatrix} 0 & 1 \\ -6.89 & -4.54 \end{bmatrix}, \quad B = \begin{bmatrix} 0 \\ 6.89 \end{bmatrix}, \quad C = \begin{bmatrix} 0 & -1.32 \end{bmatrix}, \quad D = 1$$

The state-space form is given by

$$\dot{\mathbf{z}} = A\mathbf{z} + Bx$$
$$y = C\mathbf{z} + Dx$$

Taking the Laplace transform and finding the relationship between the input x and the output y, we obtain the transfer function as follows:

$$sZ(s) = AZ(s) + BX(s) \Rightarrow Z = (sI - A)^{-1}BX(s)$$
$$\Rightarrow Y(s) = \left[C(sI - A)^{-1}B + D\right] X(s)$$

where $\mathbf{z}(0) = \mathbf{0}$. The transfer function, $C(sI - A)^{-1} + D$, is the same for the above two sets of (A, B, C, D). It implies that there is no unique state-space realization for a given transfer function. Any linear transformation such as

$$\mathbf{q} = T\mathbf{z}$$

where T is a non-singular matrix, i.e. T^{-1} exists, provides the following state-space form:

$$\dot{\mathbf{q}} = T\dot{\mathbf{z}} = TA\mathbf{z} + TBx = TA(T^{-1}T\mathbf{z}) + TBx$$
$$= (TAT^{-1})\mathbf{q} + (TB)x = \tilde{A}\mathbf{q} + \tilde{B}x$$
$$y = C(T^{-1}T\mathbf{z}) + Dx = (CT^{-1})\mathbf{q} + Dx = \tilde{C}\mathbf{q} + Dx$$

and $\tilde{C}(sI - \tilde{A})^{-1}\tilde{B} + D$ provides the same transfer function.

For a transfer function given by

$$Y(s) = \frac{b_0 s^n + b_1 s^{n-1} + \cdots + b_{n-1}s + b_n}{s^n + a_1 s^{n-1} + \cdots + a_{n-1}s + a_n} X(s)$$

the controllable canonical form is as follows (The MathWorks, 2021):

$$A = \begin{bmatrix} 0 & 1 & 0 & \dots & 0 \\ 0 & 0 & 1 & \dots & 0 \\ \vdots & \vdots & \vdots & \ddots & \vdots \\ -a_n & -a_{n-1} & \dots & \dots & -a_1 \end{bmatrix}, \quad B = \begin{bmatrix} 0 \\ 0 \\ \vdots \\ 0 \\ 1 \end{bmatrix},$$

$$C = \begin{bmatrix} b_n - a_n b_0 & b_{n-1} - a_{n-1} b_0 & \dots & b_2 - a_2 b_0 & b_1 - a_1 b_0 \end{bmatrix}, \quad D = b_0$$

Divide the numerator and the denominator of (2,2)-Padé approximation by $\tau^2/12$

$$Y(s) = \frac{s^2 - (6/\tau)s + (12/\tau^2)}{s^2 + (6/\tau)s + (12/\tau^2)} X(s)$$

The controllable canonical form of (2,2) Padé approximation is

$$A = \begin{bmatrix} 0 & 1 \\ -12/\tau^2 & -6/\tau \end{bmatrix}, \quad B = \begin{bmatrix} 0 \\ 1 \end{bmatrix}, \quad C = \begin{bmatrix} 0 & -12/\tau \end{bmatrix}, \quad D = 1$$

Using this state-space realization, the time delays of O_F and M_F are implemented. For example, $O_F(t - \tau_p)$ is obtained by solving the following first-order differential equation:

$$\dot{\mathbf{z}}_p = \begin{bmatrix} 0 & 1 \\ -12/\tau_p^2 & -6/\tau_p \end{bmatrix} \mathbf{z}_p + \begin{bmatrix} 0 \\ 1 \end{bmatrix} O_F(t) = A(\tau_p)\mathbf{z}_p + B_p O_F(t)$$

$$O_F(t - \tau_p) = \begin{bmatrix} 0 & -12/\tau_p \end{bmatrix} \mathbf{z}_p + O_F(t) = C(\tau_p)\mathbf{z}_p + D_p O_F(t)$$

where $\mathbf{z}_p(0) = \mathbf{0}$. Similarly, the realizations for $O_F(t - \tau_m)$, $M_F(t - \tau_\rho)$, and $M_F(t - \tau_e)$ are implemented.

Define the state vector for the governing differential equations

$$\mathbf{x} = \begin{bmatrix} O_F(t) & M_F(t) & E(t) & T(t) & \mathbf{z}_p^T & \mathbf{z}_m^T & \mathbf{z}_\rho^T & \mathbf{z}_e^T \end{bmatrix}^T$$

Programs 4.1 and 4.2 implement the original four non-linear differential equations for O_F, M_F, $E(t)$, and $T(t)$ and additional eight linear differential equations for the four delayed states in MATLAB and Python, respectively.

Line 5 in Program 4.1 and line 7 in Program 4.2 reset negative state values to zero. All quantities in biological networks are positive values, e.g. molecular concentration. Numerical integrators do not concern if the solution includes negative or positive values. The reset is one of the two safeguards to prevent negative molecular concentrations. Another safeguard at the end of the programs, starting from line 112 in Program 4.1 and line 121 in Program 4.2, respectively, resets the derivatives of $O_F(t)$, $M_F(t)$, $E(t)$, and $T(t)$ equal to zero if the current concentrations are negative to prevent the concentrations decreasing further.

The functions receive all 23 model parameters. The six dependent parameters, K_i, K_t, K_r, k_ρ, k_p, and $k_d D$, are calculated from the provided parameters. As the steady state of $T(t)$ is equal to K_i, K is calculated accordingly. To calculate K for $\bar{T} = K_i$, $\bar{G}$, and $\bar{E}$ are required. In addition, to calculate $\bar{E}$, $\bar{M}_F$ and $\bar{O}_F$ are required. The functions return $\mathbf{dx}/dt$ as the 12×1 column vector.

```
 1 %% Santillan 's model delayed differential equation
 2 function dxdt = Santillan_E_coli_Tryptophan(time, state_all ,
       parameters , T_ext)
 3
 4     state_org = state_all;
 5     state_all(state_all<0) = 0.0;
 6
 7     %------------------------------------------------------------
 8     % Uncertain parameters
 9     %------------------------------------------------------------
10     tau_p            = parameters(1);
11     tau_m            = parameters(2);
12     tau_rho          = parameters(3);
13     tau_e            = parameters(4);
14     R                = parameters(5);
15     n_H              = parameters(6);
16     b                = parameters(7);
17     e                = parameters(8);
18     f                = parameters(9);
19     O                = parameters(10);
20     k_mr             = parameters(11);
21     k_pr             = parameters(12);
22     k_mi             = parameters(13);
23     k_pi             = parameters(14);
24     k_mt             = parameters(15);
25     k_pt             = parameters(16);
26     c                = parameters(17);
27     d                = parameters(18);
28     gama             = parameters(19);
29     T_consume_rate   = parameters(20);
30     P                = parameters(21);
31     rho              = parameters(22);
32     mu               = parameters(23);
33
34     %----------------------------------------
35     % Dependent variables
36     %----------------------------------------
37     K_i          = k_mi/k_pi;
38     K_t          = k_mt/k_pt;
39     K_r          = k_mr/k_pr;
40
41     k_rho        = 1/(tau_rho*rho);
42     k_p          = 1/(tau_p*P);
43     kdD          = rho*k_rho/30;
44
```

```
45      %------------------------------------
46      % Steady-state
47      %------------------------------------
48      T_SS = K_i;
49      K_g  = T_SS/20;
50      g_SS = T_consume_rate*(K_i + K_g)/K_i;
51      G_SS = g_SS*K_i/(K_i+K_g);
52
53      R_A_SS =  T_SS/(T_SS+K_t)*R;
54      O_F_SS = (K_r*mu*O)/(K_r*k_p*(1-exp(-mu*tau_p))+mu*(K_r+R_A_SS)
           );
55      M_F_SS = k_p*P*O_F_SS*exp(-mu*tau_m)*(1-b*(1-exp(-K_i/c))) ...
56               /(k_rho*rho*(1-exp(-mu*tau_rho))+kdD+mu);
57      E_SS = (k_rho*rho*M_F_SS*exp(-mu*tau_e))/(2*(gama+mu));
58
59      K = 2*(G_SS + mu*K_i)/E_SS;
60
61      % state
62      O_F = state_all(1);
63      M_F = state_all(2);
64      E   = state_all(3);
65      T   = state_all(4);
66
67      % delayed state
68      state_tau_p     = state_all(5:6);
69      state_tau_m     = state_all(7:8);
70      state_tau_rho   = state_all(9:10);
71      state_tau_e     = state_all(11:12);
72
73      A_tau_p     = [0 1; -12/tau_p^2      -6/tau_p];
74      A_tau_m     = [0 1; -12/tau_m^2      -6/tau_m];
75      A_tau_rho   = [0 1; -12/tau_rho^2    -6/tau_rho];
76      A_tau_e     = [0 1; -12/tau_e^2      -6/tau_e];
77      B_tau       = [0; 1];
78      C_tau_p     = [0 -12/tau_p];
79      C_tau_m     = [0 -12/tau_m];
80      C_tau_rho   = [0 -12/tau_rho];
81      C_tau_e     = [0 -12/tau_e];
82      D_tau       = 1;
83
84      % dxdt = Ax + Bu
85      dO_F_tau_p      = A_tau_p*state_tau_p(:)          + B_tau*O_F;
86      dO_F_tau_m      = A_tau_m*state_tau_m(:)          + B_tau*O_F;
87      dM_F_tau_rho    = A_tau_rho*state_tau_rho(:)      + B_tau*M_F;
88      dM_F_tau_e      = A_tau_e*state_tau_e(:)          + B_tau*M_F;
89
90      % y = Cx + Du
91      O_F_tau_p   = C_tau_p*state_tau_p(:)          + D_tau*O_F;
92      O_F_tau_m   = C_tau_m*state_tau_m(:)          + D_tau*O_F;
93      M_F_tau_rho = C_tau_rho*state_tau_rho(:)      + D_tau*M_F;
94      M_F_tau_e   = C_tau_e*state_tau_e(:)          + D_tau*M_F;
95
```

```
96      d_delay_dt = [dO_F_tau_p(:); dO_F_tau_m(:); dM_F_tau_rho(:);
            dM_F_tau_e(:)];
97
98      % auxilary variables
99      A_T = b*(1-exp(-T/c));
100     E_A = K_i^n_H/(K_i^n_H + T^n_H)*E;
101     R_A = T/(T+K_t)*R;
102     G   = g_SS*T/(T+K_g);
103     F   = d*T_ext/(e + T_ext*(1+T/f));
104
105     % kinetics
106     dOF_dt = K_r/(K_r + R_A)*(mu*O - k_p*P*(O_F - O_F_tau_p*exp(-mu
            *tau_p))) - mu*O_F;
107     dMF_dt = k_p*P*O_F_tau_m*exp(-mu*tau_m)*(1-A_T) ...
108         - k_rho*rho*(M_F - M_F_tau_rho*exp(-mu*tau_rho)) - (kdD +
                mu)*M_F;
109     dE_dt = 0.5*k_rho*rho*M_F_tau_e*exp(-mu*tau_e) - (gama + mu)*E;
110     dT_dt = K*E_A - G + F - mu*T;
111
112     if state_org(1) < 0 && dOF_dt < 0
113         dOF_dt = 0;
114     end
115     if state_org(2) < 0 && dMF_dt < 0
116         dMF_dt = 0;
117     end
118     if state_org(3) < 0 && dE_dt < 0
119         dE_dt = 0;
120     end
121     if state_org(4) < 0 && dT_dt < 0
122         dT_dt = 0;
123     end
124
125     dOF_MF_E_T_dt = [dOF_dt dMF_dt dE_dt dT_dt]';
126
127     % return all state
128     dxdt = [dOF_MF_E_T_dt; d_delay_dt];
129
130 end
```

Program 4.1 (MATLAB) Santillan's model for *E. coli* tryptophan operon regulation with Padé approximation for the time delay

```
1  import numpy as np
2
3  # Santillan's model delayed differential equation
4  def Santillan_E_coli_Tryptophan(time, state_all, parameters, T_ext)
       :
5
6      state_org = state_all
7      state_all[state_all<0] = 0.0
8
9      #-----------------------------------------------------------
10     # Uncertain parameters
11     #-----------------------------------------------------------
12     tau_p            = parameters[0]
13     tau_m            = parameters[1]
```

```
14    tau_rho          = parameters[2]
15    tau_e            = parameters[3]
16    R                = parameters[4]
17    n_H              = parameters[5]
18    b                = parameters[6]
19    e                = parameters[7]
20    f                = parameters[8]
21    O                = parameters[9]
22    k_mr             = parameters[10]
23    k_pr             = parameters[11]
24    k_mi             = parameters[12]
25    k_pi             = parameters[13]
26    k_mt             = parameters[14]
27    k_pt             = parameters[15]
28    c                = parameters[16]
29    d                = parameters[17]
30    gama             = parameters[18]
31    T_consume_rate   = parameters[19]
32    P                = parameters[20]
33    rho              = parameters[21]
34    mu               = parameters[22]
35
36    #----------------------------------------
37    # Dependent variables
38    #----------------------------------------
39    K_i          = k_mi/k_pi
40    K_t          = k_mt/k_pt
41    K_r          = k_mr/k_pr
42
43    k_rho        = 1/(tau_rho*rho)
44    k_p          = 1/(tau_p*P)
45    kdD          = rho*k_rho/30
46
47    #----------------------------------------
48    # Steady-state
49    #----------------------------------------
50    T_SS = K_i
51    K_g  = T_SS/20
52    g_SS = T_consume_rate*(K_i + K_g)/K_i
53    G_SS = g_SS*K_i/(K_i+K_g)
54
55    R_A_SS =  T_SS/(T_SS+K_t)*R
56    O_F_SS = (K_r*mu*O)/(K_r*k_p*(1-np.exp(-mu*tau_p))+mu*(K_r+
           R_A_SS))
57    M_F_SS = k_p*P*O_F_SS*np.exp(-mu*tau_m)*(1-b*(1-np.exp(-K_i/c))
           ) \
58         /(k_rho*rho*(1-np.exp(-mu*tau_rho))+kdD+mu)
59    E_SS = (k_rho*rho*M_F_SS*np.exp(-mu*tau_e))/(2*(gama+mu))
60
61    K = 2*(G_SS + mu*K_i)/E_SS
62
63    # state
64    O_F = state_all[0]
```

```
65      M_F = state_all[1]
66      E   = state_all[2]
67      T   = state_all[3]
68
69      # delayed state
70      state_tau_p     = state_all[4:6];    state_tau_p.resize((2,1))
71      state_tau_m     = state_all[6:8];    state_tau_m.resize((2,1))
72      state_tau_rho   = state_all[8:10];   state_tau_rho.resize((2,1))
73      state_tau_e     = state_all[10::];   state_tau_e.resize((2,1))
74
75      A_tau_p      = np.array([[0,1], [-12/tau_p**2,     -6/tau_p]])
76      A_tau_m      = np.array([[0,1], [-12/tau_m**2,     -6/tau_m]])
77      A_tau_rho    = np.array([[0,1], [-12/tau_rho**2,   -6/tau_rho]])
78      A_tau_e      = np.array([[0,1], [-12/tau_e**2,     -6/tau_e]])
79      B_tau        = np.array([[0], [1]])
80      C_tau_p      = np.array([[0, -12/tau_p]])
81      C_tau_m      = np.array([[0, -12/tau_m]])
82      C_tau_rho    = np.array([[0, -12/tau_rho]])
83      C_tau_e      = np.array([[0, -12/tau_e]])
84      D_tau        = np.array([[1]])
85
86      # dxdt = Ax + Bu
87      dO_F_tau_p   = A_tau_p@state_tau_p + B_tau@np.array([[O_F]])
88      dO_F_tau_m   = A_tau_m@state_tau_m + B_tau@np.array([[O_F]])
89      dM_F_tau_rho= A_tau_rho@state_tau_rho + B_tau@np.array([[M_F]])
90      dM_F_tau_e   = A_tau_e@state_tau_e + B_tau@np.array([[M_F]])
91
92      # y = Cx + Du
93      O_F_tau_p    = C_tau_p@state_tau_p      + D_tau@np.array([[O_F]])
94      O_F_tau_m    = C_tau_m@state_tau_m      + D_tau@np.array([[O_F]])
95      M_F_tau_rho  = C_tau_rho@state_tau_rho+ D_tau@np.array([[M_F]])
96      M_F_tau_e    = C_tau_e@state_tau_e      + D_tau@np.array([[M_F]])
97
98      # make 1x1 array to scalar
99      O_F_tau_p=O_F_tau_p[0][0]
100     O_F_tau_m=O_F_tau_m[0][0]
101     M_F_tau_rho=M_F_tau_rho[0][0]
102     M_F_tau_e=M_F_tau_e[0][0]
103
104     d_delay_dt = np.vstack((dO_F_tau_p,dO_F_tau_m,dM_F_tau_rho,
            dM_F_tau_e))
105     d_delay_dt = d_delay_dt.squeeze()
106
107     # auxilary variables
108     A_T = b*(1-np.exp(-T/c))
109     E_A = K_i**n_H/(K_i**n_H + T**n_H)*E
110     R_A = T/(T+K_t)*R
111     G   = g_SS*T/(T+K_g)
112     F   = d*T_ext/(e + T_ext*(1+T/f))
113
114     # kinetics
115     dOF_dt = K_r/(K_r + R_A)*(mu*O - k_p*P*(O_F - O_F_tau_p*np.exp
            (-mu*tau_p))) - mu*O_F
```

```
116        dMF_dt = k_p*P*O_F_tau_m*np.exp(-mu*tau_m)*(1-A_T) \
117            - k_rho*rho*(M_F - M_F_tau_rho*np.exp(-mu*tau_rho)) - (kdD
                 + mu)*M_F
118        dE_dt = 0.5*k_rho*rho*M_F_tau_e*np.exp(-mu*tau_e) - (gama + mu)
             *E;
119        dT_dt = K*E_A - G + F - mu*T;
120
121        if state_org[0] < 0 and dOF_dt < 0:
122            dOF_dt = 0
123        if state_org[1] < 0 and dMF_dt < 0:
124            dMF_dt = 0;
125        if state_org[2] < 0 and dE_dt < 0:
126            dE_dt = 0
127        if state_org[3] < 0 and dT_dt < 0:
128            dT_dt = 0
129
130        dOF_MF_E_T_dt = np.array([dOF_dt, dMF_dt, dE_dt, dT_dt])
131
132        # return all state
133        dxdt = np.hstack((dOF_MF_E_T_dt,d_delay_dt))
134
135        return dxdt
```

Program 4.2 (Python) Santillan's model for *E. coli* tryptophan operon regulation with Padé approximation for the time delay

4.2.2.4 Python

Program 4.2 has several points related to matrix operations in constructing the state-space form for the delay terms. We introduce one-dimensional arrays in Python on page 47. Additional cares for matrices in Python stem from how one-dimensional arrays or scalar values interact with two-dimensional arrays in Python.

Consider the following Python code generating a 2×1 matrix, B:

```
1 In [1]: import numpy np
2 In [2]: B = np.array([[1],[2]])
3 In [3]: B.shape
4 Out[3]: (2,1)
```

Multiple B by 3 in four ways:

```
1 In [4]: y = B*3
2 In [5]: y.shape
3 Out[5]: (2,1)
```

returns what we expect.

```
1 In [6]: y = B@3
```

returns the dimension mismatch error. Python interprets *B@3* as a 2 ×1 matrix multiplies a 0 ×0 matrix.

```
1 In [7]: y = B@np.array([[3]])
2 In [8]: y.shape
3 Out[8]: (2,1)
```

returns what we expect in the matrix multiplication as Python interprets it as the product of a 2 ×1 matrix and a 1 × 1 matrix. In addition,

```
1 In [7]: y = B@np.array([3])
2 In [8]: y.shape
3 Out[8]: (2,)
```

returns the same expected result, but the result is in one-dimensional array. This will be confusing when it adds to two-dimensional array. For example,

```
1 In [9]: y = B@np.array([3])
2 In [10]: x = np.array([[1],[2]])
3 In [11]: x+y
```

does not return the result of $[1, 2]^T + [3, 6]^T = [4, 8]^T$ but each element of x adds to y as follows:

$$\mathrm{x+y} = \begin{bmatrix} 1 \\ 2 \end{bmatrix} + y = \begin{bmatrix} 1+y \\ 2+y \end{bmatrix} = \begin{bmatrix} 4 & 7 \\ 5 & 8 \end{bmatrix}$$

> **Matrix operations in Python**: When two-dimensional arrays and one-dimensional arrays are mixed in the matrix operations in Python, unexpected results could occur. It is safe to use only two-dimensional arrays in Python matrix operations.

4.2.2.5 Model Parameter Ranges

All 23 independent parameters are given in Table 4.2, where some are experimentally measured and some are estimated (Santillán and Mackey, 2001). For the wild-type *E. coli*, three different nutritional shift experiments are performed in Yanofsky and Horn (1994):

- Experiment A: Tryptophan → the same media without tryptophan
- Experiment B: Tryptophan, phenylalanine, tyrosine → the same media without tryptophan. This is considered to be the harshest condition
- Experiment C: Tryptophan, acid-hydrolyzed casein → the same media without tryptophan

Table 4.2 The nominal values of the model parameters in the tryptophan regulation model by Santillán and Mackey (2001).

	Unit	Value	Index		Unit	Value	Index
τ_p	(min)	0.1	1	k_{-i}	(1/min)	0.072	13
τ_m	(min)	0.1	2	k_{+i}	$(\mu\text{M min})^{-1}$	0.0176	14
τ_ρ	(min)	0.05	3	k_{-t}	(1/min)	2.1×10^4	15
τ_e	(min)	0.66	4	k_{+t}	$(\mu\text{M min})^{-1}$	348	16
R	(μM)	0.8	5	c	$(\cdot)$	0.04	17
n_H	$(\cdot)$	1.2	6	d	$(\cdot)$	23.5	18
b	$(\cdot)$	0.85	7	γ	(1/min)	0.0	19
e	$(\cdot)$	0.9	8	T_{cr}	(μM/min)	22.7	20
f	$(\cdot)$	380	9	P	(μM)	2.6	21
O	(μM)	0.0033	10	ρ	(μM)	2.9	22
k_{-r}	(1/min)	0.012*	11	μ	(1/min)	0.01	23
k_{+r}	$(\mu\text{M min })^{-1}$	4.6*	12				

* One hundred times different values are found in Santillán and Mackey (2001) and its supplementary material. The ratio, $K_r = k_{-r}/k + r$, remains the same for both cases, however.

Source: Based on Santillán and Mackey (2001).

Table 4.1 provides the time history measurements of the active enzyme concentration for each experiment in Yanofsky and Horn (1994). For three different experimental sets in Table 4.1, to solve the model fitting optimization problem, (4.5), the search space for the parameters in Table 4.2 needs to be set. A larger search space would provide parameter combinations that better fit the experimental measurements. However, we potentially have parameter combinations that extend beyond the range of biologically meaningful values. Before solving the model-fitting optimization problem, we establish the parameter search ranges.

The last four parameters in Table 4.2 have the experimentally known ranges. Table 4.3 provides the ranges of these four parameters and their uncertainty formula, where δ_{20}, δ_{21}, δ_{22}, and δ_{23} are the real numbers between -1 and $+1$.

The uncertainties of the other parameters in Table 4.2 are modelled as follows:

$$p_i = \bar{p}_i(1 + \delta_i)$$

where p_i is the perturbed parameter, $\bar{p}_i$ is the nominal value, and δ_i is the uncertainty in $[-1, +1]$ for $i = 1, 2, 3, 4, 5, 8, 9, \ldots, 18, 19$. For $i = 6$, the uncertainty is modelled as follows:

$$n_H = 2 + \delta_6$$

Table 4.3 Experimentally known ranges of the parameters from the Supplementary of (Santillán and Mackey, 2001).

Parameter	Unit	Uncertainty range	Uncertainty formula
T_{cr}	(μM/min)	[14.0, 29.0]	$21.5 + 7.5\delta_{20}$
P	(μM)	[2.11, 3.46]	$2.785 + 0.675\delta_{21}$
ρ	(μM)	[2.37, 3.87]	$3.12 + 0.75\delta_{22}$
μ	(μM)	[0.01, 0.0418]	$0.0259 + 0.0159\delta_{23}$

Source: Based on Santillán and Mackey (2001).

where $\delta_6 \in [-1, 1]$. The nominal value of the Hill coefficient is equal to 1.2, and the uncertain range covers from 1 to 3. The lower bound restricts the minimum Hill coefficient equal to 1. The upper bound allows us to consider up to 2.5 times larger values from the nominal value. For $i = 7$, the uncertainty is modelled as follows:

$$b = 0.65 + 0.35\delta_7$$

where $\delta_7 \in [-1, 1]$. The nominal value of b is equal to 0.85, and the uncertain range covers from 0.3 to 1.0. The term, $b(1 - e^{T(t)/c})$, in (4.9) is the probability of premature transcription termination. b restricts the maximum possible probability between 30% and 100%.

The MATLAB and the Python scripts for the uncertainty modelling are implemented in Program 4.3 and 4.4, respectively.

```
1  %% uncertain parameters
2  %
3  % [Ref] Moises Santillan and Michael C. Mackey. Dynamic reguiation
       of the tryptophan
4  % operon: A modeling study and comparison with experimental data.
5  % Proceedings of the National Academy of Sciences, 98(4):1364-1369,
       February 2001.
6  %
7  function [perturbed_model_para] =
       Santillans_Tryptophan_Model_constants(delta)
8
9      %-----------------------------------------------------------
10     % Uncertain ranges without experimental evidences
11     %-----------------------------------------------------------
12     Santillan_tau_p   = 0.1*(1 + delta(1));              % 1
13     Santillan_tau_m   = 0.1*(1 + delta(2));              % 2
14     Santillan_tau_rho = 0.05*(1 + delta(3));             % 3
15     Santillan_tau_e   = 0.66*(1 + delta(4));             % 4
16
17     Santillan_R = 0.8*(1 + delta(5));                    % 5
18
```

```
19     Santillan_n_H = 2 + delta(6);                                % 6
20                           % nominal = 1.2
21                           % delta_nominal = -0.8
22
23     Santillan_b =  0.65 + 0.35*delta(7);              % 7  [0.3, 1.0]
24                           % nominal = 0.85
25                           % delta_nominal = 0.5714
26
27     Santillan_e = 0.9*(1 + delta(8));                            % 8
28     Santillan_f = 380*(1 + delta(9));                            % 9
29
30     Santillan_O = 3.32e-3*(1 + delta(10));                       % 10
31
32     Santillan_k_mr = 1.2e-2*(1 + delta(11));                     % 11
33                          % value in [Ref] & its supplementary is
                                 different
34     Santillan_k_pr = 4.6*(1 + delta(12));
35                          % value in [Ref] & its supplementary is
                                 different
36                          % but the ratio, kmr/kpr is the same
37
38     Santillan_k_mi = 7.2e-2*(1 + delta(13));                     % 13
39     Santillan_k_pi = 1.76e-2*(1 + delta(14));                    % 14
40
41     Santillan_k_mt = 2.1e4*(1 + delta(15));                      % 15
42     Santillan_k_pt = 348*(1 + delta(16));                        % 16
43
44     Santillan_c = 4e-2*(1 + delta(17));                          % 17
45     Santillan_d = 23.5*(1 + delta(18));                          % 18
46
47     Santillan_gama = 0.01*(1 + delta(19));                       % 19
48                         % nominal value 0
49                         % delta nominal = -1
50
51     %---------------------------------------
52     % Uncertain ranges from experiments
53     %---------------------------------------
54     Santillan_T_consume_rate = 21.5 + 7.5*delta(20);             % 20
55                         % range 14 ~ 29
56                         % nominal 22.7 -> 0.16
57
58     Santillan_P = 2.785 + 0.675*delta(21);
59                                                                  % 21
60     % range 2.11 - 3.46 micro-Molar,
61     % nominal 2.6 -> -0.2741
62     % 1250 molecule per cell, cell average volume 6.0e-16 - 9.8e-16
63     % liters, average volumn = (6.0 + 9.8)/2*1e-16 = 7.9e-16 liters
64     % 1250 molecule = 1250/6.022e23 = 2.0757e-21 mole
65     % 2.0757e-21/7.9e-16 = 2.62e-6 Molar = 2.62 micro-Molar
66
67     Santillan_rho = 3.12 + 0.75*delta(22);
68                                                                  % 22
69     % range 2.37 - 3.87 micro-Molar,
```

```
70      % nominal 2.9 -> -0.2933
71      % 1400 molecule per cell , cell average volume 6.0e-16 - 9.8e-16
72      % liters , average volumn = (6.0 + 9.8)/2*1e-16 = 7.9e-16 liters
73      % 1400 molecule = 1400/6.022e23 = 2.3248e-21 mole
74      % 2.3248e-21/7.9e-16 = 2.94e-6 Molar = 2.94 micro-Molar
75
76      Santillan_mu = 0.0259 + 0.0159*delta(23);
77                                                                  % 23
78      % range 0.01 ~ 0.0418 [min^-1],
79      % nominal 0.01 -> -1
80      % actual range from 0.6 h^-1 ~ 2.5 h^-1
81
82      %% return values
83      num_para = 23;
84      perturbed_model_para = zeros(1,num_para);
85      perturbed_model_para(1) = Santillan_tau_p;
86      perturbed_model_para(2) = Santillan_tau_m;
87      perturbed_model_para(3) = Santillan_tau_rho;
88      perturbed_model_para(4) = Santillan_tau_e;
89      perturbed_model_para(5) = Santillan_R;
90      perturbed_model_para(6) = Santillan_n_H;
91      perturbed_model_para(7) = Santillan_b;
92      perturbed_model_para(8) = Santillan_e;
93      perturbed_model_para(9) = Santillan_f;
94      perturbed_model_para(10) = Santillan_O;
95      perturbed_model_para(11) = Santillan_k_mr;
96      perturbed_model_para(12) = Santillan_k_pr;
97      perturbed_model_para(13) = Santillan_k_mi;
98      perturbed_model_para(14) = Santillan_k_pi;
99      perturbed_model_para(15) = Santillan_k_mt;
100     perturbed_model_para(16) = Santillan_k_pt;
101     perturbed_model_para(17) = Santillan_c;
102     perturbed_model_para(18) = Santillan_d;
103     perturbed_model_para(19) = Santillan_gama;
104     perturbed_model_para(20) = Santillan_T_consume_rate;
105     perturbed_model_para(21) = Santillan_P;
106     perturbed_model_para(22) = Santillan_rho;
107     perturbed_model_para(23) = Santillan_mu;
108
109 end
```

Program 4.3 (MATLAB) Santillan's model uncertain parameters

```
1 # uncertain parameters
2 #
3 # [Ref] Moises Santillan and Michael C. Mackey. Dynamic regulation
      of the tryptophan
4 # operon: A modeling study and comparison with experimental data.
5 # Proceedings of the National Academy of Sciences, 98(4):1364-1369,
      February 2001.
6 #
7 def Santillans_Tryptophan_Model_constants(delta):
8
```

```
 9      #------------------------------------------------------------
10      # Uncertain ranges without experimental evidences
11      #------------------------------------------------------------
12      Santillan_tau_p   = 0.1*(1 + delta[0])                    # 1
13      Santillan_tau_m   = 0.1*(1 + delta[1])                    # 2
14      Santillan_tau_rho = 0.05*(1 + delta[2])                   # 3
15      Santillan_tau_e   = 0.66*(1 + delta[3])                   # 4
16
17      Santillan_R = 0.8*(1 + delta[4])                          # 5
18
19      Santillan_n_H = 2 + delta[5]                              # 6
20                     # nominal = 1.2
21                     # delta_nominal = -0.8
22
23      Santillan_b = 0.65 + 0.35*delta[6]             # 7  [0.3, 1.0]
24                     # nominal = 0.85
25                     # delta_nominal = 0.5714
26
27      Santillan_e = 0.9*(1 + delta[7])                          # 8
28      Santillan_f = 380*(1 + delta[8])                          # 9
29
30      Santillan_O = 3.32e-3*(1 + delta[9])                      # 10
31
32      Santillan_k_mr = 1.2e-2*(1 + delta[10])                   # 11
33          # value in [Ref] & its supplementary is different
34      Santillan_k_pr = 4.6*(1 + delta[11])                      # 12
35          # value in [Ref] & its supplementary is different
36          # but the ratio, kmr/kpr is the same
37
38      Santillan_k_mi = 7.2e-2*(1 + delta[12])                   # 13
39      Santillan_k_pi = 1.76e-2*(1 + delta[13])                  # 14
40
41      Santillan_k_mt = 2.1e4*(1 + delta[14])                    # 15
42      Santillan_k_pt = 348*(1 + delta[15])                      # 16
43
44      Santillan_c = 4e-2*(1 + delta[16])                        # 17
45      Santillan_d = 23.5*(1 + delta[17])                        # 18
46
47      Santillan_gama = 0.01*(1 + delta[18])                     # 19
48                # nominal value 0
49                # delta nominal = -1
50
51      #---------------------------------------
52      # Uncertain ranges from experiments
53      #---------------------------------------
54      Santillan_T_consume_rate = 21.5 + 7.5*delta[19]           # 20
55                # range 14 ~ 29
56                # nominal 22.7 -> 0.16
57
58      Santillan_P = 2.785 + 0.675*delta[20]
59                                                                # 21
60      # range 2.11 - 3.46 micro-Molar,
61      # nominal 2.6 -> -0.2741
```

```
62      # 1250 molecule per cell , cell average volume 6.0e-16 - 9.8e-16
63      # liters , average volumn = (6.0 + 9.8)/2*1e-16 = 7.9e-16 liters
64      # 1250 molecule = 1250/6.022e23 = 2.0757e-21 mole
65      # 2.0757e-21/7.9e-16 = 2.62e-6 Molar = 2.62 micro-Molar
66
67      Santillan_rho = 3.12 + 0.75*delta[21]
68                                                               # 21
69      # range 2.37 - 3.87 micro-Molar,
70      # nominal 2.9 -> -0.2933
71      # 1400 molecule per cell , cell average volume 6.0e-16 - 9.8e-16
72      # liters , average volumn = (6.0 + 9.8)/2*1e-16 = 7.9e-16 liters
73      # 1400 molecule = 1400/6.022e23 = 2.3248e-21 mole
74      # 2.3248e-21/7.9e-16 = 2.94e-6 Molar = 2.94 micro-Molar
75
76      Santillan_mu = 0.0259 + 0.0159*delta[22]
77                                                               # 22
78      # range 0.01 ~ 0.0417 [min^-1],
79      # nominal 0.01 -> -1
80      # actual range from 0.6 h^-1 ~ 2.5 h^-1
81
82      # return values
83      num_para = 23
84      perturbed_model_para = np.zeros(num_para)
85      perturbed_model_para[0] = Santillan_tau_p
86      perturbed_model_para[1] = Santillan_tau_m
87      perturbed_model_para[2] = Santillan_tau_rho
88      perturbed_model_para[3] = Santillan_tau_e
89      perturbed_model_para[4] = Santillan_R
90      perturbed_model_para[5] = Santillan_n_H
91      perturbed_model_para[6] = Santillan_b
92      perturbed_model_para[7] = Santillan_e
93      perturbed_model_para[8] = Santillan_f
94      perturbed_model_para[9] = Santillan_O
95      perturbed_model_para[10] = Santillan_k_mr
96      perturbed_model_para[11] = Santillan_k_pr
97      perturbed_model_para[12] = Santillan_k_mi
98      perturbed_model_para[13] = Santillan_k_pi
99      perturbed_model_para[14] = Santillan_k_mt
100     perturbed_model_para[15] = Santillan_k_pt
101     perturbed_model_para[16] = Santillan_c
102     perturbed_model_para[17] = Santillan_d
103     perturbed_model_para[18] = Santillan_gama
104     perturbed_model_para[19] = Santillan_T_consume_rate
105     perturbed_model_para[20] = Santillan_P
106     perturbed_model_para[21] = Santillan_rho
107     perturbed_model_para[22] = Santillan_mu
108
109     return perturbed_model_para
```

Program 4.4 (Python) Santillan's model uncertain parameters

4.2.2.6 Model Fitting Optimization

All the three experiments provide the external tryptophan to the *E. coli* cell cultures, wait until they reach the steady states, and shift them into no external tryptophan conditions. We simulate these scenarios: Firstly, set the external tryptophan equal to 400 times the steady state of the tryptophan, $\bar{T}$ (Santillán and Mackey, 2001), and the initial condition equal to zero. Secondly, solve the differential equation until it reaches a steady state, where the simulation time sets to 1200 minutes, which would be long enough to reach to a steady state for the parameters in the uncertain space. Thirdly, set the external tryptophan equal to zero and the initial condition equal to the steady state found in the previous step. Finally, solve the differential equation, compare the normalized active enzyme concentration with the normalized experiment data using the data in Table 4.1, and calculate the cost function value to be minimized in (4.5). The pseudo-code to implement the cost function is given in Algorithm 4.1.

Algorithm 4.1 *E. coli* model fitting cost function

1: Input: δ_i for $i = 1, 2, \ldots, 23$

2: Set the initial condition and the external tryptophan equal to 400 times the steady state of T as follows:

$$O_F(0) = 0,\ M_F(0) = 0,\ E(0) = 0,\ T(0) = 0,$$
$$T_{\text{ext}} = 400\bar{T} = 400K_i = 400k_{-i}/k_{+i}$$

3: Solve (4.7), (4.9), (4.10), and (4.11) for $t \in [0, 600]$ minutes.

4: Set the initial condition equal to the final values of the simulation and the external tryptophan equal to zero

$$O_F(0) = O_F(1200),\ M_F(0) = M_F(1200),$$
$$E(0) = E(1200),\ T(0) = T(1200), T_{\text{ext}} = 0.0$$

5: Solve (4.7), (4.9), (4.10), and (4.11) for $t \in [0, 1200]$ minutes

6: Calculate E_A using (4.12) at the time points given in Table 4.1

7: Calculate J using (4.5)

8: Return J

After implementing Algorithm 4.1 and testing it with δ combinations for calculating J, we would notice one possible issue in solving the non-linear differential equations. It is discussed earlier in the implementation of the differential equations that they are not generic differential equations but the ones representing the kinetics of how the molecular concentrations evolve. As molecular concentrations are non-negative quantities, they should not be allowed to decrease below zero. Although we device two safeguards in the implementations, we keep monitoring the state values, and if they are below a threshold, the numerical integrator stops. Both the ODE solvers, *ode45()* in MATLAB and *solve_ivp()* in Scipy, Python, have the event detection capability. It can be used to detect if any states go below a threshold.

The negative event detection function in MATLAB or Python is implemented as follows:

```
function [value, isterminal, direction] = negativeConcentration(~,
    state)
    tol = -0.1;
    OF_MF_E_T = state(1:4);
    delay_output = state(5:2:11);
    all_positive_state = [OF_MF_E_T(:)' delay_output(:)'];
    value = any(all_positive_state<tol)-1;
    isterminal = 1;
    direction = 0;
end
```

Program 4.5 (MATLAB) Negative event detection function for *ode45()*

```
# check negative states to stop the integrator
def negativeConcentration(time,state,parameters, T_ext):
    tol = -1e-1;
    OF_MF_E_T = state[0:4]
    delay_output = state[4::2]
    all_positive_state = np.hstack((OF_MF_E_T,delay_output))
    value = 1-float(any(all_positive_state<tol))
    return value
```

Program 4.6 (Python) Negative event detection function for *solve_ivp()*

For both event detection functions, the threshold equal to –0.1 allows the concentrations to become small negative numbers between 0 and –0.1. The arguments of the MATLAB event detection functions must be the same as the functions for implementing differential equations, i.e. the time and the states. When some mandatory arguments are unused inside the functions, they can be replaced by '~'. The requirement to match the arguments in Python is stricter than in MATLAB. The arguments of the negative detection function must be exactly the same as the ones of the differential equation function in Program 4.2. Hence, *parameters* and *Text* in the arguments of the negative event detection are included, although they are not used inside the function. The ODE solver in Algorithm 4.1 is implemented as follows:

```
1  %% Cost function for the model fitting
2  function J_cost = Santillan_Model_Fit_Cost(delta, tspan, time_exp,
       Act_Enzy_exp)
3
4      try
5          num_state = 12;
6          model_para = Santillans_Tryptophan_Model_constants(delta);
7
8          % Initially the culture in the medium with presence of the
               external tryptophan
9          T_ext = 400*(model_para(13)/model_para(14));  % 400 times
               of T(t) steady state
10         ode_option = odeset('RelTol',1e-3,'AbsTol',1e-6,'Events',
               @negativeConcentration);
11         state_t0 = zeros(1,num_state);
12         sol = ode45(@(time, state)Santillan_E_coli_Tryptophan(time,
                state, ...
               model_para, T_ext),tspan, state_t0, ode_option);
14         OF_MF_E_T_IC = mean(sol.y(:,end-50:end),2); % it reaches to
                the steady state
15
16         sol2 = sol;
17
18         % No external tryptophan medium shift experiment
19         T_ext = 0;
20         state_t0 = OF_MF_E_T_IC(:); % the steady state becomes the
                initial condition
21         tspan_sim = [0 time_exp(end)];
22         sol = ode45(@(time, state)Santillan_E_coli_Tryptophan(time,
                state, ...
               model_para, T_ext), tspan_sim, state_t0, ode_option);
24
25         % evaluate the Enzyme and the Tryptophan at the given
               measurent time
26         state_at_time_exp = deval(sol,time_exp);
27         E_at_time_exp = state_at_time_exp(3,:);
28         T_at_time_exp = state_at_time_exp(4,:);
29
30         % calculate the active enzyme using the model
31         n_H = model_para(6);
32         K_i = model_para(13)/model_para(14);
33         EA_model = (K_i^n_H./(K_i^n_H + T_at_time_exp.^n_H)).*
               E_at_time_exp;
34
35         % normalize the active enzyme
36         y_bar = EA_model/EA_model(end);
37         y_tilde = Act_Enzy_exp/Act_Enzy_exp(end);
38
```

```
39          % calculate the cost
40          J_cost = sum((y_bar-y_tilde).^2);
41
42      catch
43          J_cost = 1e3;
44      end
45
46  end
```

Program 4.7 (MATLAB) The cost function, J, with the event detection

```
1  # Cost function for the model fitting
2  def Santillan_Model_Fit_Cost(delta, tspan, time_exp, Act_Enzy_exp):
3
4      try:
5          num_state = 12;
6          model_para = Santillans_Tryptophan_Model_constants(delta);
7
8          negativeConcentration.terminal = True
9          negativeConcentration.direction = 0
10
11         # Initially the culture in the medium with presence of the
               external tryptophan
12         T_ext = 400*(model_para[12]/model_para[13]); # 400 times of
                T(t) steady state
13         time_eval = np.linspace(tspan[0],tspan[1],1000)
14         state_t0 = np.zeros(num_state)
15
16         sol = solve_ivp(Santillan_E_coli_Tryptophan, tspan,
17                         state_t0, events=negativeConcentration, args=(
                                model_para, T_ext),
18                         t_eval=time_eval, rtol=1e-3, atol=1e-6)
19         OF_MF_E_T_IC = np.mean(sol.y[:,-50:-1],axis=1) # it reaches
                to the steady state
20
21         # No external tryptophan medium shift experiment
22         T_ext = 0
23         state_t0=OF_MF_E_T_IC # the steady state becomes the
               initial condition
24         sol = solve_ivp(Santillan_E_coli_Tryptophan, (tspan[0],
               time_exp[-1]),
25                         state_t0, args=(model_para, T_ext),
26                         t_eval=time_exp, rtol=1e-3, atol=1e-6)
27
28         # evaluate the Enzyme and the Tryptophan at the given
               measurement time
29         state_at_time_exp = sol.y[0:4,:]
30         E_at_time_exp = state_at_time_exp[2,:]
31         T_at_time_exp = state_at_time_exp[3,:]
32
33         # calculate the active enzyme using the model
34         n_H = model_para[5]
35         K_i = model_para[12]/model_para[13]
```

```
36          EA_model = (K_i**n_H/(K_i**n_H + T_at_time_exp**n_H))*
                E_at_time_exp
37
38          # normalize the active enzyme
39          y_bar = EA_model/EA_model[-1]
40          y_tilde = Act_Enzy_exp/Act_Enzy_exp[-1]
41
42          # calculate the cost
43          J_cost = np.sum((y_bar-y_tilde)**2)
44
45      except:
46          J_cost = 1e3
47
48      return J_cost
```

Program 4.8 (Python) The cost function, J, with the event detection

> **Error or exception handling**: *try-catch* in MATLAB and *try-except* in Python are extremely useful methods to make programs robust.

The implementations of the cost function in Programs 4.7 and 4.8 use the error handling method using *try-catch* in MATLAB or *try-except* in Python. Consider the case illustrated in Figure 4.4, where δ_i and δ_j are the optimization parameters,

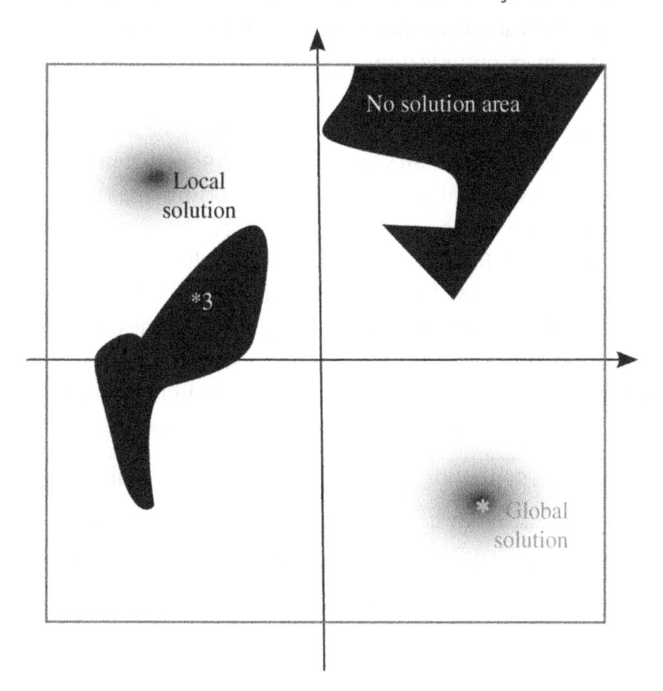

Figure 4.4 Nonlinear optimization problem.

and *1, *2, *3, and *4 are the initial point of the optimization algorithm. The optimization starting from *2 would have a higher chance to converge to the global solution than the others. Whereas the one from *4 would have a higher chance to arrive at the local solution. The optimizations starting from *1 or *3 have a distinct issue. They are inside or closer to the regions where no solution exists, i.e. the ODE solvers would stop by the negative concentration checking function. When the ODE stops by the event checking function, it might not reach the final simulation time, and the cost function does not have all the states to calculate its value. For these erroneous or exceptional cases, *try* command detects the event and passes the program flow to the error/exception handling parts, which are *catch* in MATLAB and *except* in Python. The cost function values are set to an arbitrarily chosen large number, 1000, for the exceptional cases.

Finally, the main part of the optimization is to be implemented using the genetic algorithm and the differential evolution algorithm. Mathematical optimization problems naturally arise in science or engineering problems. The optimization algorithms are categorized into two broad types, i.e. the local and the global ones. The local algorithms are mostly relying on the gradient of the cost function to be minimized. There are many theoretical results about the convergence of the local algorithms while the solutions depend highly on the initial guess. The global algorithms attempt to find the global solution(s). Details on optimization algorithms are found in Spall (2005) about stochastic global optimization algorithms, Boyd et al. (2004) about convex optimization and Fletcher (2013) about local optimizations.

Program 4.9 uses the genetic algorithm in MATLAB, which is a global optimization algorithm. The genetic algorithm mimics the natural evolution that improves survivability by adapting to the environment. The genetic algorithm uses a group of sampling points called the population in the search space. The cost function value for each sample is evaluated. Based on the cost function values, each sample is kept, removed, or moved into a new location in the search space for the next step or called the next generation. The genetic algorithm minimizes the cost function through the evolutionary process, namely, selection, crossover, and mutation of the population (Menon et al., 2006). To use the genetic algorithm optimization function, *ga()*, in MATLAB, the cost function is defined by the following:

```
FitnessFunction = @(delta)Santillan_Model_Fit_Cost(delta,time_span,
    time_exp, Enzy_exp);
```

delta in front of the cost function, i.e. *@(delta)*, is the optimization variable to be identified in the optimizer, *ga()*. The search space is constrained between ±0.99

instead of ±1 to avoid some of the parameters becoming zero, which does not make any biological sense, as follows:

```
lb = -0.99*ones(1,delta_dim);
ub = 0.99*ones(1,delta_dim);
opt_opts = optimoptions('ga','Display','iter');
```

where the optimization option indicates that the algorithm to be used is the genetic algorithm and the intermediate result for each iteration is to be displayed. In addition, the following call executes the optimization:

```
[delta_best,fval] = ga(FitnessFunction,delta_dim,[],[],[],[],lb,ub
    ,[],opt_opts);
```

where the default population size, 200, is used.

```
1  clear
2
3  time_A = [0 20 38 59 89 119 149];
4  Enzy_A = [25 657 617 618 577 577 567];
5
6  time_B = [0 29 60 89 179];
7  Enzy_B = [0 1370 1362 1291 913];
8
9  time_C = [0 29 58 88 118 178];
10 Enzy_C = [0 754 888 763 704 683];
11
12 %% Main Part: Parameter Identification
13 experiment_num = 2; % 1(A), 2(B), 3(C)
14
15 % choose experiment
16 switch experiment_num
17     case 1
18         time_exp = time_A;
19         Enzy_exp = Enzy_A;
20     case 2
21         time_exp = time_B;
22         Enzy_exp = Enzy_B;
23     case 3
24         time_exp = time_C;
25         Enzy_exp = Enzy_C;
26 end
27
28 delta_dim = 23;
29
30 % time span for obtaining the steady state
31 time_span = [0 1200]; % [minutes]
32
```

```
33 % model fitting optimization
34 FitnessFunction = @(delta)Santillan_Model_Fit_Cost(delta ,time_span ,
        time_exp , Enzy_exp);
35 lb = -0.99*ones(1,delta_dim);
36 ub = 0.99*ones(1,delta_dim);
37 opt_opts = optimoptions('ga','Display','iter');
38
39 [delta_best ,fval] = ga(FitnessFunction ,delta_dim ,[] ,[] ,[] ,[] ,lb ,ub
       ,[] ,opt_opts);
```

Program 4.9 (MATLAB) The model-fitting optimization using genetic algorithm

Program 4.10 uses the differential evolution algorithm in scipy Python, which is another global optimization algorithm (Storn and Price, 1997). The differential evolution is an evolutionary global optimization algorithm the same as the genetic algorithm. The main idea is to use the difference between two points in the search space. The update part of each point in the population points uses the difference vectors to direct the search direction.

```
1  experiment_num = 2 # 1(A), 2(B), 3(C)
2
3  time_A = np.array([0, 20, 38, 59, 89, 119, 149])
4  Enzy_A = np.array([25, 657, 617, 618, 577, 577, 567])
5
6  time_B = np.array([0, 29, 60, 89, 179])
7  Enzy_B = np.array([0, 1370, 1362, 1291, 913])
8
9  time_C = np.array([0, 29, 58, 88, 118, 178])
10 Enzy_C = np.array([0, 754, 888, 763, 704, 683])
11
12 # choose experiment
13 if experiment_num==1:
14         time_exp = time_A
15         Enzy_exp = Enzy_A
16 elif experiment_num==2:
17         time_exp = time_B
18         Enzy_exp = Enzy_B
19 elif experiment_num==3:
20         time_exp = time_C
21         Enzy_exp = Enzy_C
22
23 #-----------------------------------------------------------
24 # Main Model Fitting Optimization
25 #-----------------------------------------------------------
26 delta_dim = 23;
27
28 # time span for obtaining the steady state
29 time_span = np.array([0, 1200]) # [minutes]
30
31 state_all = np.random.randn(12)
32 delta = 0.99*(2*np.random.rand(23)-1)
33
```

```
34 Act_Enzy_exp = Enzy_exp
35 plot_sw = False
36 bounds = [(-0.99,0.99)]*delta_dim
37
38 from scipy import optimize
39 result = optimize.differential_evolution(Santillan_Model_Fit_Cost,
40                 bounds,
41                 args=(time_span, time_exp, Act_Enzy_exp, plot_sw),
42                 updating='deferred', disp=True, popsize=200,
                       maxiter=100, workers=4)
```

Program 4.10 (Python) The model-fitting optimization using genetic algorithm

4.2.2.7 Optimal Solution (MATLAB)

The genetic algorithm in MATLAB solves the model fitting problem and obtains the following optimal solution for each experiment:

$$\delta^* = \begin{bmatrix} \delta_1^* \\ \delta_2^* \\ \vdots \\ \delta_{22}^* \\ \delta_{23}^* \end{bmatrix} \rightarrow \begin{bmatrix} \delta_1^* & \delta_2^* & \delta_3^* & \delta_4^* & \delta_5^* \\ \delta_6^* & \delta_7^* & \delta_8^* & \delta_9^* & \delta_{10}^* \\ \delta_{11}^* & \delta_{12}^* & \delta_{13}^* & \delta_{14}^* & \delta_{15}^* \\ \delta_{16}^* & \delta_{17}^* & \delta_{18}^* & \delta_{19}^* & \delta_{20}^* \\ \delta_{21}^* & \delta_{22}^* & \delta_{23}^* & & \end{bmatrix} \tag{4.16}$$

$$\rightarrow \begin{bmatrix} -0.2810 & -0.3898 & 0.7068 & 0.0111 & 0.6425 \\ -0.8315 & -0.4280 & 0.7985 & 0.1770 & 0.1069 \\ -0.9174 & 0.6862 & 0.7437 & -0.6154 & 0.8466 \\ -0.5668 & -0.9273 & 0.2702 & -0.8495 & -0.7187 \\ 0.8848 & 0.3515 & 0.6146 & & \end{bmatrix}_A \tag{4.17}$$

$$\rightarrow \begin{bmatrix} 0.6415 & -0.8862 & 0.7426 & 0.2918 & 0.8398 \\ -0.9839 & -0.2078 & 0.0135 & 0.9869 & -0.7969 \\ -0.9889 & 0.9796 & -0.5617 & -0.9464 & 0.3877 \\ -0.9095 & -0.9177 & 0.8251 & -0.9529 & -0.7319 \\ -0.1656 & 0.5099 & 0.2271 & & \end{bmatrix}_B \tag{4.18}$$

$$\rightarrow \begin{bmatrix} -0.9895 & 0.9009 & 0.9616 & 0.9898 & 0.6706 \\ -0.9865 & -0.6868 & 0.1184 & 0.6514 & 0.7565 \\ -0.9780 & 0.9749 & 0.9785 & -0.4989 & 0.9878 \\ -0.6734 & -0.4203 & 0.9420 & -0.2586 & -0.7786 \\ -0.6590 & -0.5503 & 0.0357 & & \end{bmatrix}_C \tag{4.19}$$

where $[\cdot]_A$, $[\cdot]_B$, and $[\cdot]_C$ are the optimal solutions for Experiments A, B, and C, respectively. The trajectory of the optimal model for each experiment is compared with the measurements in Figure 4.5. All trajectories are reasonably close to the experimental measurements.

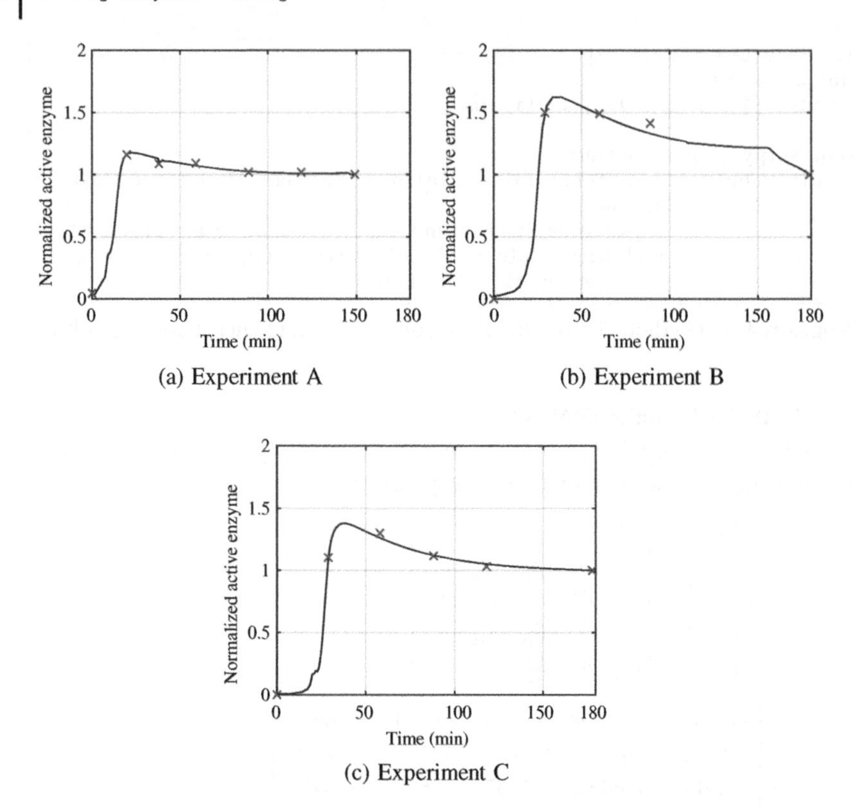

(a) Experiment A

(b) Experiment B

(c) Experiment C

Figure 4.5 (MATLAB) Model fitting results for Experiments A, B, and C, where the experimental data indicated by the crosses is normalized by the last data for each experiment and the normalized model output of the optimal fitted model is in the solid line.

The noise strength, (4.6b), for each parameter, δ_i for $i = 1, 2, \ldots, 23$, is calculated as follows:

$$\begin{bmatrix} \varphi_1 & \varphi_2 & \varphi_3 & \varphi_4 & \varphi_5 \\ \varphi_6 & \varphi_7 & \varphi_8 & \varphi_9 & \varphi_{10} \\ \varphi_{11} & \varphi_{12} & \varphi_{13} & \varphi_{14} & \varphi_{15} \\ \varphi_{16} & \varphi_{17} & \varphi_{18} & \varphi_{19} & \varphi_{20} \\ \varphi_{21} & \varphi_{22} & \varphi_{23} & & \end{bmatrix} = \begin{bmatrix} 3.1900 & \mathbf{6.8072} & 0.0237 & 0.5894 & 0.0159 \\ 0.0084 & 0.1304 & 0.5857 & 0.2737 & \mathbf{27.4299} \\ 0.0015 & 0.0321 & 1.7800 & 0.0785 & 0.1329 \\ 0.0429 & 0.1114 & 0.1896 & 0.2042 & 0.0013 \\ \mathbf{30.9861} & 3.1538 & 0.2974 & & \end{bmatrix}$$

where δ_2, δ_{10}, and δ_{21} have significantly larger values compared to the others. The changes of these three parameters for the experiments are summarized in Table 4.4. The optimization results show that these three parameters τ_m, O, and P belong to the adaptation parameter $\mathbf{p}_E$.

Table 4.4 Three optimal parameters changed the most by the genetic algorithm in MATLAB.

	Experiment A	Experiment B	Experiment C
τ_m	0.0610	0.0114	0.1901
O	0.0037	0.0007	0.0058
P	3.3822	2.6732	2.3402

4.2.2.8 Optimal Solution (Python)

The differential evolution in scipy Python solves the model fitting problem and obtains the following solution for each experiment:

$$\begin{bmatrix} \delta_1^* & \delta_2^* & \delta_3^* & \delta_4^* & \delta_5^* \\ \delta_6^* & \delta_7^* & \delta_8^* & \delta_9^* & \delta_{10}^* \\ \delta_{11}^* & \delta_{12}^* & \delta_{13}^* & \delta_{14}^* & \delta_{15}^* \\ \delta_{16}^* & \delta_{17}^* & \delta_{18}^* & \delta_{19}^* & \delta_{20}^* \\ \delta_{21}^* & \delta_{22}^* & \delta_{23}^* & & \end{bmatrix} \rightarrow \begin{bmatrix} 0.4139 & -0.650 & -0.5078 & 0.6556 & 0.8028 \\ -0.6978 & -0.014 & -0.7477 & -0.6088 & -0.4281 \\ -0.9512 & 0.984 & 0.4909 & -0.6512 & 0.1610 \\ -0.4746 & -0.064 & 0.4921 & -0.5255 & -0.4847 \\ -0.6853 & 0.266 & 0.1039 & & \end{bmatrix}_A \tag{4.20}$$

$$\begin{bmatrix} \delta_1^* & \delta_2^* \\ \delta_3^* & \delta_4^* \\ \delta_5^* & \delta_6^* \\ \delta_7^* & \delta_8^* \\ \delta_9^* & \delta_{10}^* \\ \delta_{11}^* & \delta_{12}^* \\ \delta_{13}^* & \delta_{14}^* \\ \delta_{15}^* & \delta_{16}^* \\ \delta_{17}^* & \delta_{18}^* \\ \delta_{19}^* & \delta_{20}^* \\ \delta_{21}^* & \delta_{22}^* \\ \delta_{23}^* & \end{bmatrix} \rightarrow \begin{bmatrix} -0.8164609821840342 & -0.9377737042517797 \\ -0.2790376682215831 & 0.9639037189678477 \\ 0.9899716285333048 & -0.9403394600209112 \\ 0.7598769991256485 & 0.3748592645823837 \\ 0.9659493912171094 & 0.005719318898349843 4 \\ -0.9890715372649793 & 0.5645345097583515 \\ 0.3565540724489650 3 & -0.6388522834084576 \\ -0.4876497370209529 & -0.4666376552759634 \\ -0.3606741179703174 & 0.8618508223375989 \\ -0.6752330337194928 & -0.0357508444106379 7 \\ 0.2271734961513452 4 & 0.4169772469116487 \\ 0.3175712292813526 & \end{bmatrix}_B \tag{4.21}$$

$$\begin{bmatrix} \delta_1^* & \delta_2^* & \delta_3^* & \delta_4^* \\ \delta_5^* & \delta_6^* & \delta_7^* & \delta_8^* \\ \delta_9^* & \delta_{10}^* & \delta_{11}^* & \delta_{12}^* \\ \delta_{13}^* & \delta_{14}^* & \delta_{15}^* & \delta_{16}^* \\ \delta_{17}^* & \delta_{18}^* & \delta_{19}^* & \delta_{20}^* \\ \delta_{21}^* & \delta_{22}^* & \delta_{23}^* & \end{bmatrix} \rightarrow \begin{bmatrix} -0.9529 & -0.1960 & 0.8548 & 0.7683 \\ -0.0008 & -0.9602 & -0.0880 & -0.7117 \\ 0.9453 & 0.0330 & -0.9786 & 0.0622 \\ 0.2542 & -0.7535 & 0.6106 & -0.8708 \\ 0.2591 & 0.8535 & -0.0101 & -0.8994 \\ 0.8292 & 0.2678 & 0.2170 & \end{bmatrix}_C \tag{4.22}$$

where the solution in Experiment B is presented with 16 decimal places.

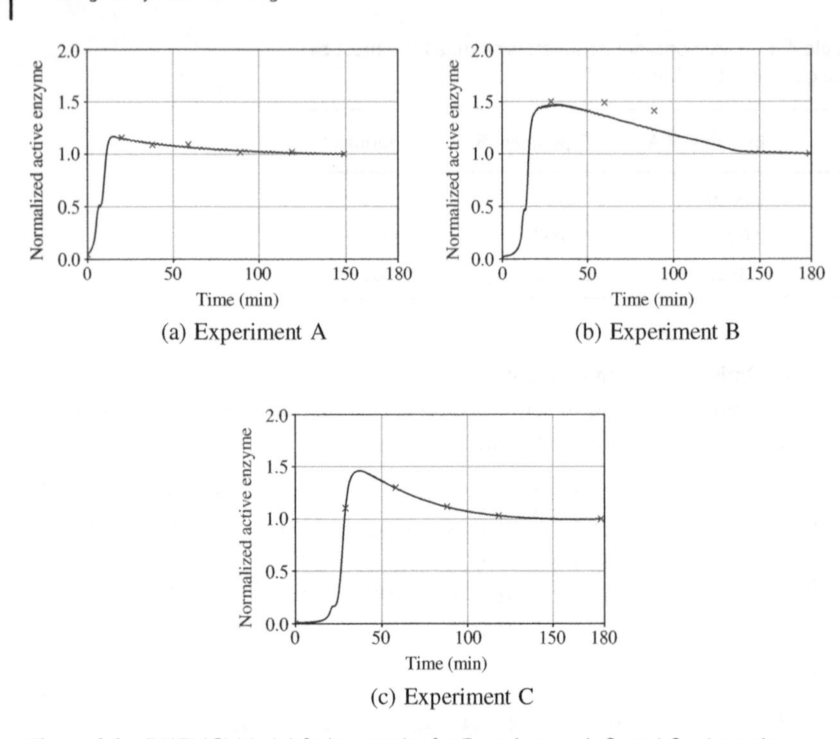

Figure 4.6 (MATLAB) Model fitting results for Experiments A, B, and C, where the experimental data indicated by the crosses is normalized by the last data for each experiment and the normalized model output of the optimal fitted model is in the solid line.

The trajectory of the optimal model for each experiment is compared with the measurements in Figure 4.6. All trajectories are reasonably close to the experimental measurements.

The cost function values of Experiment B with the optimal δ is about 0.05, while it is about 10 times larger, 0.4, with the approximated δ to 4 decimal places. The question is whether 0.4 is the acceptable accuracy of the model. Figure 4.7 shows the normalized active enzyme trajectories for Experiment B with two different delta precisions. The optimal δ with 16 or 4 decimal points produces significantly different histories from each other after around 20 minutes from the initial time. Although they have significant quantitative differences, it might be acceptable differences in terms of the unknown measurement errors. If the hypothetical error bars shown in the figure are true, then two trajectories are equally acceptable in terms of model fitting.

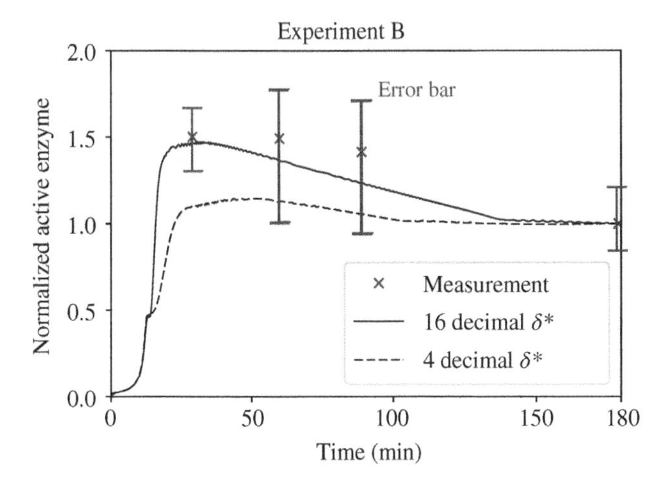

Figure 4.7 (Python) Sensitivity to the decimal points of the optimal δ.

The noise strength, (4.6b), for each parameter, δ_i for $i = 1, 2, \ldots, 23$, is calculated as follows:

$$\begin{bmatrix} \varphi_1 & \varphi_2 & \varphi_3 & \varphi_4 & \varphi_5 \\ \varphi_6 & \varphi_7 & \varphi_8 & \varphi_9 & \varphi_{10} \\ \varphi_{11} & \varphi_{12} & \varphi_{13} & \varphi_{14} & \varphi_{15} \\ \varphi_{16} & \varphi_{17} & \varphi_{18} & \varphi_{19} & \varphi_{20} \\ \varphi_{21} & \varphi_{22} & \varphi_{23} & & \end{bmatrix} = \begin{bmatrix} 0.836 & 0.157 & \mathbf{15.7} & 0.0204 & 0.309 \\ 0.0164 & 0.671 & 0.751 & 1.25 & 0.344 \\ 0.000261 & 0.265 & 0.0256 & 0.00388 & \mathbf{2.15} \\ 0.0589 & \mathbf{2.87} & 0.0404 & 0.201 & 0.263 \\ \mathbf{3.13} & 0.0158 & 0.0358 & & \end{bmatrix}$$

where δ_3, δ_{15}, δ_{17}, and δ_{21} have significantly larger values compared to the others. The changes of these three parameters for the experiments are summarized in Table 4.5. Based on the optimization results, these four parameters, τ_ρ, k_{-t}, c, and P would belong to the adaptive parameters, $\mathbf{p}_E$.

Table 4.5 Four optimal parameters changed the most by the differential evolution algorithm in Python.

	Experiment A	Experiment B	Experiment C
τ_ρ	0.0246	0.0360	0.0927
k_{-t}	24,382.5	10,759.4	33,823.2
c	0.0374	0.0256	0.0636
P	2.3224	2.9383	3.3448

4.2.2.9 Adaptive Parameters

Tables 4.4 and 4.5 provide two groups of the adaptive parameters, $\mathbf{p}_E$. All adaptive parameters in the groups directly affect the free operon, O_F, and the free mRNA productions, M_F. The only common element found in both sets is the mRNA polymerase, P. The mRNA polymerase binds to the free operon and produces the mRNA. Interestingly, the second messenger nucleotide ppGpp (Guanosine tetraphosphate) is known to directly bind to the mRNA polymerase and alter the transcription rate to adapt to environmental fluctuations (Sanchez-Vazquez et al., 2019, Zuo et al., 2013).

To check the robustness of the optimal parameters, perturb the parameter uncertainties as follows:

$$\delta_i = \delta_i^*(1 + 0.05 * \epsilon_i) \tag{4.23}$$

for $i = 1, 2, \ldots, 22, 23$, where δ_i^* is the optimal parameter perturbation (4.16) to fit the model to the measurements, ϵ_i is a random perturbation, and the parameter is perturbed by $\pm 5\%$. Take 10,000 random samples of ϵ, which is the same dimension as $\boldsymbol{\delta}^*$, the i-th element, ϵ_i, is a random number from the uniform distribution in $[-1, 1]$. The robustness of (4.16) is shown in Figure 4.8. The changes of the cost function, J, with respect to the norm of ϵ are presented.

4.2.2.10 Limitations

There are only three experiments, which are far from enough experiments in the statistical sense. There is no information about the errors in the measurements. The numerical integrator in MATLAB or scipy Python would have different ways to control numerical errors, which might affect the overall performance of the optimization. Note that the scipy integrator is not as fast as the MATLAB integrator, and the differential evolution in scipy requires a significant computation time. The principal purpose of model fitting in *systems biology* is to establish hypotheses

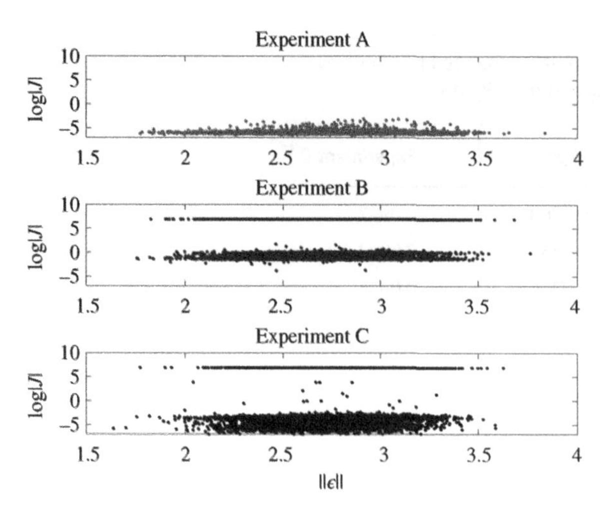

Figure 4.8 The cost function, J, variations with respect to the optimal parameter perturbation, $\|\epsilon\|$.

and design experiments, which would, in turn, save the time and resources wasted by unnecessary or poorly designed experiments.

4.3 Biological Oscillation

Periodic oscillations of biomolecular concentrations are crucial to keeping functionalities of the living cells. For example, circadian rhythm, the 24 hours periodic oscillation, exists in many life forms such as *Drosophila*, the fruit fly (Goldbeter, 1995), *Neurospora*, a fungus (Smolen et al., 2001), and the mammalian (Leloup and Goldbeter, 2003), which are evolving on the Earth provided the 24 hours day and night switching environment.

Dictyostelium discoideum is an amoeba, unicellular life form commonly found in forest soil. A group of amoeba aggregates to form a spore when there is no food available in the environment. The wave of the molecular concentration, 3′,5′-cyclic adenosine monophosphate (cyclic AMP or cAMP, in short), initiates the aggregation, where each cell moves towards the higher concentration, called *chemotaxis*. During the aggregation, an individual amoeba secretes cAMP from inside the cell and reacts to the external concentration changes of cAMP. The concentration changes periodically with a period of 5–10 minutes (Laub and Loomis, 1998).

cAMP is an essential intracellular messenger molecule triggering various responses inside the cell. To understand the meaning of the naming, 3′,5′-cAMP, see the structure of the cAMP molecule in Figure 4.9. The pentagon structure

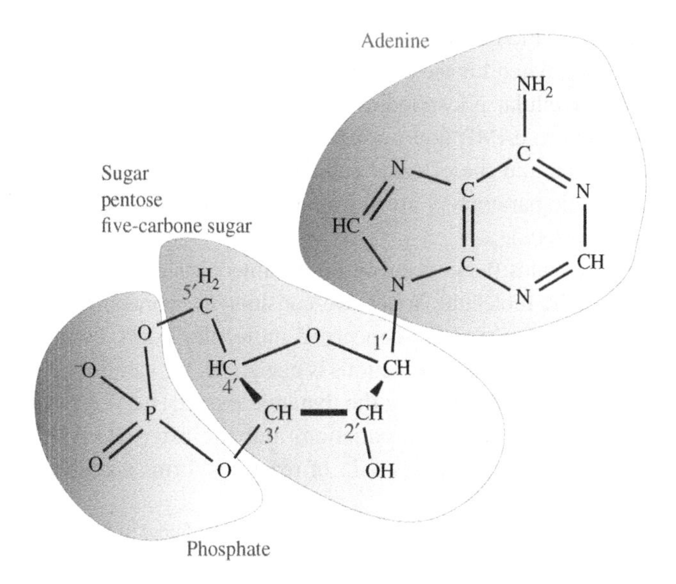

Figure 4.9 3′,5′-cyclic adenosine monophosphate (cAMP or cyclic AMP, in short).

in the centre is the sugar or ribose, where the five carbons number from 1′ (one-prime) to 5′ (five prime). In the upper-right part of the molecule, the nitrogen and the 1′ carbon link the adenine and the sugar molecule. The lower left part of the molecule is the phosphate. While adenosine triphosphate (ATP), i.e. the energy storage and transferring molecule, has a chain of three phosphates, the single (mono) phosphate is in the cyclic AMP holding two carbons, 3′ and 5′ carbons, in the sugar and forms the cyclic structure. Hence, the name of the molecule is 3′,5′-cAMP.

Laub and Loomis (1998) have proposed the oscillation network model for *Dictyostelium* cAMP concentration changes during the aggregation phase of the amoeba.[1] The mathematical model is given by

$$\frac{d[\text{ACA}]}{dt} = k_1[\text{CAR1}] - k_2[\text{ACA}][\text{PKA}] \tag{4.24a}$$

$$\frac{d[\text{PKA}]}{dt} = k_3[\text{i-cAMP}] - k_4[\text{PKA}] \tag{4.24b}$$

$$\frac{d[\text{ERK2}]}{dt} = k_5[\text{CAR1}] - k_6[\text{PKA}][\text{ERK2}] \tag{4.24c}$$

$$\frac{d[\text{REG A}]}{dt} = k_7 - k_8[\text{ERK2}][\text{REG A}] \tag{4.24d}$$

$$\frac{d[\text{i-cAMP}]}{dt} = k_9[\text{ACA}] - k_{10}[\text{REG A}][\text{i-cAMP}] \tag{4.24e}$$

$$\frac{d[\text{e-cAMP}]}{dt} = k_{11}[\text{ACA}] - k_{12}[\text{e-cAMP}] \tag{4.24f}$$

$$\frac{d[\text{CAR1}]}{dt} = k_{13}[\text{e-cAMP}] - k_{14}[\text{CAR1}] \tag{4.24g}$$

where ACA is adenylyl cyclase, PKA is the cAMP-dependent protein kinase, ERK2 is the extracellular signal-regulated kinase 2, a mitogen-activated protein kinase (MAPK), REG A is an intercellular phosphodiesterase, CAR1 is the cell surface receptor with a high affinity to cAMP, and i-cAMP and e-cAMP are the cAMP concentrations of the internal and the external cellular space, respectively. The nominal values of the kinetic parameters are summarized in Table 4.6 (Maeda et al., 2004, Ma and Iglesias, 2002).

Recall the bracket, [·], indicates the concentration of molecules, e.g. [ACA] is the concentration of the ACA molecule. When we consider the concentration of molecules, it assumes that plenty of the concerned molecules exists, i.e. the stochasticity from individual molecular interactions is negligible. We can infer all detailed logic behind the model by examining the dynamic model, (4.24). CAR1 activates ACA, ACA makes more cAMP production, and the external cAMP binding to CAR1 activates more ACA. This chain of reactions forms a positive

1 The model in Laub and Loomis (1998) includes typos. The correct model is found in Ma and Iglesias (2002) or Maeda et al. (2004).

Table 4.6 Laub–Loomis model kinetic parameters.

Parameter	Value	Unit	Parameter	Value	Unit
k_1	2.0	min^{-1}	k_8	1.3	$\mu M^{-1}\ min^{-1}$
k_2	0.9	$\mu M^{-1}\ min^{-1}$	k_9	0.3	min^{-1}
k_3	2.5	min^{-1}	k_{10}	0.8	$\mu M^{-1}\ min^{-1}$
k_4	1.5	min^{-1}	k_{11}	0.7	min^{-1}
k_5	0.6	min^{-1}	k_{12}	4.9	min^{-1}
k_6	0.8	$\mu M^{-1}\ min^{-1}$	k_{13}	23.0	min^{-1}
k_7	1.0	$\mu M/ min$	k_{14}	4.5	min^{-1}

feedback loop. Another chain between ACA, the internal cAMP, PKA, and ACA forms a negative feedback loop to reduce the production of cAMP via inhibiting ACA. These negative and positive feedback loops are known to have an important role in biological oscillations (Tsai et al., 2008). It is convenient to have a graphical visualization of these interactions. Figure 4.10 is the interaction network of the cAMP oscillation model, where → indicates activation and ⊣ indicates inhibition. For two molecules connected by the arrow, if the molecule at the tail of the arrow increases the change rate of the molecule at the arrowhead, it is *activation*. If the change rate decreases, it is *inhibition*.

For the parameter values given in Table 4.6, Figure 4.11 shows the time histories of the internal cAMP concentration and the CAR1 receptor concentration, i.e. the solution of the ODE, (4.24). The oscillation has a stable period of seven minutes.

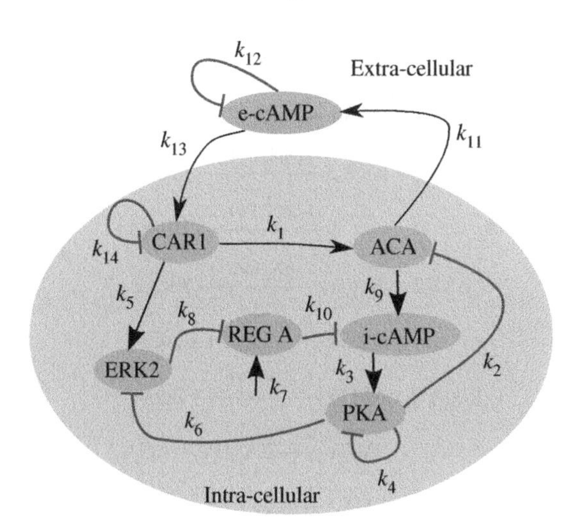

Figure 4.10 *Dictyostelium* cAMP oscillation network.

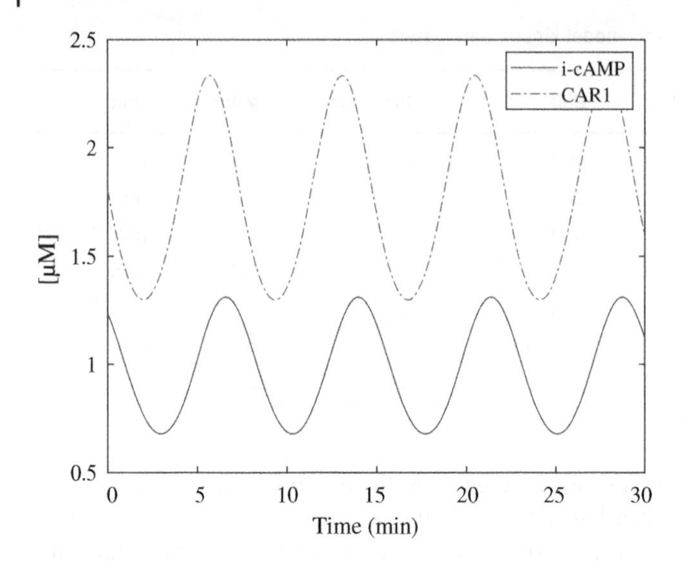

Figure 4.11 The concentration oscillations of the internal cAMP molecules and CAR1 receptors simulated using the ODE model.

The reverse procedure to the one shown in Section 1.2.2 obtains the 14 elementary biological interactions from the ODEs (4.24) as follows:

$$\text{CAR1} \xrightarrow{k_1} \text{ACA} + \text{CAR1} \tag{4.25a}$$

$$\text{ACA} + \text{PKA} \xrightarrow{k_2/(N_A V 10^{-6})} \text{PKA} \tag{4.25b}$$

$$\text{cAMPi} \xrightarrow{k_3} \text{PKA} + \text{cAMPi} \tag{4.25c}$$

$$\text{PKA} \xrightarrow{k_4} \varnothing \tag{4.25d}$$

$$\text{CAR1} \xrightarrow{k_5} \text{ERK2} + \text{CAR1} \tag{4.25e}$$

$$\text{PKA} + \text{ERK2} \xrightarrow{k_6/(N_A V 10^{-6})} \text{PKA} \tag{4.25f}$$

$$1 \xrightarrow{k_7 \times (N_A V 10^{-6})} [\text{RegA}] \tag{4.25g}$$

$$\text{ERK2} + \text{RegA} \xrightarrow{k_8/(N_A V 10^{-6})} \text{ERK2} \tag{4.25h}$$

$$\text{ACA} \xrightarrow{k_9} \text{cAMPi} + \text{ACA} \tag{4.25i}$$

$$\text{RegA} + \text{cAMPi} \xrightarrow{k_{10}/(N_A V 10^{-6})} \text{RegA} \tag{4.25j}$$

$$\text{ACA} \xrightarrow{k_{11}} \text{cAMPe} + \text{ACA} \tag{4.25k}$$

$$\text{cAMPe} \xrightarrow{k_{12}} \varnothing \tag{4.25l}$$

$$\text{cAMPe} \xrightarrow{k_{13}} \text{CAR1} + \text{cAMPe} \tag{4.25m}$$

$$\text{CAR1} \xrightarrow{k_{14}} \varnothing \tag{4.25n}$$

where V is the cell volume equal to $3.672 \times 10^{-14}\ell$ (Kim et al., 2007a) and 10^{-6} is multiplied because of changing the unit μM of k_2, k_6, k_7, k_8, and k_{10} to M. The Molar (M) is the unit of the number of molecules per volume divided by Avogadro's number as follows:

$$1[\text{M}] = 1\frac{[\#\text{ of molecules}]}{N_A V}$$

For k_2, k_6, k_8, and k_{10}, the unit change from 1 per minute-Molar to 1 per minute-the number of molecules is as follows:

$$\left(k_i\frac{1}{[\text{min}][\mu\text{M}]}\right) \times \frac{1}{N_A V} = \frac{k_i/(N_A V 10^{-6})}{[\text{min}]\,[\text{M}]}$$
$$= k_i/(N_A V 10^{-6})\ \frac{1}{[\text{min}]\,[\#\text{ of molecules}]}$$

for $i = 2, 6, 8$, and 10. Similarly, for k_7,

$$\left(k_7\frac{[\mu\text{M}]}{[\text{min}]}\right) \times (N_A V) = \frac{k_7(N_A V 10^{-6})[\text{M}]}{[\text{min}]}$$
$$= k_7 \times (N_A V 10^{-6})\ \frac{[\#\text{ of molecules}]}{[\text{min}]}$$

4.3.1 Gillespie's Direct Method

The method provides the exact simulation result for the molecular interactions, where the principal assumption is *the well-mixed condition*, i.e. all the molecules in the interactions distribute uniformly in the space (Gillespie, 1976).

Gillespie's method answers the following two questions to simulate stochastic molecular interactions:

- When does the next reaction occur?
- What reaction does occur?

When does the next reaction occur? The probability that the first reaction, (4.25a), occurs during the time interval, δt, is proportional to (the current number of CAR1 molecules)×(the reaction rate) as follows:

$$p_1(\delta t) = k_1\ \text{CAR1}\ \delta t = a_1 \delta_t$$

where a_1 is equal to ($k_1\times$ CAR1) and is called the *propensity function*. Similarly, the probability for the second reaction is

$$p_2(\delta t) = \frac{k_2}{N_A V 10^{-6}}\ \text{ACA PKA}\ \delta t = a_2 \delta t$$

where a_2 is equal to $(k_2 \times \text{ACA PKA}/(N_A V 10^{-6})$. Obtaining the probabilities for the rest of the reaction, from p_3 to p_{14}, is left as an exercise for the reader.

The probability that none of the 14 reactions occurs for δt is

$$p_{\text{no reaction}}(\delta t) = (1 - p_1)(1 - p_2) \ldots (1 - p_{14}) \approx 1 - \sum_{i=1}^{14} p_i = 1 - \sum_{i=1}^{14} a_i \delta t$$

where $p_1, p_2, \ldots, p_{14}$ are significantly smaller than 1 as δt is short, and the higher order terms are neglected in the approximation. For $\tau > 0$ and $d\tau > 0$, if none of the reactions occurs from the current time, t, to $t + \tau$ and one of the reaction occurs between $t + \tau$ and $t + \tau + d\tau$, what is the probability distribution of τ? Note that $d\tau$ is significantly smaller than τ, i.e. $d\tau \ll \tau$. By the definition, τ is the time interval between any two reactions occurring. Let τ be equal to N number of δt, i.e. $\tau = N\delta t$. The probability that no reaction occurs for τ is equal to

$$p_{\text{no reaction}}(\tau) = p_{\text{no reaction}}(N\delta t) = \left(1 - \sum_{i=1}^{14} \frac{a_i \tau}{N}\right)^N$$

Take the limitation by $\delta t \to 0$ or equivalently by $N \to \infty$

$$p_{\text{no reaction}}(\tau) = \lim_{N \to \infty} \left[1 + \frac{-\sum_{i=1}^{14} (a_i \tau)}{N}\right]^N$$

The term on the right-hand side is the definition of the exponential, i.e.

$$p_{\text{no reaction}}(\tau) = e^{-\sum a_i \tau}$$

The probability that no reaction occurs between t and $t + \tau$ and the i-th reaction will occur between $t + \tau$ and $t + \tau + d\tau$ is

$$p(i, \tau)d\tau = p_i(d\tau) \times p_{\text{no reaction}}(\tau) = a_i e^{-\sum a_i \tau} d\tau$$

for $i = 1, 2, \ldots, 14$. As a result, the probability density function for τ is given by

$$p(\tau) = \sum_{i=1}^{14} p(i, \tau) = \left(\sum_{i=1}^{14} a_i\right) e^{-\sum a_i \tau} \tag{4.26}$$

for $\tau > 0$ and $p_\tau = 0$ otherwise. p_τ satisfies the condition for the function to be the probability density function as follows (Shanmugan and Breipohl, 1988):

$$p(\tau) \geq 0$$

$$\int_{-\infty}^{\infty} p(\tau)d\tau = \int_{0^+}^{\infty} \left(\sum_{i=1}^{14} a_i\right) e^{-\sum a_i \tau} d\tau = -e^{-\sum a_i \tau}\Big|_{\tau=0}^{\tau \to \infty} = 1$$

$$\int_b^a p(\tau)d\tau \geq 0 \text{ for any } b < a$$

The time length from the current time to the next reaction time, τ, follows the exponential distribution. Hence, the answer to the first question of Gillespie's direct method is *generating a random number, τ, whose distribution is given by the exponential distribution*, (4.26).

What reaction does occur? If one of the 14 reactions occurs, which one does occur? As we know that the probability of each reaction will happen for the given time interval, δt, normalize the propensity function by the sum of the propensity function as follows:

$$\bar{a}_i = \frac{a_i}{\sum_{i=1}^{14} a_i}$$

for $i = 1, 2, \ldots, 14$. Generate a random number, x, from the uniform distribution between 0 and 1 and determine the reaction to occur depending on the random number and update the number of molecules as follows:

$$\text{reaction \#1 occurs if } 0 \le x < \bar{a}_1 \tag{4.27a}$$

$$\text{reaction \#2 occurs if } \bar{a}_1 \le x < \bar{a}_1 + \bar{a}_2 \tag{4.27b}$$

$$\text{reaction \#3 occurs if } \bar{a}_1 + \bar{a}_2 \le x < \bar{a}_1 + \bar{a}_2 + \bar{a}_3 \tag{4.27c}$$

$$\vdots \tag{4.27d}$$

$$\text{reaction \#13 occurs if } \sum_{i=1}^{12} a_i \le x < \sum_{i=1}^{13} a_i \tag{4.27e}$$

$$\text{reaction \#14 occurs if } \sum_{i=1}^{13} a_i \le x \le 1 \tag{4.27f}$$

For each reaction, update the number of molecules as follows:

$$\text{reaction \#1 ACA} \leftarrow \text{ACA} + 1 \tag{4.28a}$$

$$\text{reaction \#2 ACA} \leftarrow \text{ACA} - 1 \tag{4.28b}$$

$$\text{reaction \#3 PKA} \leftarrow \text{PKA} + 1 \tag{4.28c}$$

$$\text{reaction \#4 PKA} \leftarrow \text{PKA} - 1 \tag{4.28d}$$

$$\text{reaction \#5 ERK2} \leftarrow \text{ERK2} + 1 \tag{4.28e}$$

$$\text{reaction \#6 ERK2} \leftarrow \text{ERK2} - 1 \tag{4.28f}$$

$$\text{reaction \#7 RegA} \leftarrow \text{RegA} + 1 \tag{4.28g}$$

$$\text{reaction \#8 RegA} \leftarrow \text{RegA} - 1 \tag{4.28h}$$

$$\text{reaction \#9 cAMPi} \leftarrow \text{cAMPi} + 1 \tag{4.28i}$$

$$\text{reaction \#10 cAMPi} \leftarrow \text{cAMPi} - 1 \tag{4.28j}$$

$$\text{reaction \#11 cAMPe} \leftarrow \text{cAMPe} + 1 \tag{4.28k}$$

$$\text{reaction } \#12 \; \text{cAMPe} \leftarrow \text{cAMPe} - 1 \tag{4.28l}$$

$$\text{reaction } \#13 \; \text{CAR1} \leftarrow \text{CAR1} + 1 \tag{4.28m}$$

$$\text{reaction } \#14 \; \text{CAR1} \leftarrow \text{CAR1} - 1 \tag{4.28n}$$

The pseudo-code of the direct method for (4.25) is summarized in Algorithm 4.2.

Algorithm 4.2 Gillespie's direct method

1: Set the initial number of molecules: ACA, PKA, ERK2, RegA, cAMPi, cAMPe, CAR1
2: Set the initial time, $t = 0$, and the final time, t_f
3: **while** $t < t_f$ **do**
4: Generate the random number τ from the pdf, (4.26)
5: $t \leftarrow t + \tau$
6: Generate x from the uniform distribution between 0 and 1
7: Determine the reaction to occur using x and (4.27)
8: Update the number of molecules for the chosen reaction using (4.28)
9: **end while**

4.3.2 Simulation Implementation

Let the initial number of molecules for the seven molecular species be equal to

$$\begin{aligned} &\text{ACA} = 35{,}403, \quad \text{PKA} = 32{,}888, \quad \text{ERK2} = 11{,}838, \quad \text{RegA} = 27{,}348, \\ &\text{cAMPi} = 15{,}489, \quad \text{cAMPe} = 4980, \quad \text{CAR1} = 25{,}423 \end{aligned} \tag{4.29}$$

This particular set of initial conditions is on the oscillation trajectory. Beginning the simulation at arbitrary positive integer value,[2] simulating the reactions until they converge to the oscillation, and extracting the number of molecules at the end of the simulation. This procedure provides the initial values given in (4.29).

Figure 4.12 shows the probability density function of τ, (4.26), at the initial condition. The histogram uses τ generated from the exponential distribution during one minute of the simulation time. Initially, more than 1 million molecules are evenly distributed in the cell volume, and many reactions occur in one minute as shown in the histogram. As the number of molecules increases, the chance that two molecules collide with each other increases, and the reaction time, τ, becomes shorter. These short τ will slow down the simulation progress as the time increment is small. Gillespie's direct method is mainly for the low number of molecules simulations. Some other methods, such as τ-leap or Langevin

2 Use small integer numbers for the initial condition. Why?

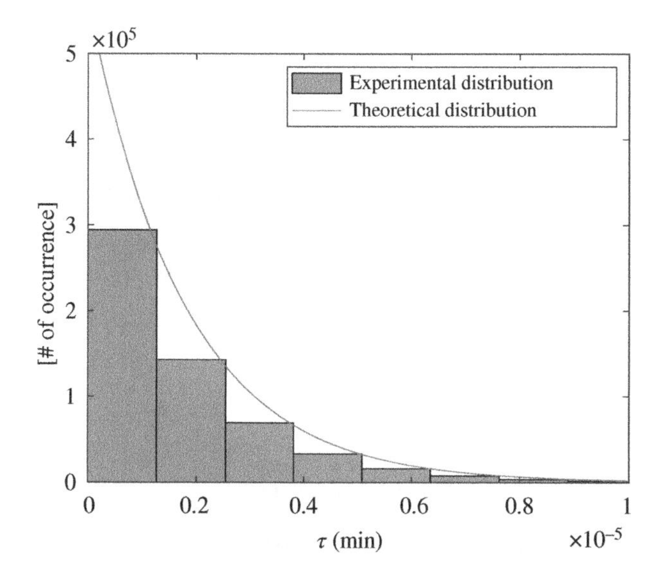

Figure 4.12 Experimental and theoretical distributions of τ in Gillespie's direct method, where the experimental distribution is from τ generated for one minute from the beginning of the simulation.

approximation, address slow simulation progress (Cao et al., 2006, Kim et al., 2018).

Figure 4.12 shows that the average reaction time interval is around 0.4 ms. That is, for every 0.4 ms, one of the reactions occurs. Storing all molecule number changes by the reactions requires lots of computer memory space. To reduce the memory size, we save the molecule numbers with a lot longer sampling time, e.g. 0.1 seconds.

Program 4.11 is the MATLAB implementation of the cAMP oscillation network using Gillespie's direct method. For the robustness analysis part from line 17, which we shall discuss later, set *p_delta* equal to zero and simulate it for the nominal kinetic parameter case. The initial molecular numbers are found by a priori simulation with a simulation time longer than 60 minutes so that the trajectory converges to the oscillation cycle close enough and sets to the molecular numbers at the end of the simulation. To generate τ, whose distribution is given by (4.26), the inverse of $\sum a_i$ is given to the MATLAB *exprnd()* function in line 55. *exprnd(a)* generates a random number, x, from the following distribution:

$$p(x) = \frac{1}{a}e^{-x/a}$$

In the last parts of the program, the simulation history saves the simulation time and the corresponding numbers of the molecules every 0.1 seconds.

```
1  clear
2
3  % simulation time values
4  time_current = 0;     % initial time
5  time_final   = 60.0; % final time [min]
6  time_record  = time_current; % data record time
7  dt_record    = 0.1;  % minimum time interval for data recording
8  max_num_data = floor((time_final-time_current)/dt_record+0.5);
9
10 % kinetic parameters for the Laub-Loomis Dicty cAMP oscillation
11 % network model from k1 to k14
12 ki_para_org = [2.0; 0.9; 2.5; 1.5; 0.6; 0.8; 1.0; 1.3; 0.3; 0.8;
       0.7; 4.9; 23.0; 4.5];
13 Cell_Vol = 3.672e-14; % [litre]
14 NA = 6.022e23;            % Avogadro's number
15 num_molecule_species = 7;
16
17 % robustness
18 delta_worst = [-1 -1 1 1 -1 1 1 -1 1 1 -1 1 -1 1]';
19 p_delta = 0;
20 ki_para=ki_para_org.*(1+(p_delta/100)*delta_worst);
21
22 % initial number of molecules
23 ACA   = 35403; % [# of molecules]
24 PKA   = 32888; % [# of molecules]
25 ERK2  = 11838; % [# of molecules]
26 REGA  = 27348; % [# of molecules]
27 icAMP = 15489; % [# of molecules]
28 ecAMP = 4980;  % [# of molecules]
29 CAR1  = 25423; % [# of molecules]
30
31 % storing data
32 species_all = zeros(max_num_data, num_molecule_species+1);
33 species_all(1,:) = [time_current ACA PKA ERK2 REGA icAMP ecAMP CAR1
       ];
34 data_idx = 1;
35
36 while data_idx < max_num_data
37
38     propensity_a(1) = ki_para(1)*CAR1;
39     propensity_a(2) = ki_para(2)*ACA*PKA/(NA*Cell_Vol*1e-6);
40     propensity_a(3) = ki_para(3)*icAMP;
41     propensity_a(4) = ki_para(4)*PKA;
42     propensity_a(5) = ki_para(5)*CAR1;
43     propensity_a(6) = ki_para(6)*PKA*ERK2/(NA*Cell_Vol*1e-6);
44     propensity_a(7) = ki_para(7)*(NA*Cell_Vol*1e-6);
45     propensity_a(8) = ki_para(8)*ERK2*REGA/(NA*Cell_Vol*1e-6);
46     propensity_a(9) = ki_para(9)*ACA;
47     propensity_a(10) = ki_para(10)*REGA*icAMP/(NA*Cell_Vol*1e-6);
48     propensity_a(11) = ki_para(11)*ACA;
49     propensity_a(12) = ki_para(12)*ecAMP;
50     propensity_a(13) = ki_para(13)*ecAMP;
51     propensity_a(14) = ki_para(14)*CAR1;
```

```
52
53      % determine the reaction time tau
54      sum_propensity_a = sum(propensity_a);
55      tau = exprnd(1/sum_propensity_a);
56
57      % determine the reaction
58      normalized_propensity_a = propensity_a/sum_propensity_a;
59      cumsum_propensity_a = cumsum(normalized_propensity_a);
60      which_reaction = rand(1);
61      reaction_idx = cumsum((cumsum_propensity_a-which_reaction)<0);
62      reaction = reaction_idx(end)+1;
63
64      % update number of molecules
65      switch reaction
66          case 1
67              ACA = ACA + 1;
68          case 2
69              ACA = ACA - 1;
70          case 3
71              PKA = PKA + 1;
72          case 4
73              PKA = PKA - 1;
74          case 5
75              ERK2 = ERK2 + 1;
76          case 6
77              ERK2 = ERK2 - 1;
78          case 7
79              REGA = REGA + 1;
80          case 8
81              REGA = REGA - 1;
82          case 9
83              icAMP = icAMP + 1;
84          case 10
85              icAMP = icAMP - 1;
86          case 11
87              ecAMP = ecAMP + 1;
88          case 12
89              ecAMP = ecAMP - 1;
90          case 13
91              CAR1 = CAR1 + 1;
92          case 14
93              CAR1 = CAR1 - 1;
94          otherwise
95              error('Wrong reaction number!');
96      end
97
98      time_current = time_current + tau;
99
100     if time_record < time_current
101        data_idx = data_idx + 1;
102        species_all(data_idx,:) = [time_current ACA PKA ERK2 REGA
               icAMP ecAMP CAR1];
103        time_record = time_record + dt_record;
```

```
104             disp(time_record);
105         end
106 
107 end
```

Program 4.11 (MATLAB) cAMP oscillation network simulation using Gillespie's direct method

Figure 4.13 shows the stochastic simulation results of the *Dictyostelium* cAMP oscillation network using Gillespie's direct method. Compare the results with the ones in Figure 4.11, which is from the deterministic simulation, their trajectories show several quantitative measures well matched to each other. The periods are about seven minutes, the phase difference between the internal cAMP and the CAR1 receptor is about one minute, and the average value of the internal cAMP is about twice bigger than the one of the CAR1 receptor. We conclude that both simulations in the deterministic and the stochastic settings provide the same results.

The Python implementation of Gillespie's direct method is given in Program 4.12 and the simulation results are shown in Figure 4.14. Unlike MATLAB, Python does not have a *switch-case* statement,[3] and the program implements it using *if-elif-else*.

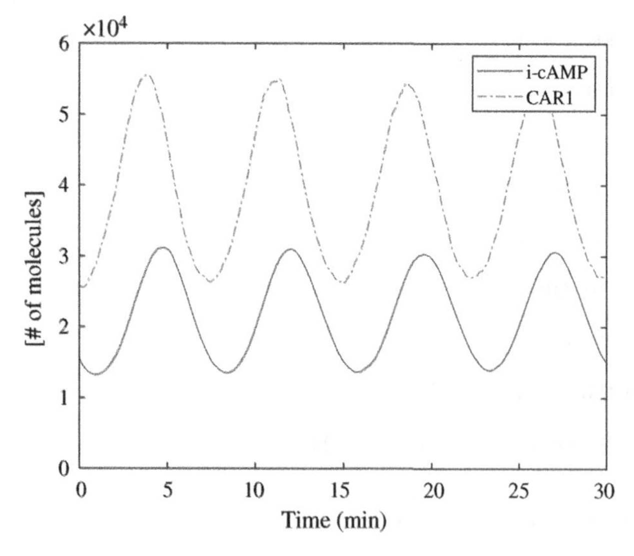

Figure 4.13 (MATLAB) The concentration oscillations of the internal cAMP molecules and CAR1 receptors simulated using Gillespie's direct method.

3 Python 3.10 introduces *match-case* to provide the same functionality of switch-case in other languages. https://www.python.org/dev/peps/pep-0622/.

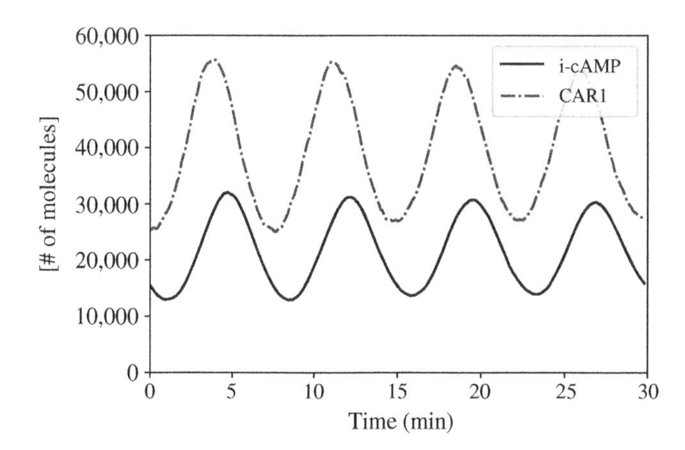

Figure 4.14 (Python) The concentration oscillations of the internal cAMP molecules and CAR1 receptors simulated using Gillespie's direct method.

```
1  import numpy as np
2
3  # simulation time values
4  time_current = 0     # initial time
5  time_final   = 60.0 # final time [min]
6  time_record  = time_current # data record time
7  dt_record    = 0.1  # minimum time interval for data recording
8  max_num_data = np.floor((time_final-time_current)/dt_record+0.5);
9
10 # kinetic parameters for the Laub-Loomis Dicty cAMP oscillation
11 # network model from k1 to k14
12 ki_para_org = np.array([2.0, 0.9, 2.5, 1.5, 0.6, 0.8, 1.0, 1.3,
       0.3, 0.8, 0.7, 4.9, 23.0, 4.5])
13 Cell_Vol = 3.672e-14; # [litre]
14 NA = 6.022e23;          # Avogadro's number
15 num_molecule_species = 7
16 num_reactions = 14
17
18 # robustness
19 delta_worst = np.array([-1, -1, 1, 1, -1, 1, 1, -1, 1, 1, -1, 1,
       -1, 1])
20 p_delta = 0;
21 ki_para=ki_para_org*(1+(p_delta/100)*delta_worst)
22
23 # initial number of molecules
24 ACA   = 35403  # [# of molecules]
25 PKA   = 32888  # [# of molecules]
26 ERK2  = 11838  # [# of molecules]
27 REGA  = 27348  # [# of molecules]
28 icAMP = 15489  # [# of molecules]
29 ecAMP = 4980   # [# of molecules]
30 CAR1  = 25423  # [# of molecules]
```

```
31
32 # storing data
33 species_all = np.zeros((int(max_num_data), num_molecule_species+1))
34 species_all[0,:] = np.array([time_current, ACA, PKA, ERK2, REGA,
       icAMP, ecAMP, CAR1])
35 data_idx = 0
36
37 propensity_a = np.zeros(num_reactions)
38
39 while data_idx < max_num_data-1:
40
41     propensity_a[0] = ki_para[0]*CAR1
42     propensity_a[1] = ki_para[1]*ACA*PKA/(NA*Cell_Vol*1e-6)
43     propensity_a[2] = ki_para[2]*icAMP
44     propensity_a[3] = ki_para[3]*PKA
45     propensity_a[4] = ki_para[4]*CAR1
46     propensity_a[5] = ki_para[5]*PKA*ERK2/(NA*Cell_Vol*1e-6)
47     propensity_a[6] = ki_para[6]*(NA*Cell_Vol*1e-6)
48     propensity_a[7] = ki_para[7]*ERK2*REGA/(NA*Cell_Vol*1e-6)
49     propensity_a[8] = ki_para[8]*ACA
50     propensity_a[9] = ki_para[9]*REGA*icAMP/(NA*Cell_Vol*1e-6)
51     propensity_a[10] = ki_para[10]*ACA
52     propensity_a[11] = ki_para[11]*ecAMP
53     propensity_a[12] = ki_para[12]*ecAMP
54     propensity_a[13] = ki_para[13]*CAR1
55
56     # determine the reaction time tau
57     sum_propensity_a = np.sum(propensity_a)
58     tau = np.random.exponential(1/sum_propensity_a)
59
60     # determine the reaction
61     normalized_propensity_a = propensity_a/sum_propensity_a
62     cumsum_propensity_a = np.cumsum(normalized_propensity_a)
63     which_reaction = np.random.rand(1)
64     reaction_idx = np.cumsum((cumsum_propensity_a-which_reaction)
           <0)
65     reaction = reaction_idx[-1]
66
67     # update number of molecules
68     if reaction==0:
69         ACA = ACA + 1
70     elif reaction==1:
71         ACA = ACA - 1
72     elif reaction==2:
73         PKA = PKA + 1
74     elif reaction==3:
75         PKA = PKA - 1
76     elif reaction==4:
77         ERK2 = ERK2 + 1
78     elif reaction==5:
79         ERK2 = ERK2 - 1
80     elif reaction==6:
81         REGA = REGA + 1
```

```
82        elif reaction==7:
83            REGA = REGA - 1
84        elif reaction==8:
85            icAMP = icAMP + 1
86        elif reaction==9:
87            icAMP = icAMP - 1
88        elif reaction==10:
89            ecAMP = ecAMP + 1
90        elif reaction==11:
91            ecAMP = ecAMP - 1
92        elif reaction==12:
93            CAR1 = CAR1 + 1
94        elif reaction==13:
95            CAR1 = CAR1 - 1
96        else:
97            print(reaction, 'Wrong reaction number!')
98
99        time_current = time_current + tau
100
101       if time_record < time_current:
102           data_idx = data_idx + 1
103           species_all[data_idx,:] = np.array([time_current, ACA, PKA,
                  ERK2, REGA, icAMP, ecAMP, CAR1])
104           time_record = time_record + dt_record
105           print(time_record)
```

Program 4.12 (Python) cAMP oscillation network simulation using Gillespie's direct method

4.3.3 Robustness Analysis

Even within the same species, each cell is different. The kinetic parameters for the cAMP network of *Dictyostelium* would vary from cell to cell. The robustness evaluation towards parametric perturbations in the network models is a way to provide the plausibility or verification tests of biomolecular networks.

Consider the parametric perturbations given by

$$k_i = \bar{k}_i \left(1 + \frac{p_\delta}{100}\delta_i\right)$$

for $i = 1, 2, \ldots, 14$, $\bar{k}_i$ is the nominal values of the kinetic parameters of the cAMP oscillation network given in Table 4.6, p_δ is the percentage perturbation greater than or equal to zero, and δ_i is the normalized perturbation between -1 and 1. The worst perturbation, $\boldsymbol{\delta}^*$, is the smallest magnitude perturbation *destroying the oscillation*, where

$$\boldsymbol{\delta} = \begin{bmatrix} \delta_1 & \delta_2 & \ldots & \delta_{14} \end{bmatrix}^T$$

and the magnitude would be a vector norm, typically the 2-norm or the ∞-norm.

Destroying the oscillation is not trivial to define as an optimization problem. If there is no oscillation, it is the case that the trajectories of the states are converged or diverged. As the number of molecules in biological networks is finite, we exclude diverging cases and only consider the converging ones. When the state converges to constants, the time derivative converges to zero. The integral of the time derivative can be a measure of how vivid the oscillations are. The cost function to be minimized is defined by

$$\begin{aligned} \underset{\|\delta\|\leq 1}{\text{Minimize}} J &= \frac{1}{2}\int_{t_0}^{t_f} \left(\frac{d[\text{ACA}]}{dt}\right)^2 dt \\ &= \frac{1}{2}\int_{t_0}^{t_f} (k_1\,[\text{CAR1}] - k_2\,[\text{ACA}]\,[\text{PKA}])^2\,dt \\ &\approx \frac{1}{2}\sum_{i=0}^{N} \{k_1\,[\text{CAR1}(t_i)] - k_2\,[\text{ACA}(t_i)]\,[\text{PKA}(t_i)]\}^2 \Delta t \end{aligned} \tag{4.30}$$

where t_0 must be large enough to reduce the effect of the initial conditions to the integration, t_f is long enough to include several oscillations if they exist, $\Delta t = t_i - t_{i-1}$, $t_f = t_N$, and N is the number of intervals. The last expression with the summation is an approximation of the integral with N-Δt intervals. For a fixed p_δ, solve the minimization problem and check if the worst perturbation destroys the oscillation manually. Repeat this procedure by reducing p_δ. Write pseudo-code for the robustness analysis of the oscillation is left as an exercise.

The MATLAB and the Python codes for the cost function are given in Programs 4.13 and 4.14.

```
1  function J_cost = Dicty_x1_square_integral(delta)
2
3      ki_para_org = [2.0; 0.9; 2.5; 1.5; 0.6; 0.8; 1.0; 1.3; 0.3;
           0.8; 0.7; 4.9; 23.0; 4.5];
4      p_delta = 2; % [percents]
5      ki_para=ki_para_org.*(1+(p_delta/100)*delta(:));
6
7      x0 = rand(7,1);
8      dt = 0.1;
9      time_interval = 0:dt:1200; % [min]
10
11     [~,xout] = ode45(@(time,state) Dicty_cAMP(time,state,ki_para),
           time_interval, x0);
12
13     ACA = xout(6000:end,1);
14     PKA = xout(6000:end,2);
15     CAR1 = xout(6000:end,7);
16     J_cost = sum((ki_para(1)*CAR1 - ki_para(2)*(ACA.*PKA)).^2)*dt
           *0.5;
17
18 end
```

Program 4.13 (MATLAB) The cost function for the oscillation robustness analysis

```
1  # Cost function to be minimized for robustness analysis
2  def Dicty_x1_square_integral(delta):
3
4      ki_para_org = np.array([2.0, 0.9, 2.5, 1.5, 0.6, 0.8, 1.0, 1.3,
           0.3, 0.8, 0.7, 4.9, 23.0, 4.5])
5      p_delta = 2 # [percents]
6      ki_para=ki_para_org*(1+(p_delta/100)*delta)
7
8      init_cond = np.random.rand(7)
9      dt = 0.1
10     tf = 1200
11     time_interval = np.linspace(0,tf,int(tf/dt)) # [min]
12
13     sol_out = solve_ivp(Dicty_cAMP, (0, tf), init_cond, t_eval=
          time_interval, args=(ki_para,))
14     xout = sol_out.y
15
16     ACA = xout[0,5999::]
17     PKA = xout[2,5999::]
18     CAR1 = xout[6,5999::]
19     J_cost = np.sum((ki_para[0]*CAR1 - ki_para[1]*(ACA*PKA))**2)*dt
          *0.5
20
21     return J_cost
```

Program 4.14 (Python) The cost function for the oscillation robustness analysis

The worst perturbation found in Kim et al. (2006) is

$$\delta^* = \begin{bmatrix} -1 & -1 & 1 & 1 & -1 & 1 & 1 & -1 & 1 & 1 & -1 & 1 & -1 & 1 \end{bmatrix}^T \tag{4.31}$$

and $p_\delta^* = 0.6$. As shown in the time history of two molecular species concentration changes in Figure 4.15, it needs only 0.6% perturbation to the nominal kinetic parameters to annihilate the oscillation. It is essential to consider the biological interpretation or meaning in biological network models. The cAMP oscillation takes place during the aggregation phase of *Dictyostelium* cells. The oscillation disappears after the slug, a tight mould of the cells, forms (Hashimura et al., 2019). The oscillation lasts less than 10 hours (600 minutes) before disappearing. Hence, the fragility of oscillation in 50 hours (3000 minutes) could be acceptable.

p_δ^* increases to 2% with the same δ^* to have a better biological interpretation. Figure 4.16 shows that the oscillations diminish significantly in 6 hours (300 minutes). If this happens to the cells during the aggregation phase, it has damaging effects. From the robustness analysis results showing the biologically acceptable fragility, we could conclude that the model might not be correct. Is the conclusion acceptable? We cannot reject the model immediately now. Multiple wet-lab experiments support most of the network connections shown in Figure 4.10 and the kinetic parameters in Table 4.6. Recall that the simulations shown in Figures 4.15 and 4.16 are based on the deterministic ODEs.

Figure 4.17 shows the stochastic simulation with the same 2% worst perturbation used for the deterministic case shown in Figure 4.16. The difference between

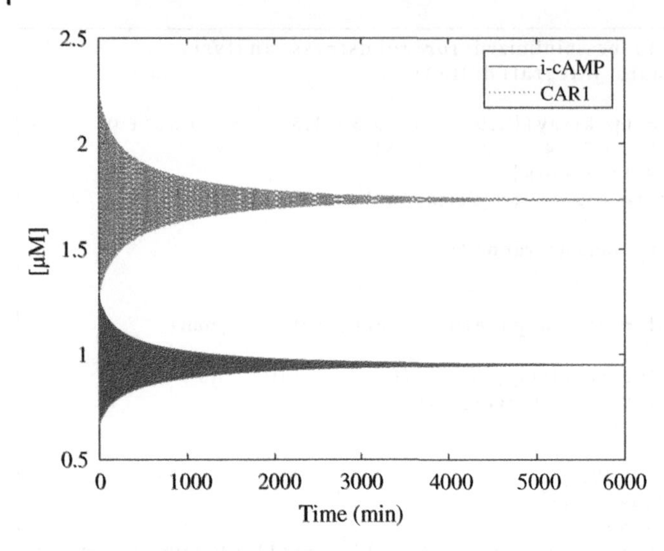

Figure 4.15 (Deterministic ODE model) The concentration change time history with the 0.6% worst perturbation (4.31).

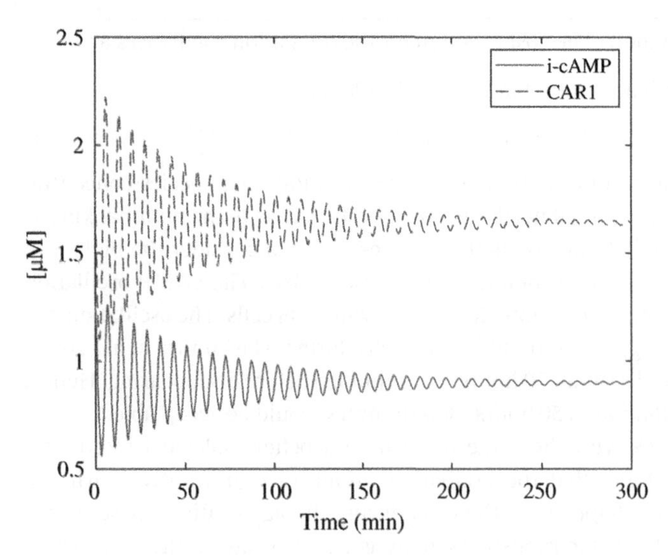

Figure 4.16 (Deterministic ODE model) The concentration change time history with the 2% worst perturbation (4.31).

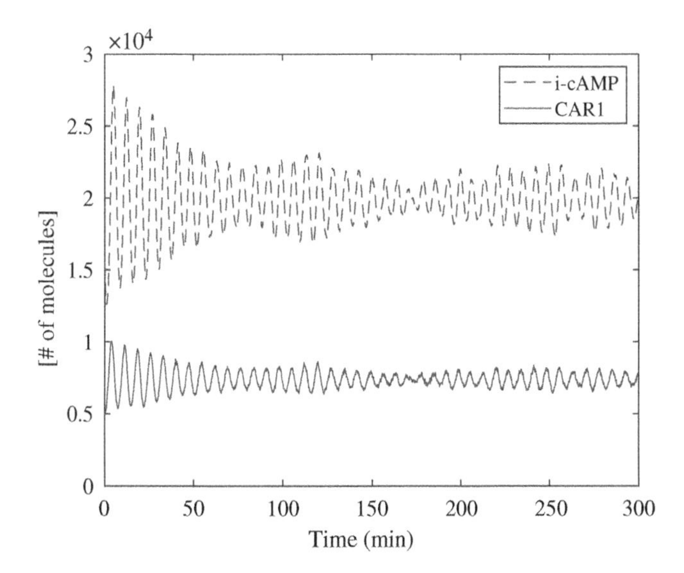

Figure 4.17 (Stochastic simulation) The concentration change time history with the 2% worst perturbation (4.31).

these two simulations is dramatic. While the oscillation disappears for the deterministic case, it sustains for the stochastic case, where the kinetic parameters are the same for both. It tells us that the importance of stochastic fluctuations in biological interactions as it would make qualitative differences. The deterministic model shows extreme sensitivity to the parametric perturbation. The stochastic model shows resilience to the same perturbation. Stochastic noise, which degrades the performance of engineering systems, in general, should be minimized, provides robustness to the oscillation.

Vilar et al. (2002) have presented the theoretical results of how the stochastic nature of molecular interactions becomes the source of genetic oscillation. Kim et al. (2008) have shown the necessary condition for the noise intensity to cause vibration. Kim et al. (2007b) show that synchronization between multiple *Dictyostelium* cells via waves of external cAMP plays an important role in improving the robustness of the oscillation.

Exercises

Exercise 4.1 Using the definition of Laplace transform,

$$Y(s) = \int_{t=0}^{t=\infty} y(t)e^{-st}\, dt$$

show that

$$y(t) = x(t - \tau) \Rightarrow Y(s) = e^{-\tau s}X(s)$$

where $x(t - \tau) = 0$ for $t \in [0, \tau)$.

Exercise 4.2 Using the definition of Laplace transform,

$$Y(s) = \int_{t=0}^{t=\infty} y(t)e^{-st}\, dt$$

show that

$$sY(s) - y(0) = \int_{t=0}^{t=\infty} \dot{y}(t)e^{-st}\, dt$$

Exercise 4.3 (MATLAB/Python) Implement Algorithm 4.1 and test for random parameter combinations to calculate the cost value, J, in (4.5). Find cases that the function fails and discuss the main cause of the fails.

Exercise 4.4 Obtain the probability for each reaction in (4.25) occurs.

Exercise 4.5 (Python) Perturb the optimal parameters in (4.20), (4.21), or (4.22) using (4.23). Plot the robustness of the parameters as shown in Figure 4.8 using 10,000 random samples of ϵ.

Exercise 4.6 (MATLAB/Python) Implement the ODE model for the cAMP oscillation in (4.24) and perform a simulation to produce Figure 4.11.

Exercise 4.7 Write pseudo-code for the robustness analysis algorithm of the *Dictyostelium* cAMP network using (4.30) and the descriptions below (4.30).

Exercise 4.8 (MATLAB/Python) Using the cost functions in Program 4.13 or Program 4.14, implement the pseudo-code in Exercise 4.7.

Exercise 4.9 (MATLAB/Python) Modify the 14 elementary reactions for a single cell *Dictyostelium* cAMP oscillation in (4.28) for 20 cells and perform the simulations for the worst perturbation given in (4.31) with $p_\delta = 2$.

Bibliography

B. Alberts, A. Johnson, J. Lewis, D. Morgan, M. Raff, K. Roberts, and P. Walter. *Molecular Biology of the Cell (J. Wilson, & T. Hunt, Eds.) (6th ed.)*. W. W. Norton & Company, 2015.

Stephen Boyd, Stephen P. Boyd, and Lieven Vandenberghe. *Convex Optimization*. Cambridge University Press, 2004.

Yang Cao, Daniel T. Gillespie, and Linda R. Petzold. Efficient step size selection for the tau-leaping simulation method. *The Journal of Chemical Physics*, 124(4):044109, 2006.

J. M. Carlson and John Doyle. Highly optimized tolerance: robustness and design in complex systems. *Physical Review Letters*, 84(11):2529–2532, 2000.

Alejandro Colman-Lerner, Andrew Gordon, Eduard Serra, Tina Chin, Orna Resnekov, Drew Endy, C. Gustavo Pesce, and Roger Brent. Regulated cell-to-cell variation in a cell-fate decision system. *Nature*, 437(29):699–706, 2005.

Michael B. Elowitz, Arnold J. Levine, Eric D. Siggia, and Peter S. Swain. Stochastic gene expression in a single cell. *Science*, 297(16):1183–1186, 2002.

Roger Fletcher. *Practical Methods of Optimization*. John Wiley & Sons, 2013.

Gene F. Franklin, J. David Powell, and Abbas Emami-Naeini. *Feedback Control of Dynamic Systems*. Pearson, London, 2015.

Daniel T. Gillespie. A general method for numerically simulating the stochastic time evolution of coupled chemical reactions. *Journal of Computational Physics*, 22(4):403–434, 1976.

Albert Goldbeter. A model for circadian oscillations in the *Drosophila* period protein (PER). *Proceedings of the Royal Society of London. Series B: Biological Sciences*, 261(1362):319–324, 1995. https://doi.org/10.1098/rspb.1995.0153. https://royalsocietypublishing.org/doi/abs/10.1098/rspb.1995.0153.

Hidenori Hashimura, Yusuke V. Morimoto, Masato Yasui, and Masahiro Ueda. Collective cell migration of *Dictyostelium* without camp oscillations at multicellular stages. *Communications Biology*, 2(1):1–15, 2019.

Mads Kærn, Timothy C. Elston, William J. Blake, and James J. Collins. Stochasticity in gene expression: from theories to phenotypes. *Nature Reviews Genetics*, 6:451–464, 2005.

Jongrae Kim, Declan G. Bates, Ian Postlethwaite, Lan Ma, and Pablo A. Iglesias. Robustness analysis of biochemical network models. *IEE Proceedings-Systems Biology*, 153(3):96–104, 2006.

Jongrae Kim, Pat Heslop-Harrison, Ian Postlethwaite, and Declan G. Bates. Stochastic noise and synchronisation during *Dictyostelium* aggregation make camp oscillations robust. *PLOS Computational Biology*, 3(11):1–9, 11 2007a. https://doi.org/10.1371/journal.pcbi.0030218.

Jongrae Kim, Pat Heslop-Harrison, Ian Postlethwaite, and Declan G. Bates. Stochastic noise and synchronisation during *Dictyostelium* aggregation make camp oscillations robust. *PLoS Computational Biology*, 3(11):e218, 2007b.

Jongrae Kim, Declan G. Bates, and Ian Postlethwaite. Evaluation of stochastic effects on biomolecular networks using the generalized nyquist stability criterion. *IEEE Transactions on Automatic Control*, 53(8):1937–1941, 2008.

Jongrae Kim, Mathias Foo, and Declan G. Bates. Computationally efficient modelling of stochastic spatio-temporal dynamics in biomolecular networks. *Scientific Reports*, 8(1):1–7, 2018.

Michael T. Laub and William F. Loomis. A molecular network that produces spontaneous oscillations in excitable cells of *D ictyostelium*. *Molecular Biology of the Cell*, 9(12):3521–3532, 1998.

Jean-Christophe Leloup and Albert Goldbeter. Toward a detailed computational model for the mammalian circadian clock. *Proceedings of the National Academy of Sciences of the United States of America*, 100(12):7051–7056, 2003. ISSN 0027-8424. https://doi.org/10.1073/pnas.1132112100. https://www.pnas.org/content/100/12/7051.

Lan Ma and Pablo A. Iglesias. Quantifying robustness of biochemical network models. *BMC Bioinformatics*, 3(1):1–13, 2002.

Mineko Maeda, Sijie Lu, Gad Shaulsky, Yuji Miyazaki, Hidekazu Kuwayama, Yoshimasa Tanaka, Adam Kuspa, and William F. Loomis. Periodic signaling controlled by an oscillatory circuit that includes protein kinases ERK2 and PKA. *Science*, 304(5672):875–878, 2004.

Prathyush P. Menon, Jongrae Kim, Declan G. Bates, and Ian Postlethwaite. Clearance of nonlinear flight control laws using hybrid evolutionary optimization. *IEEE Transactions on Evolutionary Computation*, 10(6):689–699, 2006. https://doi.org/10.1109/TEVC.2006.873220.

Mineo Morohashi, Amanda E. Winnz, Mark T. Borisuk, Hamid Bolouri, John Doyle, and Hiroaki Kitano. Robustness as a measure of plausibility in models of biochemical networks. *Journal of Theoretical Biology*, 216(1):19–30, 2002.

Matteo Osella, Eileen Nugent, and Marco Cosentino Lagomarsino. Concerted control of *Escherichia coli* cell division. *Proceedings of the National Academy of Sciences of the United States of America*, 111(9):3431–3435, 2014. ISSN 0027-8424. https://doi.org/10.1073/pnas.1313715111. https://www.pnas.org/content/111/9/3431.

Evangelia Papadimitriou and Peter I. Lelkes. Measurement of cell numbers in microtiter culture plates using the fluorescent dye Hoechst 33258. *Journal of Immunological Methods*, 162(1):41–45, 1993. ISSN 0022-1759. https://doi.org/10.1016/0022-1759(93)90405-V. https://www.sciencedirect.com/science/article/pii/002217599390405V.

Adithya Kumar Pediredla and Chandra Sekhar Seelamantula. Active-contour-based automated image quantitation techniques for western blot analysis. In *2011 7th International Symposium on Image and Signal Processing and Analysis (ISPA)*, pages 331–336, 2011.

Patricia Sanchez-Vazquez, Colin N. Dewey, Nicole Kitten, Wilma Ross, and Richard L. Gourse. Genome-wide effects on *Escherichia coli* transcription from ppGpp binding to its two sites on RNA polymerase. *Proceedings of the National Academy of Sciences of the United States of America*, 116(17):8310–8319, 2019. ISSN 0027-8424.

https://doi.org/10.1073/pnas.1819682116. https://www.pnas.org/content/116/17/8310.

Moisés Santillán and Michael C. Mackey. Dynamic reguiation of the tryptophan operon: a modeling study and comparison with experimental data. *Proceedings of the National Academy of Sciences of the United States of America*, 98(4):1364–1369, 2001.

Rajesh Babu Sekar and Ammasi Periasamy. Fluorescence resonance energy transfer (FRET) microscopy imaging of live cell protein localizations. *The Journal of Cell Biology*, 160(5):629–633, 2003.

K. S. Shanmugan and A. M. Breipohl. *Random Signals: Detection, Estimation and Data Analysis*. John Wiley & Sons, 1988. ISBN 978-0471815556.

Paul Smolen, Douglas A. Baxter, and John H. Byrne. Modeling circadian oscillations with interlocking positive and negative feedback loops. *Journal of Neuroscience*, 21(17):6644–6656, 2001.

James C. Spall. *Introduction to Stochastic Search and Optimization: Estimation, Simulation, and Control*, volume 65. John Wiley & Sons, 2005.

Rainer Storn and Kenneth Price. Differential evolution–a simple and efficient Heuristic for global optimization over continuous spaces. *Journal of Global Optimization*, 11(4):341–359, 1997.

The MathWorks. Canonical state-space realizations. https://uk.mathworks.com/help/control/ug/canonical-state-space-realizations.html, 2021. Accessed: 2021-07-28.

Tony Yu-Chen Tsai, Yoon Sup Choi, Wenzhe Ma, Joseph R. Pomerening, Chao Tang, and James E. Ferrell. Robust, tunable biological oscillations from interlinked positive and negative feedback loops. *Science*, 321(5885):126–129, 2008.

José M. G. Vilar, Hao Yuan Kueh, Naama Barkai, and Stanislas Leibler. Mechanisms of noise-resistance in genetic oscillators. *Proceedings of the National Academy of Sciences of the United States of America*, 99(9):5988–5992, 2002.

Charles Yanofsky and Virginia Horn. Role of regulatory features of the *trp* operon of *Escherichia coli* in mediating a response to a nutritional shift. *Journal of Bacteriology*, 176(20):6245–6254, 1994.

Yuhong Zuo, Yeming Wang, and Thomas A. Steitz. The mechanism of *E. coli* RNA polymerase regulation by ppGpp is suggested by the structure of their complex. *Molecular Cell*, 50(3):430–436, 2013.

5

Biological System Control

5.1 Control Algorithm Implementation

Recall the enzyme–substrate interactions, (4.8),

$$\mathrm{E} + \mathrm{S} \underset{k_{\mathrm{off}}}{\overset{k_{\mathrm{on}}}{\rightleftharpoons}} \mathrm{ES}$$

$$\mathrm{ES} \xrightarrow{k_{\mathrm{cat}}} \mathrm{P} + \mathrm{E}$$

$$\mathrm{P} \xrightarrow{k_{\mathrm{dg}}} \varnothing$$

where the product, P, degrades with the rate of k_{dg}. These four elementary reactions represent the system whose input and output are S and P, respectively. Figure 5.1 shows the block diagram for the reactions.

As we have the input/output system, we can set the standard control problem with the feedback loop shown in Figure 5.2. By the definition given in Springer Nature Limited (2021), 'Synthetic biology is the design and construction of new biological parts, devices, and systems, and the re-design of the existing, natural biological systems for useful purposes.' Elementary molecular interactions implement the desired P, the subtraction operator for calculating the error, and the controller for providing the input S in Figure 5.2.

5.1.1 PI Controller

Consider the control to be the proportional integral (PI) controller

$$[\mathrm{S}_{\mathrm{true}}] = k_P[\Delta\mathrm{P}] + k_I \int_0^t [\Delta\mathrm{P}(\tau)]d\tau$$

Dynamic System Modelling and Analysis with MATLAB and Python: For Control Engineers,
First Edition. Jongrae Kim.

Companion Website: www.wiley.com/go/kim/dynamicmodeling

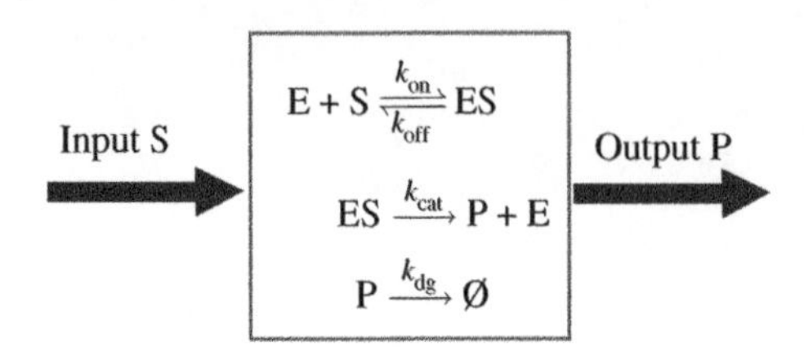

Figure 5.1 Enzyme–substrate reactions as an input (S) and output (P) system.

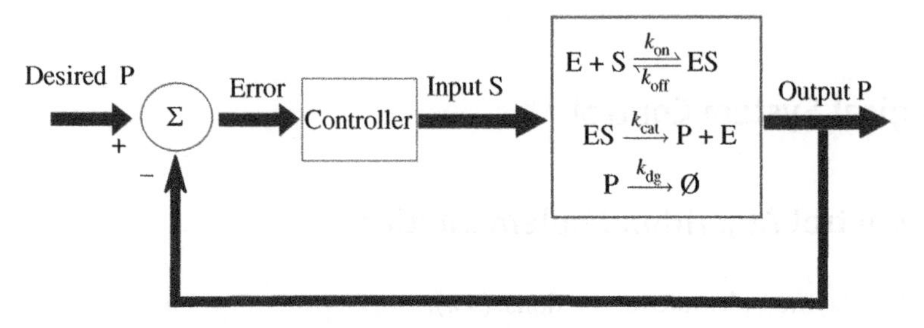

Figure 5.2 Feedback control structure for the enzyme–substrate system.

where $[\cdot]$ is the concentration of the molecules, k_P and k_I are the proportional gain and integral gain, respectively, and ΔP is the error calculated by the subtraction operator. The state-space form of the PI controller is given by

$$\frac{dz}{dt} = [\Delta P] = 0z + 1 \times [\Delta P] = A_{\text{true}}z + B_{\text{true}}[\Delta P]$$
$$[S_{\text{true}}] = k_I z + k_P[\Delta P] = C_{\text{true}}z + D_{\text{true}}[\Delta P]$$

and the corresponding transfer function is given by

$$K_{\text{true}}(s) = C_{\text{true}}\left(s - A_{\text{true}}\right)^{-1}B_{\text{true}} + D_{\text{true}} = \frac{k_I}{s} + k_P \tag{5.1}$$

Direct implementations of the PI controller require two multiplications, k_p and k_I, to the error and the integration of the error, respectively, one integration, and the summation. These operations are not available immediately in biological networks. Specifically designed multiple elementary molecular interactions implement the operations.

5.1.1.1 Integral Term

Consider the following elementary chemical reaction (Foo et al., 2016):

$$\Delta P \xrightarrow{k_I} \Delta P + X_1$$

where the ΔP molecule produces an intermediate molecular species, X_1, and the reaction rate is k_I. The corresponding differential equation for X_1 is

$$\frac{d[X_1]}{dt} = k_I[\Delta P]$$

X_1 becomes the integration term multiplied by the integration gain as follows:

$$[X_1(t)] = k_I \int_0^t [\Delta P(\tau)] d\tau$$

5.1.1.2 Proportional Term

Consider the following two elementary chemical reactions (Foo et al., 2016):

$$\Delta P \xrightarrow{\gamma_G k_P} \Delta P + X_2$$
$$X_2 \xrightarrow{\gamma_G} \varnothing$$

where the ΔP molecule produces another intermediate molecular species, X_2, with the rate equal to $\gamma_G k_P$, and X_2 degrades with the rate of γ_G. The corresponding differential equation for X_2 is

$$\frac{d[X_2]}{dt} = -\gamma_G[X_2] + \gamma_G k_P[\Delta P]$$

Once it reaches the steady-state, i.e. $[X_2(t)] = [X_2^{ss}] = \text{const}$, then

$$0 = -\gamma_G[X_2^{ss}] + \gamma_G k_P[\Delta P] \Rightarrow [X_2^{ss}] = k_P[\Delta P]$$

For a constant or slowly varying ΔP, the steady-state of $[X_2(t)]$ becomes the proportional term.

5.1.1.3 Summation of the Proportional and the Integral Terms

Consider the following three elementary chemical reactions (Foo et al., 2016):

$$X_1 \xrightarrow{k_{s2}} X_1 + S$$
$$X_2 \xrightarrow{k_{s2}} X_2 + S$$
$$S \xrightarrow{k_{s2}} \varnothing$$

where two molecular species X_1 and X_2 produce the substrate, S, with the rate of k_{s2}, and the substrate degrades at the same rate. The corresponding differential equation for S is

$$\frac{d[S]}{dt} = k_{s2}\left([X_1] + [X_2] - [S]\right)$$

Similar to before, consider the steady-state of S as follows:

$$0 = k_{s2}\left([X_1^{ss}] + [X_2^{ss}] - [S^{ss}]\right) \Rightarrow [S^{ss}] = [X_1^{ss}] + [X_2^{ss}]$$

5.1.1.4 Approximated PI Controller

Therefore, the steady-state of the substrate is given by

$$[S^{ss}] = [X_2^{ss}] + [X_3^{ss}] = k_P[\Delta P] + k_I \int_0^t [\Delta P](\tau) d\tau$$

and it provides the approximated PI control input. Note that X_1 reaches its steady-state only if the error, ΔP, converges to zero, and X_2 reaches its steady-state only if the error, ΔP, is constant. S cannot converge to a steady-state unless X_1 and X_2 converge, and the above analysis based on approximations cannot be exact. As long as all states change slowly, however, the approximation error would be in the acceptable ranges. The linear time-invariant (LTI) model for the PI controller is given by

$$\frac{d}{dt}\begin{bmatrix}[X_1]\\ [X_2]\\ [S]\end{bmatrix} = \begin{bmatrix}0 & 0 & 0\\ 0 & -\gamma_G & 0\\ k_{s2} & k_{s2} & -k_{s2}\end{bmatrix}\begin{bmatrix}[X_1]\\ [X_2]\\ [S]\end{bmatrix} + \begin{bmatrix}k_I\\ \gamma_G k_P\\ 0\end{bmatrix}[\Delta P] = A_c\mathbf{x}_c + B_c[\Delta P]$$

$$[S] = \begin{bmatrix}0 & 0 & 1\end{bmatrix}\begin{bmatrix}[X_1]\\ [X_2]\\ [S]\end{bmatrix} + 0[\Delta P] = C_c\mathbf{x}_c + D_c[\Delta P]$$

The transfer function of the approximated PI controller is given by

$$K_{\text{approx}}(s) = C_c(sI_4 - A_c)^{-1}B_c + D_c = \frac{(\gamma_G k_{s2}k_p + k_{s2}k_I)\,s + \gamma_G k_{s2}k_I}{s^3 + (k_{s2} + \gamma_G)\,s^2 + \gamma_G k_{s2}s} \tag{5.2}$$

For $|s| \ll 1$, the higher order terms are negligible and the transfer function is approximated by

$$\begin{aligned} K_{\text{approx}}(s) &\approx \frac{(\gamma_G k_{s2}k_p + k_{s2}k_I)\,s + \gamma_G k_{s2}k_I}{\gamma_G k_{s2}s} \\ &= \frac{\gamma_G k_{s2}k_p + k_{s2}k_I}{\gamma_G k_{s2}} + \frac{k_I}{s} = \left(k_p + \frac{k_I}{\gamma_G}\right) + \frac{k_I}{s} \end{aligned} \tag{5.3}$$

5.1.1.5 Comparison of PI Controller and the Approximation

Assume that the values of the reaction rates for the PI controller are given as follows: Foo et al. (2016)

$$k_P = 20, \quad k_I = 2.5\times10^{-4}, \quad \gamma_G = 8\times10^{-4}, \quad k_{s2} = 4\times10^{-4} \tag{5.4}$$

where the units are arbitrary. Substitute these into (5.1) and (5.3)

$$K_{\text{true}}(s) = \frac{2.5\times10^{-4}}{s} + 20$$

$$K_{\text{approx}}(s) = \left(20 + \frac{2.5\times10^{-4}}{8\times10^{-4}}\right) + \frac{2.5\times10^{-4}}{s} = \frac{2.5\times10^{-4}}{s} + 20.31$$

We confirm that K_{approx} is approximately close to K_{true} in low frequency, i.e. $|s| = |j\omega| = \omega \ll 1$. The bode plots of the true PI controller, (5.1), and the approximation, (5.2), are shown in Figure 5.3. As shown in the above derivations, the bode plots of the approximated and the true PI controller are well matched up to ω equal to 10^{-5}.

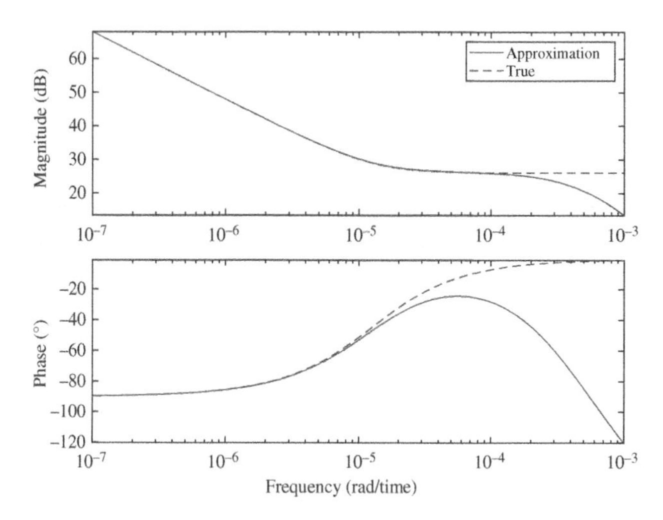

Figure 5.3 Comparison of the frequency responses between the true PI controller and the approximation.

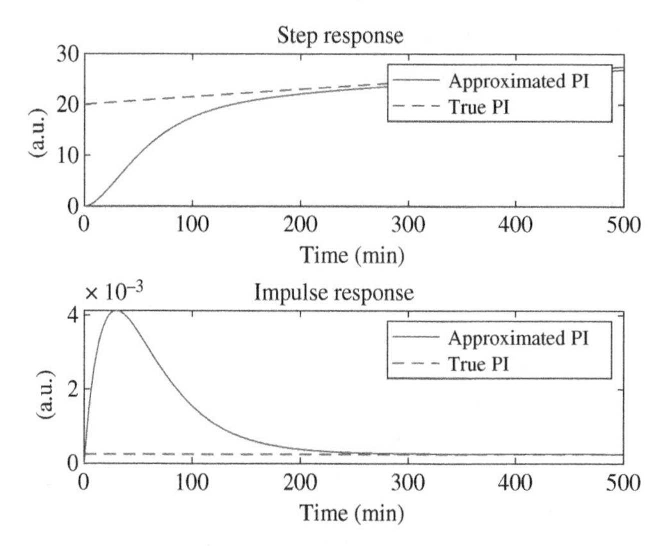

Figure 5.4 Comparison of the step and the impulse responses between the true PI controller and the approximation.

Programs 5.1 and 5.2 produce the bode plots given in Figure 5.3 and two time responses shown in Figure 5.4. In line 21, the LTI system given by

$$\dot{\mathbf{x}} = A\mathbf{x} + B\,[\Delta \mathrm{P}]$$
$$[\mathrm{S}] = C\mathbf{x} + D\,[\Delta \mathrm{P}]$$

is packed in MATLAB using *ss()*. In line 25, the magnitude and the phase values are returned by the function *bode()*. The unit for the magnitude is not in the dB unit, and they should be manually converted to the decibel unit using $20\log_{10}()$ as shown in line 30, where $\log_{10}()$ is the common logarithm function. The phase angle values returned by *bode()* is in degrees.

The time responses shown in Figure 5.4 are for the step and the impulse inputs produced by the MATLAB commands, *step()* and *impulse()*, respectively. For the following two cases of ΔP,

$$\text{Step input: } [\Delta\text{P}] = \begin{cases} 1 & \text{for } t \geq 0 \\ 0 & \text{for } t < 0 \end{cases}$$

$$\text{Impulse input: } [\Delta\text{P}] = \delta(t)$$

where $\delta(t)$ is the Dirac delta function given by Franklin et al. (2015)

$$\delta(t) = 0 \text{ for } t \neq 0, \quad \text{and} \quad \int_{-\infty}^{\infty} \delta(t)dt = 1$$

As shown in the middle plot of Figure 5.5, the Dirac delta function is the operator constructed by taking the limit of Δt approaching zero while the total area remains equal to 1. Consequently, the following integral

$$\int_{-\infty}^{\infty} f(t)\delta(t-a)dt = f(a)$$

provides the value of the function, $f(t)$, at t equal to the instance that the delta function is not zero, i.e. $t = a$.

While the step response measures how fast the system output reaches the desired response and how the transient of the output response behaves, the Dirac delta function expresses the impulse response, where the impulsive type input excites the system. Ideally, the impulse input excites all frequencies in the bode plot, and we observe the time response of the system when all frequency signals are injected into the system at the same time, whereas the step input will excite the system with the signal, which is mostly in the low-frequency range.

The step response in Figure 5.4 shows the approximated PI controller approaches to the true PI controller response at around 300 minutes, where there seems to have a constant bias error between them. The error could be reduced by adjusting the kinetic parameters. It is not always possible to realize the exact desired kinetic parameters, and the current setting producing the bias error for the step input is left to see whether the overall response could be acceptable with this deficiency.

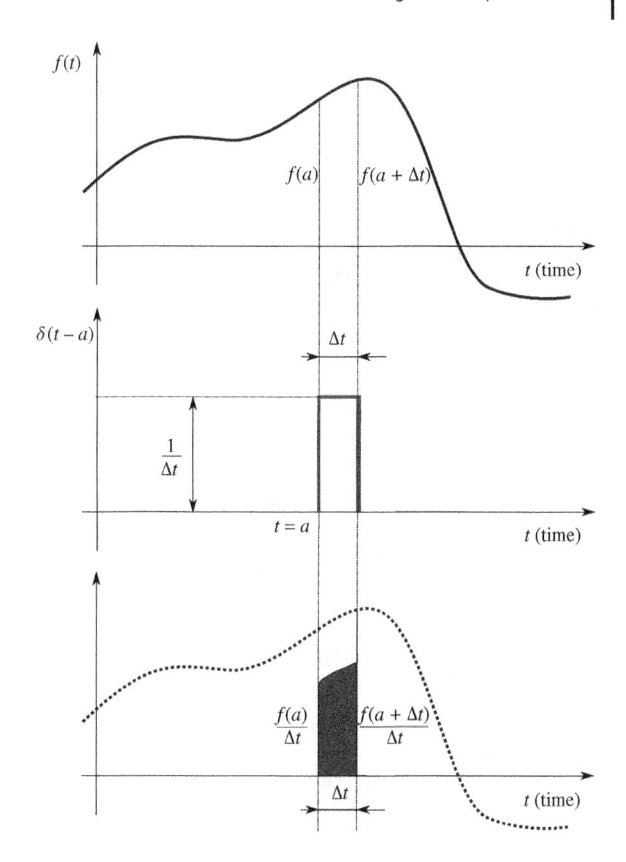

Figure 5.5 Dirac Delta function $\delta(t)$ and its property with integration.

```
1  clear
2
3  kP = 20;
4  kI = 2.5e-4;
5  gamma_G = 8e-4;
6  ks2 = 4e-4;
7
8
9  A_PI = [0 0 0; 0 -gamma_G 0; ks2 ks2 -ks2];
10 B_PI = [kI; gamma_G*kP; 0];
11 C_PI = [0 0 1];
12 D_PI = 0;
13
14 sys_PI = ss(A_PI,B_PI,C_PI,D_PI);
15
16 A_true = 0;
17 B_true = 1;
18 C_true = kI;
19 D_true = kP;
20
```

```
21 sys_true_PI = ss(A_true,B_true,C_true,D_true);
22
23 % bode plots
24 freq = logspace(-7,-3,1000); % [rad/time]
25 [mm1,pp1]=bode(sys_PI,freq);
26 [mm2,pp2]=bode(sys_true_PI,freq);
27
28 figure; clf;
29 subplot(211);
30 semilogx(freq,20*log10(squeeze(mm1)));
31 hold on;
32 semilogx(freq,20*log10(squeeze(mm2)),'r--');
33 set(gca,'FontSize',14);
34 ylabel('Magnitude [dB]');
35 legend('Approximation','True');
36 subplot(212);
37 semilogx(freq,squeeze(pp1));
38 hold on;
39 semilogx(freq,squeeze(pp2),'r--');
40 set(gca,'FontSize',14);
41 ylabel('Phase [\circ]');
42 xlabel('Frequency [rad/time]');
43
44
45 % step response and impulse response
46 time_sim = linspace(0,30000,300000);
47 [ys1,~,xs1]=step(sys_PI,time_sim);
48 [ys2,~]=step(sys_true_PI,time_sim);
49 [yp1,~,xp2]=impulse(sys_PI,time_sim);
50 [yp2,~]=impulse(sys_true_PI,time_sim);
51
52 figure; clf;
53 subplot(211);
54 plot(time_sim/60, ys1);
55 hold on;
56 plot(time_sim/60, ys2,'r--');
57 set(gca,'FontSize',14);
58 ylabel('[a.u.]');
59 xlabel('time [minutes]')
60 title('Step Response');
61 legend('approximated PI','true PI');
62 subplot(212);
63 plot(time_sim/60, yp1);
64 hold on;
65 plot(time_sim/60, yp2,'r--');
66 set(gca,'FontSize',14);
67 legend('approximated PI','true PI');
68 ylabel('[a.u.]');
69 xlabel('time [minutes]')
70 title('Impulse Response');
```

Program 5.1 (MATLAB) Compare bode plot, step response, and impulse response for the approximated PI controller and the true PI controller

The python code, Program 5.2, produces the bode plots and the time responses. The python commands for the bode plot and the time responses are under *scipy.signal*, which is imported as *spsg* at the beginning of the program. In python, the LTI system is constructed using the *lti()* command in *scipy.signal* as shown in line 22. The bode command in python returns the frequency, the magnitude in dB, and the phase angle in degrees. There are two commands for the step response in python, *step()* and *step2()*. Each uses the ordinary differential equation (ODE) solver *lsim()* or *lsim2()* in *scipy.signal*. It would be useful to check the responses using two methods if they coincide with each other.

```
import numpy as np
import matplotlib.pyplot as plt
import scipy.signal as spsg

kP = 20
kI = 2.5e-4
gamma_G = 8e-4
ks2 = 4e-4

A_PI = np.array([[0, 0, 0], [0, -gamma_G, 0], [ks2, ks2, -ks2]])
B_PI = np.array([[kI], [gamma_G*kP], [0]])
C_PI = np.array([[0, 0, 1]])
D_PI = np.array([[0]])

sys_PI = spsg.lti(A_PI,B_PI,C_PI,D_PI)

A_true = np.array([[0]])
B_true = np.array([[1]])
C_true = np.array([[kI]])
D_true = np.array([[kP]])

sys_true_PI = spsg.lti(A_true,B_true,C_true,D_true)


# bode plots
freq = np.logspace(-7,-3,1000) # [rad/time]
ww1, mm1, pp1 = spsg.bode(sys_PI,w=freq)
ww2, mm2, pp2 = spsg.bode(sys_true_PI,w=freq)

fig1, (ax1,ax2) = plt.subplots(nrows=2,ncols=1)
ax1.semilogx(ww1,mm1,'b-',ww2,mm2,'r--')
ax2.semilogx(ww1,pp1,'b-',ww2,pp2,'r--')

ax1.legend(('approximated PI','true PI'),fontsize=14)
ax2.legend(('approximated PI','true PI'),fontsize=14)

ax1.axis([1e-7,1e-3,0,65])
ax2.axis([1e-7,1e-3,-150,0.0])

ax1.set_ylabel('Magnitude [dB]',fontsize=14)
```

```
41 ax2.set_ylabel('Phase [$\circ$]',fontsize=14)
42
43 ax2.set_xlabel('Frequency [rad/time]',fontsize=14)
44
45 # step response and impulse response
46 time_sim = np.linspace(0,30000,300000)
47 ts1, ys1 = spsg.step2(sys_PI,T=time_sim)
48 ts2, ys2 = spsg.step2(sys_true_PI,T=time_sim)
49
50 tp1, yp1 = spsg.impulse(sys_PI,T=time_sim)
51 tp2, yp2 = spsg.impulse(sys_true_PI,T=time_sim)
```

Program 5.2 (Python) Compare bode plot, step response, and impulse response for the approximated PI controller and the true PI controller

5.1.2 Error Calculation: ΔP

The next component to implement the synthetic control circuit in the biological network in Figure 5.2 is the subtraction operation producing ΔP going into the PI controller. Consider the following four elementary chemical reactions (Foo et al., 2016):

$$\mathrm{P}_d \xrightarrow{k_{s1}} \mathrm{P}_d + \Delta\mathrm{P}$$

$$\Delta\mathrm{P} + \mathrm{X}_{\mathrm{sensor}} \xrightarrow{k_{s1}} \varnothing$$

$$\mathrm{P} \xrightarrow{k_{s1}} \mathrm{P} + \mathrm{X}_{\mathrm{sensor}}$$

$$\Delta\mathrm{P} \xrightarrow{k_{s1}} \varnothing$$

where the P_d concentration indicates the desired level of P concentration, which is assumed to be given. The role of X_{sensor} in the control system is the sensor, which measures P and feedbacks the measured information. Note that all the four reactions have the same reaction rate, k_{s1}.

Two ODE's for ΔP and X_{sensor} corresponding to the reactions are given by

$$\frac{d[\Delta\mathrm{P}]}{dt} = k_{s1}[\mathrm{P}_d] - k_{s1}[\Delta\mathrm{P}]\,[X_{\mathrm{sensor}}] - k_{s1}[\Delta\mathrm{P}]$$

$$\frac{d[X_{\mathrm{sensor}}]}{dt} = -k_{s1}[\Delta\mathrm{P}]\,[X_{\mathrm{sensor}}] + k_{s1}[\mathrm{P}]$$

Once they reach the steady-state

$$0 = k_{s1}[\mathrm{P}_d^{ss}] - k_{s1}[\Delta\mathrm{P}^{ss}]\,[X_{\mathrm{sensor}}^{ss}] - k_{s1}[\Delta\mathrm{P}^{ss}] \Longrightarrow [\Delta\mathrm{P}^{ss}] = [\mathrm{P}_d^{ss}] - [\Delta\mathrm{P}^{ss}]\,[X_{\mathrm{sensor}}^{ss}]$$

$$0 = -k_{s1}[\Delta\mathrm{P}^{ss}]\,[X_{\mathrm{sensor}}^{ss}] + k_{s1}[\mathrm{P}^{ss}] \Longrightarrow [\mathrm{P}^{ss}] = [\Delta\mathrm{P}^{ss}]\,[X_{\mathrm{sensor}}^{ss}]$$

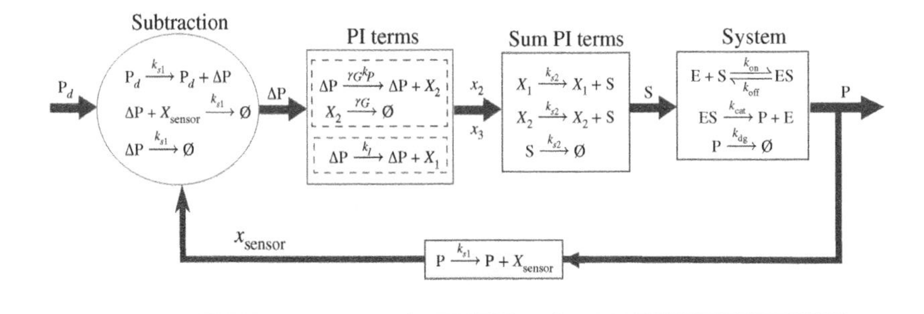

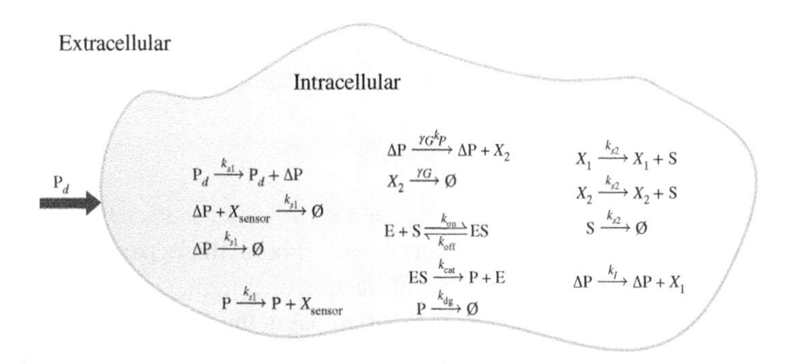

Figure 5.6 Elementary reactions for enzyme–substrate system with feedback control.

where the superscript, $(\cdot)^{ss}$, indicates the steady-state of each molecular concentration. Substituting the second equation to the first equation

$$[\Delta P^{ss}] = [P_d^{ss}] - [P^{ss}]$$

which is the error signal between the desired and the actual P.

A functional block diagram is shown in Figure 5.6. Unlike in engineering system block diagrams, each block in the diagram does not have a physical boundary to separate it from the other blocks. The block diagram shows functional separations. The 13 elementary reactions take place in the same space, i.e. the intercellular area, indicated in the lower plot of the figure.

> **Functional separation**: Elementary molecular interactions in synthetic biological circuits have functional roles distinguishable or separable from the rest. The reactions would occur in the same spatial domain.

Recall all the differential equations for the enzyme–substrate reaction and the PI controller as follows:

$$\frac{d[E]}{dt} = -k_{on}\,[E][S] + k_{cat}\,[ES]$$

$$\frac{d[\text{P}]}{dt} = k_{\text{cat}}\ [\text{ES}] - k_{\text{dg}}\ [\text{P}]$$

$$\frac{d[\text{ES}]}{dt} = k_{\text{on}}\ [\text{E}][\text{S}] - k_{\text{cat}}\ [\text{ES}]$$

$$\frac{d[X_1]}{dt} = k_I[\Delta\text{P}]$$

$$\frac{d[X_2]}{dt} = -\gamma_G[X_2] + \gamma_G k_P[\Delta\text{P}]$$

$$\frac{d[\text{S}]}{dt} = k_{s2}\left([X_1] + [X_2] - [\text{S}]\right)$$

$$\frac{d[\Delta\text{P}]}{dt} = k_{s1}[\text{P}_d] - k_{s1}[\Delta\text{P}]\ [X_{\text{sensor}}] - k_{s1}[\Delta\text{P}]$$

$$\frac{d[X_{\text{sensor}}]}{dt} = -k_{s1}[\Delta\text{P}]\ [X_{\text{sensor}}] + k_{s1}[\text{P}]$$

where $k_{\text{on}} = 5 \times 10^{-5}$, $k_{\text{cat}} = 1.6$, $k_{\text{dg}} = 8 \times 10^{-8}$, $k_P = 50$, $k_I = 5 \times 10^{-6}$, $\gamma_G = 8 \times 10^{-8}$, $k_{s1} = 3$, and $k_{s2} = 4 \times 10^{-4}$. We choose these particular values for demonstrating the limitation of the subtraction operation.

Let the desired P, P_d, be equal to 1. Program 5.3 is the MATLAB program for the simulation of the controller and the enzyme–substrate network in ODEs. The chosen set of parameters makes the ODE being stiff. The usual MATLAB function to solve ODE, *ode45()*, would struggle to solve the differential equation, has to reduce the integration step size small to satisfy the given numerical tolerance, and result in a very long computation time for solving the differential equations. The cause of the slow computation is the parameters over the very different scales from 10^{-8} to 10^1. The MATLAB has the ODE solver for stiff equations called *ode15s()*. In line 26, the function solves the enzyme–substrate equation with the PI controller and the subtractor and it solves the differential equations with a lot faster speed.

```
1  clear;
2
3  %% parameters
4  kP = 50;
5  kI = 5e-6;
6  gamma_G = 8e-4;
7  ks2 = 4e-4;
8
9  kon = 5e-5;
10 kcat = 1.6;
11 kdg = 8e-8;
12 ks1 = 3;
13
14 Pd = 1;
15
```

```
16  para = [kP kI gamma_G ks2 kon kcat kdg ks1 Pd];
17
18  %% simulation time values
19  time_current = 0;      % initial time
20  time_final   = 3600*16; % final time [min]
21  tspan = [time_current time_final];
22
23  %% simulation
24  ode_option = odeset('RelTol',1e-3,'AbsTol',1e-6);
25  state_t0 = 0.1*ones(1,8); state_t0(5)=1e-3;
26  sol = ode15s(@(time, state)ES_PI_Half_Subtraction(time, state, para
        ),tspan, state_t0, ode_option);
27
28  figure(1);
29  clf;
30  time_hr = sol.x/3600; % [hour]
31  P_history = sol.y(2,:);
32  plot(time_hr,Pd*ones(size(time_hr)),'r--');
33  hold on;
34  plot(time_hr, P_history, 'b-');
35  set(gca,'FontSize',14);
36  xlabel('time [hour]');
37  ylabel('P(t) [a.u.]');
38  legend('desired P', 'achieved P');
39
40  %% E-S PI Control Half Subtraction
41  function dxdt = ES_PI_Half_Subtraction(time,state,ki_para)
42      E   = state(1);
43      P   = state(2);
44      ES  = state(3);
45      X1  = state(4);
46      X2  = state(5);
47      S   = state(6);
48      DP  = state(7);
49      Xs  = state(8);
50
51      kP = ki_para(1);
52      kI = ki_para(2);
53      gamma_G = ki_para(3);
54      ks2 = ki_para(4);
55      kon = ki_para(5);
56      kcat = ki_para(6);
57      kdg = ki_para(7);
58      ks1 = ki_para(8);
59      Pd = ki_para(9);
60
61      dE_dt = -kon*E*S + kcat*ES;
62      dP_dt = kcat*ES - kdg*P;
63      dES_dt = kon*E*S - kcat*ES;
64      dX1_dt = kI*DP;
65      dX2_dt = -gamma_G*X2 + gamma_G*kP*DP;
66      dS_dt = ks2*X1 + ks2*X2 - ks2*S;
67      dDP_dt = ks1*Pd - ks1*DP*Xs - ks1*DP;
```

```
68        dXs_dt = -ks1*DP*Xs + ks1*P;
69
70        dxdt = [dE_dt; dP_dt; dES_dt; dX1_dt; dX2_dt; dS_dt; dDP_dt;
              dXs_dt];
71
72  end
```

Program 5.3 (MATLAB) PI controller for enzyme–substrate network

Figure 5.7 shows the trajectories of the desired P and the actual P during the first 16 hours. The P trajectory in the figure has a large offset from the desired value of 1 and appears to diverge slowly. We could reduce the offset by adjusting the control gains, k_P and k_I. In practice, it would be easier to simply tune the desired P, i.e. re-scaling P_d by multiplying K_F as follows:

$$P_d = P_{desired} \times K_F$$

where $P_{desired}$ is the desired level of P, and P_d is the injected amount to the synthetic circuit. With a few trials and errors, we find $K_F = 0.62$ adjusts the offset roughly zero as shown in Figure 5.8. The error, $P_d - P$, however, slowly diverges, and the instability is inherent in the designed closed-loop system. Scaling the desired P by K_F does not make the error zero. Why?

To answer the question, consider the right-hand side of the differential equation of $d(\Delta P)/dt$. The steady-state presumably produces the difference between P_d and P. Consider a case that the initial P_d is greater than P as the initial time shown in

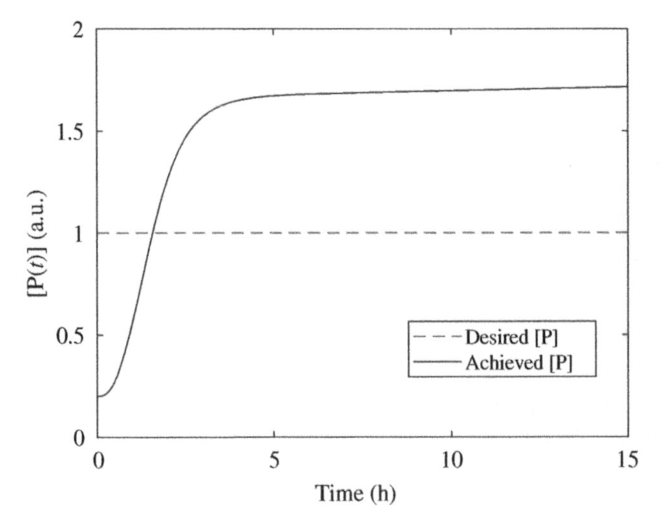

Figure 5.7 Concentration changes of the actual P with the PI controller for the desired P equal to 1.

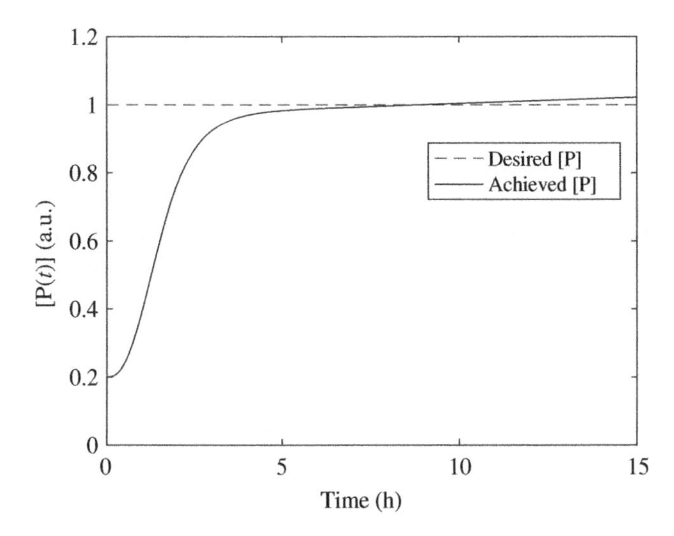

Figure 5.8 Concentration changes of the actual P with the PI controller for the desired P tuned.

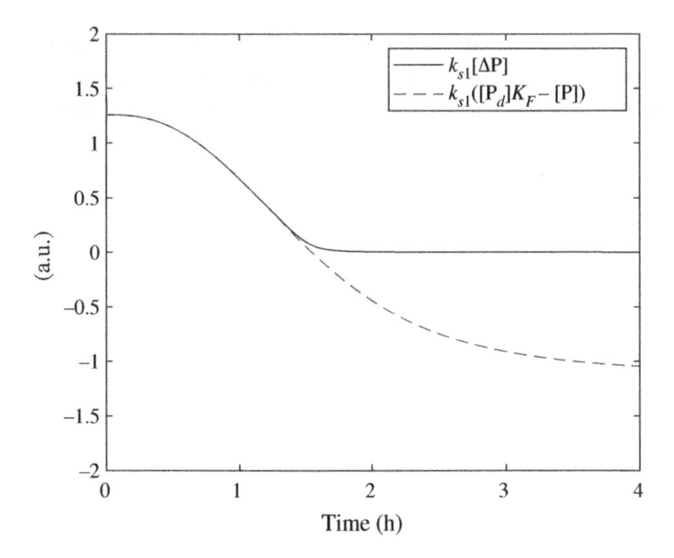

Figure 5.9 Comparison between ΔP and the true difference.

Figure 5.9, where the initial condition is given by

$$[E(0)] = 0.1, \quad [P(0)] = 0.1, \quad [ES(0)] = 0.1, \quad [X_1(0)] = 0.1,$$
$$[X_2(0)] = 0.001, \quad [S(0)] = 0.1, \quad [\Delta P(0)] = 0.1, \quad [X_{sensor}(0)] = 0.1$$

Figure 5.9 shows $k_{s1}[\Delta\mathrm{P}]$ and the true difference, $k_{s1}([\mathrm{P}_d]K_F - [\mathrm{P}])$. $k_{s1}[\Delta\mathrm{P}]$ follows the true difference well until it becomes negative, i.e. when $[\mathrm{P}_d]K_F < [\mathrm{P}]$. Initially, for $[\mathrm{P}_d]K_F > [\mathrm{P}]$, the steady-state matches the difference, $[\mathrm{P}_d]K_F - [\mathrm{P}]$, and the difference is positive. As $[\mathrm{P}_d]$ approaches [P] by the control actions, the positive difference converges to zero at around 1.4 hours. Once the difference, $[\Delta\mathrm{P}]$, becomes zero and [P] overshoots $[\mathrm{P}_d]K_F$, the difference becomes negative. As a result, the right-hand side of $d[\Delta\mathrm{P}]/dt$ becomes negative, and $[\Delta\mathrm{P}]$ remains at zero. *Recognizing that all quantities in the biomolecular network are positive is crucial. Molecular concentrations cannot express or convey negative quantities.*

After the true difference becomes negative, $[\Delta\mathrm{P}]$ stays at zero and $[X_{\mathrm{sensor}}$ cannot produce a correct measure of [P] as shown in Figure 5.10, which provides a comparison of the two terms on the right-hand side of $d[X_{\mathrm{sensor}}]/dt$. It is known to be the limitation of the one-sided subtraction operation in biological networks (Foo et al., 2016).

The limitation of the one-sided subtraction is more pronounced in the following scenario. Specify the desired P as

$$\mathrm{P}_{\mathrm{desired}} = \begin{cases} 1 & \text{for } t \leq 8\,[\text{hours}] \\ 1/2 & \text{for } t > 8\,[\text{hours}] \end{cases}$$

where the desired P is reduced to half from the initial desired P at eight hours. Figure 5.11 shows the simulation results. P seems to converge to the first step command at around four hours, but it cannot react to the lower desired P command after eight hours.

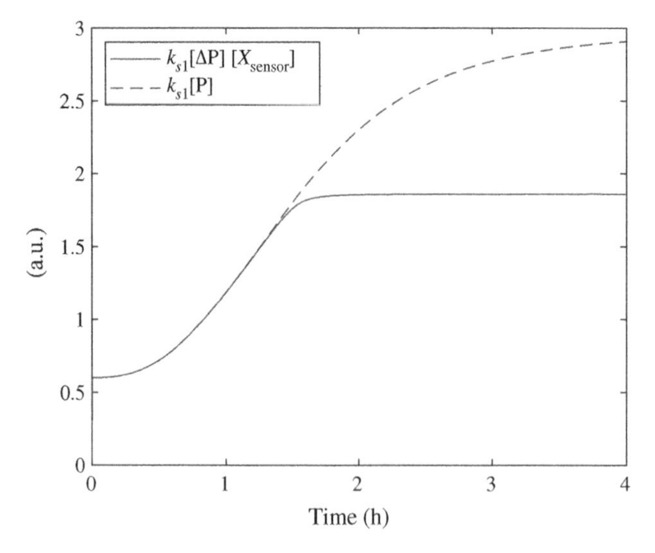

Figure 5.10 Comparison of the terms in the right-hand side of $d[X_{\mathrm{sensor}}]/dt$.

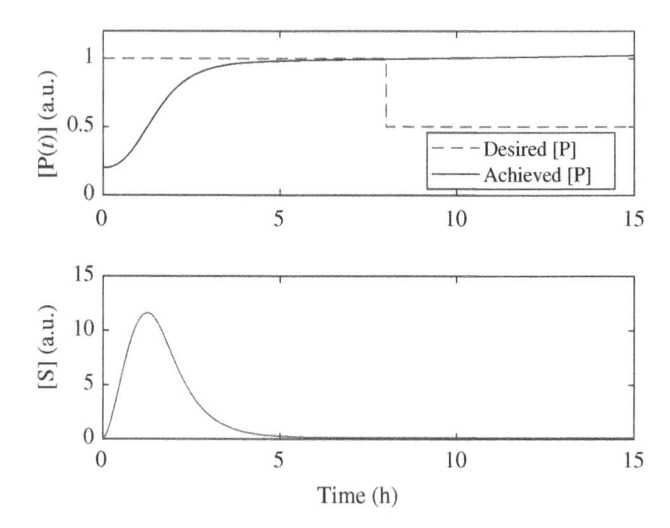

Figure 5.11 Responses of the closed-loop system with the one-sided subtraction for two-step commands.

To overcome the lack of ability to represent negative quantities using molecular concentrations, Oishi and Klavins (2011) present a method using two molecular species. In the proposed approach, each molecular species is *interpreted* as having a positive or negative value, respectively. It is not, however, implementation of negative values. Besides, as we use two molecular species to express quantities, we need at least twice as many molecular interactions.

Recall the differential equation for [P]

$$\frac{d[\mathrm{P}]}{dt} = k_{\mathrm{cat}}[\mathrm{ES}] - k_{\mathrm{dg}}[\mathrm{P}]$$

[P] is controlled by [ES], and [ES] is controlled by [S], which is the PI controller. [ES], however, has only the capability to increase [P] as $k_{\mathrm{cat}} > 0$. We need another mechanism to reduce [P]. Introduce the X_3 molecule having the capability to destroy P as follows:

$$\mathrm{X}_3 + \mathrm{P} \xrightarrow{k_{\mathrm{deg}}} \mathrm{X}_3$$

the corresponding differential equation for $X_3 = 0$ is given by

$$\frac{d[X_3]}{dt} = 0$$

and the differential equation for [P] is changed to

$$\frac{d[\mathrm{P}]}{dt} = k_{\mathrm{cat}}[\mathrm{ES}] - k_{\mathrm{dg}}[\mathrm{P}] - k_{\mathrm{deg}}[X_3][\mathrm{P}]$$

where $k_{\text{deg}}[X_3]$ is the additional degradation rate of [P]. The degradation effect must not be significant beyond the increasing speed of k_{cat}[ES] when the desired P is larger than [P], i.e. $[\Delta P] \gg 0$. For the simulation, K_{deg} is set to 0.001 and $[X_3]$ is equal to 1.

The last consideration is the integration part of the controller. $[X_1]$ integrates the error, [ΔP], through

$$\frac{d[X_1]}{dt} = k_I[\Delta P]$$

Even after the error converges to zero, the integration would drift and diverge unless the error remains exactly zero. To reject the drift, the following self-annihilation is introduced:

$$X_1 \xrightarrow{\eta} \varnothing$$

and the differential equation becomes

$$\frac{d[X_1]}{dt} = k_I[\Delta P] - \eta[X_1]$$

where η is equal to 0.0001. The comparison of the frequency response of the pure integrator, k_I/s, and the integrator with the annihilation, $k_I/(s+\eta)$, is shown in Figure 5.12. In frequency higher than 0.0001 rad/time, the integrator with the annihilation is close to the pure integrator. In the lower frequency area, it becomes a pure static gain.

Finally, the closed-loop response and the control input are shown in Figure 5.13. [P] follows the desired P within a few hours after each step command initiation.

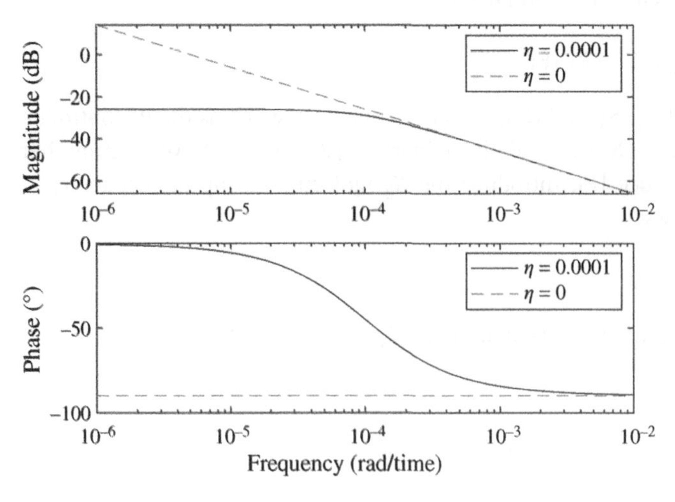

Figure 5.12 Comparison of frequency response of k_I/s and $k_I/(s+\eta)$.

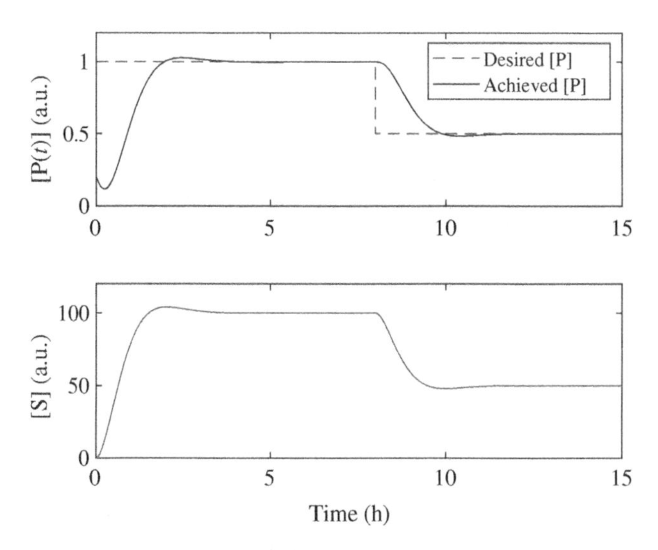

Figure 5.13 Responses of the closed-loop system with the degradation, k_{deg}, and the annihilation, η.

5.2 Robustness Analysis: μ-Analysis

Biological networks have inherent robustness towards environmental stress and internal uncertainties. One of the plausibility tests for the dynamic models of biological systems is robustness analysis. The Monte Carlo approach in Algorithm 2.3 could be one of the options for robustness analysis. We introduce a systematic approach for the robustness analysis of linear systems.

5.2.1 Simple Examples

μ-Analysis is a systematic robustness analysis method for linear systems (Doyle, 1982, Balas et al., 1993). We introduce the method using a simple example. Consider the following ODE:

$$\frac{dx}{dt} = -(2 + \delta)x \tag{5.5}$$

where δ is the parametric uncertainty, and the initial condition is $x(0) = x_0$. The solution is given by

$$x(t) = x_0 e^{-(2+\delta)t} \tag{5.6}$$

The condition for $x(t)$ to *diverge* is the following inequality:

$$\delta > 2$$

The stable and the unstable regions for $x(t)$ are divided by $\delta = 2$.

Rewrite (5.5) into the input–output format as follows:

$$\frac{dx}{dt} = -2x + w$$
$$z = x$$

where w is the input to the system and z is the system output. The transfer function for the input–output relation is obtained as

$$Z(s) = \frac{1}{s+2}W(s) = M(s)W(s)$$

where $M(s) = 1/(s+2)$. The input, w, is given by

$$w = \delta z$$

where δ is the uncertainty and is interpreted as the feedback gain, and the transfer function is

$$W(s) = \Delta(s)Z(s) = \delta Z(s)$$

where $\Delta(s) = \delta$. Two transfer functions are summarized as

$$Z(s) = M(s)W(s) \tag{5.7a}$$

$$W(s) = \Delta(s)Z(s) \tag{5.7b}$$

which is called the M–Δ form or the linear fractional transformation (LFT).

As shown in Figure 5.14, once the oscillatory signal is introduced in the loop, the amplitude of the signal would converge, diverge, or stay in the same magnitude. Each corresponds to a stable, unstable, or neutrally stable state of the system to the uncertainty, Δ. For this particular example, it is stable for $\delta < 2$, unstable for $\delta > 2$, or neutrally stable for $\delta = 2$.

Combine two transfer functions in (5.7)

$$Z(s) = M(s)W(s) = M(s)\Delta(s)Z(s) \Rightarrow Z(s) - M(s)\Delta(s)Z(s) = 0$$
$$\Rightarrow [1 - M(s)\Delta(s)]\, Z(s) = 0$$

There are two cases for $Z(s)$ in the above equation as follows:

- $1 - M(s)\Delta(s) \neq 0$, then $Z(s) = 0/[1 - M(s)\Delta(s)] = 0$
- $1 - M(s)\Delta(s) = 0$, then any $Z(s)$ satisfies the equation

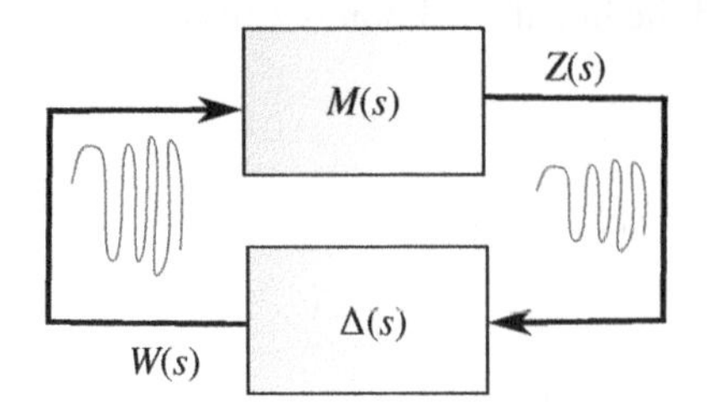

Figure 5.14 M–Δ block diagram.

for all $\omega \in [0, \infty)$, where $s = j\omega$ and $j = \sqrt{-1}$. Consider the singularity condition for the simple example as follows:

$$1 - M(j\omega)\Delta(j\omega) = 1 - \frac{1}{j\omega + 2}\delta = 0 \Rightarrow j\omega + 2 - \delta = 0$$

It becomes singular only for $\omega = 0$ and $\delta = 2$. The ODE solution, (5.6), shows that $x(t)$ remains at the initial condition, x_0, if $\delta = 2$. It makes the system neutrally stable. $\delta = 2$ is the smallest of the perturbation that makes the system not stable, $\delta \geq 2$.

μ is defined by

$$\mu(\omega) = \frac{1}{\min(|\delta|) \text{ such that } 1 - M(j\omega)\delta = 0}$$

and for the example

$$\mu(\omega) = \begin{cases} \dfrac{1}{2} \text{ for } \omega = 0 \\ 0 \text{ for } \omega \neq 0, \text{singular } \delta \text{ does not exist} \Rightarrow \lim_{\delta \to \pm\infty} \dfrac{1}{|\delta|} = 0 \end{cases}$$

is zero for non-singular cases for any $\delta \in (-\infty, \infty)$.

To make the μ-analysis problem more interesting, modify (5.5) as follows:

$$\frac{dx}{dt} = -\left(2 + 0.1\delta_1 + 0.5\delta_2 + 0.1\delta_2^2\right)x \tag{5.8}$$

where δ_1 and δ_2 are the real-valued uncertainties, and there is also the second-order term of uncertainty in δ_2. Define

$$w_1 = \delta_1 x$$
$$w_2 = \delta_2 x$$
$$w_3 = \delta_2^2 x = \delta_2\left(\delta_2 x\right) = \delta_2 w_2$$

In a compact form,

$$\mathbf{w} = \begin{bmatrix} w_1 \\ w_2 \\ w_3 \end{bmatrix} = \begin{bmatrix} \delta_1 & 0 & 0 \\ 0 & \delta_2 & 0 \\ 0 & 0 & \delta_2 \end{bmatrix} \begin{bmatrix} x \\ x \\ w_2 \end{bmatrix} = \Delta\mathbf{z}$$

and the system is written as

$$\frac{dx}{dt} = -2x + \begin{bmatrix} 0.1 & 0.5 & 0.1 \end{bmatrix} \mathbf{w} = Ax + B\mathbf{w}$$

$$\mathbf{z} = \begin{bmatrix} x \\ x \\ w_2 \end{bmatrix} = \begin{bmatrix} 1 \\ 1 \\ 0 \end{bmatrix} x + \begin{bmatrix} 0 & 0 & 0 \\ 0 & 0 & 0 \\ 0 & 1 & 0 \end{bmatrix} \mathbf{w} = Cx + D\mathbf{w}$$

where A, B, C, and D are defined appropriately. In the M–Δ form,

$$M(s) = C[s - A]^{-1}B + D = \begin{bmatrix} 1 \\ 1 \\ 0 \end{bmatrix} \frac{1}{s+2} \begin{bmatrix} 0.1 & 0.5 & 0.1 \end{bmatrix} + \begin{bmatrix} 0 & 0 & 0 \\ 0 & 0 & 0 \\ 0 & 1 & 0 \end{bmatrix} \tag{5.9}$$

$$= \begin{bmatrix} \frac{0.1}{s+2} & \frac{0.5}{s+2} & \frac{0.1}{s+2} \\ \frac{0.1}{s+2} & \frac{0.5}{s+2} & \frac{0.1}{s+2} \\ 0 & 1 & 0 \end{bmatrix}$$

For general cases, μ is defined by

$$\mu(\omega) = \frac{1}{\min(||\Delta||) \text{ such that } |I - M(j\omega)\Delta| = 0}$$

where $||\Delta||$ is typically the ∞-norm of the matrix and I is the identity matrix with the appropriate dimension.

Calculating μ is computationally expensive as increasing the computer operation exponentially with the size of $M(j\omega)$. Instead, we solve the lower and the upper bound problem:

$$\underline{\mu}(\omega) \leq \mu(\omega) \leq \overline{\mu}(\omega)$$

5.2.1.1 μ Upper Bound

There are several algorithms to provide the bounds. From the singularity condition, we deduce the following inequality for the upper bound:

$$||M(j\omega)\Delta|| \leq 1 \Rightarrow |I - M(j\omega)\Delta| \neq 0$$

As the norm of $M(j\omega)\Delta$ is smaller than 1, the determinant cannot be equal to zero. From the signal amplitude aspect in Figure 5.14, the amplitude multiplied by $M(j\omega)\Delta$ in the feedback loop, whose norm is less than 1, decreases every turn of the feedback loop and converges to zero as $t \to \infty$. With the following inequality

$$||M(j\omega)\Delta|| \leq ||M(j\omega)|| \, ||\Delta||$$

we obtain the following non-singular condition:

$$||M(j\omega)|| \, ||\Delta|| \leq 1 \Rightarrow ||M(j\omega)|| \leq \frac{1}{||\Delta||}$$

Therefore,

$$\overline{\sigma}\,[M(j\omega)] \leq \frac{1}{||\Delta||} \text{ implies } \mu(\omega) \leq \overline{\sigma}\,[M(j\omega)] \tag{5.10}$$

where the matrix norm is $\overline{\sigma}(\cdot)$, i.e. the maximum singular value norm. Efficient algorithms for calculating the singular value exist. For real parameter uncertainties, Δ is a diagonal real matrix and $||\Delta||_p$ for $p = 1,2,\infty$ are the same.

Python linear algebra package and MATLAB have the singular value decomposition function called *svd()*. In MATLAB,

```
1 >> [U,S,V] = svd(A)
```

or in python

```
1 In [13]: import numpy as np
2 In [14]: U,S,V = np.linalg.svd(A)
```

returns the following result:

$$A = USV$$

where S is the diagonal matrix whose elements are the singular values, i.e. the square root of the eigenvalues of A^*A, U, and V are unitary matrices, which satisfies $U^*U = I$ and $VV^* = I$, respectively, and the superscript $*$ indicates the complex conjugate transpose of the matrices.

Matrix norm: Matrix norms measure the magnitudes of matrices. Three of the most frequently used matrix norms are the 1-norm, the 2-norm, and the ∞-norm. They are defined by

$$\|A\|_1 = \max_{j\in[1,m]} \sum_{i=1}^{n} |a_{ij}| = (\text{maximum column sum})$$

$$\|A\|_2 = \sqrt{\bar{\lambda}(A^*A)} = \sqrt{(\text{maximum eigenvalue of } A^*A)}$$
$$= \bar{\sigma}(A^*A) = (\text{maximum singular value of } A)$$

$$\|A\|_\infty = \max_{i\in[1,n]} \sum_{j=1}^{m} |a_{ij}| = (\text{maximum row sum})$$

where A is the $n \times m$ complex matrix, whose i-th row and j-th column element is a_{ij}, and A^* is the complex conjugate transpose of A, i.e. the transpose of A, where the sign of the imaginary part of each element changes.

The upper bound in (5.10) is simply the maximum singular value at each ω. Program 5.4 calculates the upper bound between ω equal to 0.01 and 1000. The maximum upper bound over the frequency is around 1.06. Can we make it smaller? The smaller the upper bound is, the closer to μ is. Inspecting at (5.9), the cause of the maximum singular value of about 1 is the constant 1 in the matrix D. To change the constant smaller than 1, redefine $\mathbf{z}$ as follows:

$$\mathbf{w} = \begin{bmatrix} w_1 \\ w_2 \\ w_3 \end{bmatrix} = \begin{bmatrix} \delta_1 & 0 & 0 \\ 0 & \delta_2 & 0 \\ 0 & 0 & \delta_2 \end{bmatrix} \begin{bmatrix} x \\ x \\ 0.1w_2 \end{bmatrix} = \Delta\mathbf{z}$$

and the system becomes

$$\frac{dx}{dt} = -2x + \begin{bmatrix} 0.1 & 0.5 & 1 \end{bmatrix} \mathbf{w}$$

$$\mathbf{z} = \begin{bmatrix} 1 \\ 1 \\ 0 \end{bmatrix} x + \begin{bmatrix} 0 & 0 & 0 \\ 0 & 0 & 0 \\ 0 & 0.1 & 0 \end{bmatrix} \mathbf{w}$$

Construct $M(s)$ as follows:

$$M(s) = \begin{bmatrix} 1 \\ 1 \\ 0 \end{bmatrix} \frac{1}{s+2} \begin{bmatrix} 0.1 & 0.5 & 1 \end{bmatrix} + \begin{bmatrix} 0 & 0 & 0 \\ 0 & 0 & 0 \\ 0 & 0.1 & 0 \end{bmatrix} = \begin{bmatrix} \frac{0.1}{s+2} & \frac{0.5}{s+2} & \frac{1}{s+2} \\ \frac{0.1}{s+2} & \frac{0.5}{s+2} & \frac{1}{s+2} \\ 0 & 0.1 & 0 \end{bmatrix}$$

For the updated $M(s)$, the maximum upper bound over the frequency is 0.79, which is about a 25% reduction from 1.06.

```
1  import numpy as np
2  import matplotlib.pyplot as plt
3
4
5  A = np.array([[-2]])
6  B = np.array([[0.1,0.5,0.1]])
7  C = np.array([[1],[1],[0]])
8  D = np.array([[0,0,0],[0,0,0],[0,1,0]])
9
10 N_omega = 300
11 omega = np.logspace(-2,3,N_omega)
12 mu_ub = np.zeros(N_omega)
13
14 for idx in range(300):
15     jw = complex(0,omega[idx])
16     Mjw = C@np.linalg.inv(jw-A)@B+D
17     U,S,V=np.linalg.svd(Mjw)
18
19     mu_ub[idx] = S.max()
20
21
22 fig1, ax = plt.subplots(nrows=1,ncols=1)
23 ax.semilogx(omega,mu_ub)
24 ax.axis([1e-2,1e3,0,1.1])
25 ax.set_ylabel(r'$\bar{\sigma}\, [M(j\omega)]$',fontsize=14)
26 ax.set_xlabel(r'$\omega$ [rad/time]',fontsize=14)
```

Program 5.4 (Python) μ upper bound by the maximum singular value

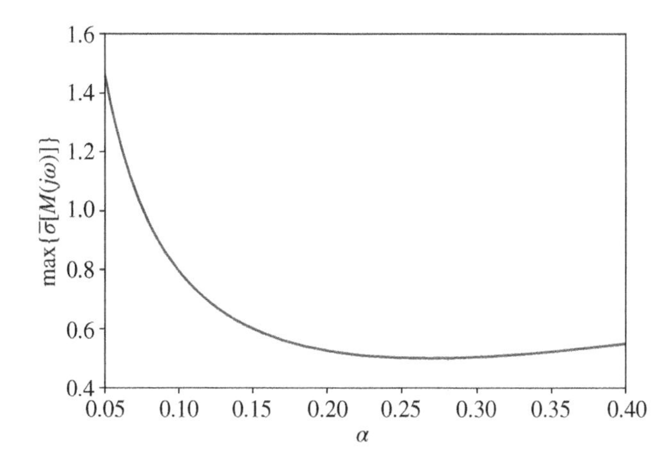

Figure 5.15 Maximum $\overline{\mu}$ with respect to α.

Based on the reduction, we would ask what is the best value to achieve the minimum value of the maximum upper bound. Consider the following generalized form of $M(s)$:

$$M(s) = \begin{bmatrix} \frac{0.1}{s+2} & \frac{0.5}{s+2} & \frac{0.1/\alpha}{s+2} \\ \frac{0.1}{s+2} & \frac{0.5}{s+2} & \frac{0.1/\alpha}{s+2} \\ 0 & \alpha & 0 \end{bmatrix}$$

Figure 5.15 shows how the maximum value of the upper bound changes with α. The minimum, 0.50, occurs at $\alpha = 0.266$, and the corresponding upper bound is shown in Figure 5.16. Depending on how we construct $M(s)$, the upper bound changes. There are advanced algorithms to obtain tighter upper bounds to the true μ (Roos, 2013, Balas et al., 1993, Young et al., 1991).

5.2.1.2 μ Lower Bound

Calculating the lower bound, i.e. searching for Δ, preferably the smallest magnitude, to make $|I - M\Delta| = 0$, becomes the original problem itself. There have been many studies of lower bound algorithms to find the minimum magnitude perturbation. Fabrizi et al. (2014) compare eight μ lower bound algorithms for various benchmark problems. Without a tight lower bound, Figure 5.16, for example, gives no information about how far or close the upper bound is from μ. If a lower bound is far away from the upper bound, we do not know either if the upper bound is too conservative or if the lower bound is far from μ. Only when two bounds are tight, we can draw a solid conclusion on system robustness.

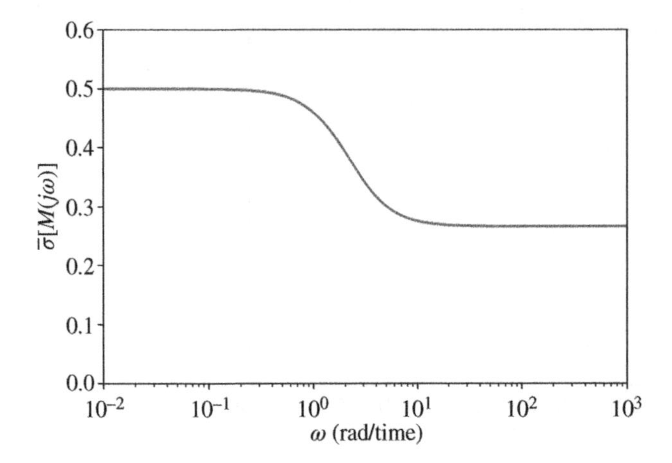

Figure 5.16 μ Upper bound for $\alpha = 0.266$.

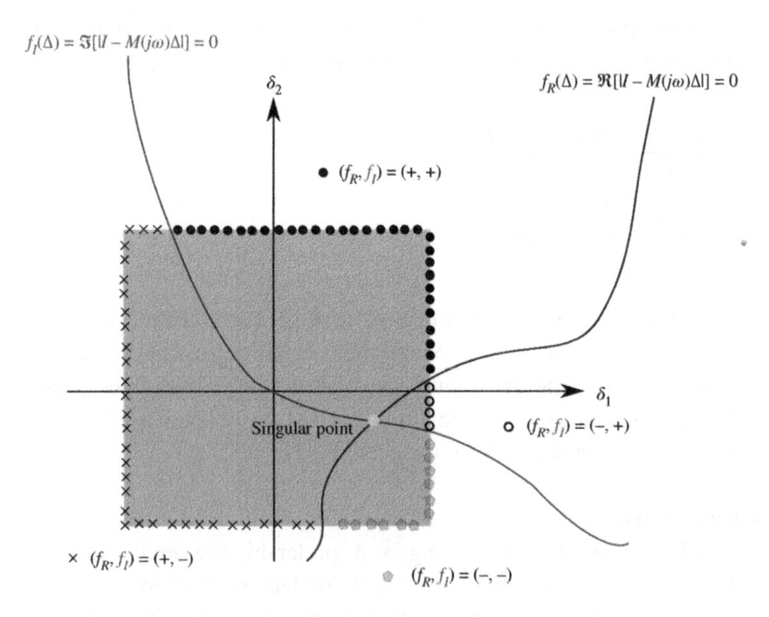

Figure 5.17 Geometric approach for μ lower bound calculation.

A geometric approach to the μ-lower bound problem has been presented first in Kim et al. (2009). A simple version of the algorithm is explained in Figure 5.17. The extension of the algorithm to the linear periodically time-varying system is shown in Zhao et al. (2011), and the improved algorithm applied to a synthetic circuit has been demonstrated its capability in Darlington et al. (2019). To provide a simple explanation of the algorithm in Figure 5.17, consider the following uncertain system:

$$\frac{dx}{dt} = \left[-2 + \delta_1 + \sin(\delta_1\delta_2)\right] x = A(\delta_1, \delta_2)x \tag{5.11}$$

where

$$A(\delta_1, \delta_2) = \left[-2 + \delta_1 + \sin(\delta_1\delta_2)\right]$$

Unlike (5.8), the system has the uncertainties, δ_1 and δ_2, which appeared in the non-polynomial form. Because of the non-linear uncertainty, the system cannot be in the LFT form, and the standard algorithms cannot calculate the bounds. Define a new variable $\delta_3 = \sin(\delta_1\delta_2)$ or approximate $\sin(\delta_1\delta_2) \approx \delta_1\delta_2$ by assuming that $|\delta_1\delta_2|$ is small. Then, we can use the standard algorithms. These introduce approximation errors in the bound calculations, however. To avoid the approximation, the system is written as

$$\frac{dx}{dt} = -2x + \left[\delta_1 + \sin(\delta_1\delta_2)\right] x = -2x + \Delta(\delta_1, \delta_2)x$$

where

$$\Delta(\delta_1, \delta_2) = \delta_1 + \sin(\delta_1\delta_2) = A(\delta_1, \delta_2) - A(0{,}0)$$

Define

$$w = \Delta(\delta_1, \delta_2)x$$

$$z = x$$

The system is written as

$$\frac{dx}{dt} = -2x + w$$

and the singularity condition is given by

$$\det\left[1 - \frac{1}{j\omega + 2}\Delta(\delta_1, \delta_2)\right] = f_R(\Delta) + f_I(\Delta)j = 0$$

where

$$f_R(\Delta) = 2 - \delta_1 - \sin(\delta_1\delta_2)$$

$$f_I(\Delta) = \omega$$

Note that in the bound algorithm illustrated in Figure 5.17, the explicit equations for $f_R(\Delta)$ and $f_I(\Delta)$ are not necessary. The evaluations of the following equations provide the values of $f_R(\Delta)$ and $f_I(\Delta)$:

$$f_R(\Delta) = \Re\left[1 - \frac{1}{j\omega + 2}\Delta(\delta_1, \delta_2)\right]$$

$$f_I(\Delta) = \Im\left[1 - \frac{1}{j\omega + 2}\Delta(\delta_1, \delta_2)\right]$$

where $\Re(\cdot)$ and $\Im(\cdot)$ take the real part and the imaginary part of the argument, respectively. As shown in Figure 5.17, $f_R(\Delta) = 0$ and $f_I(\Delta) = 0$ lines divide the uncertain space into four sections, where each section has a different sign combination of $f_R(\Delta)$ and $f_I(\Delta)$. For $\omega \neq 0$, if the singular point is inside the square box centred at the origin, we can find all four combinations of the signs along the boundary, and the square box provides a μ-lower bound. For $\omega = 0$, $f_I(\Delta)$ is zero for the whole uncertain space. Hence, if the singular point is inside the square box, i.e. the box includes or any part of $f_R(\Delta) = 0$, a singular point is inside the box. In this case, we only need to find two sign combinations, $f_R(\Delta) < 0$ and $f_R(\Delta) > 0$.

The geometric approach based on the above observation about the singularity condition provides three advantages over most of the other lower bound algorithms as follows:

- The algorithm can handle non-linear functions such as $\sin\delta$ and $\cos\delta$ without approximation.
- The algorithm can use parallel computers.
- The algorithm based on random samples improves the bounds as the algorithm evaluates more samples.

Random sample-based algorithms are powerful tools to solve many complex problems, especially when we can afford to use parallel computer architectures such as multi-core processors and parallel processing using a graphical processor unit (GPU) (NVIDIA Developer, 2021).

The pseudo-code of the lower bound algorithm in Figure 5.17 is given in Algorithm 5.1, where the bisection method is used to reduce the square box size (Press et al., 2007). As the ∞-norm of the square box is half of the side length, the μ-lower bound is equal to twice the maximum side length inversion, i.e. $2/\bar{d}$.

Figure 5.18 shows the square box contact with the unstable region shaded in the figure. The half of the side length is slightly bigger than 1. The true μ, the inversion of the half, is about 0.9. For $N_s = 1000$, $\varepsilon = 10^{-6}$, $\underline{d} = 0.001$, $\bar{d} = 10$, and $\omega = 0$, Algorithm 5.1 computes a lower bound around 0.92, which is reasonably close to the true μ.

5.2.1.3 Complex Numbers in MATLAB/Python

Both MATLAB and Python recognize *1j* as the imaginary number, $\sqrt{-1}$. Also, *1i* or *sqrt(-1)* works as the imaginary number in MATLAB, not Python. In Python, *numpy.sqrt(-1)* produces an error instead of generating the imaginary number. An alternative way to have the imaginary number in Python is *complex(0,1)*.

Algorithm 5.1 Geometric approach based μ-lower bound algorithm

1: Set the number of samples, N_s, the minimum and the maximum side length of the square box, $[\underline{d}, \bar{d}]$, the tolerance, ε, and the frequency ω
2: **while** $\bar{d} - \underline{d} > \varepsilon$ **do** the bisection search as follows:
3: Set the current side length of the square, d, equal to $(\underline{d} + \bar{d})/2$
4: Take N_s random samples on the boundary of the square box
5: **if** $\omega = 0$ **then**
6: n_{sign} equal to 2
7: **else**
8: n_{sign} equal to 4
9: **end if**
10: Evaluate $f_R(\Delta)$ and $f_I(\Delta)$
11: **if** the number of sign combinations found equal to n_{sign} **then**
12: $\bar{d} \leftarrow d$
13: **else**
14: $\underline{d} \leftarrow d$
15: **end if**
16: **end while**
17: Declare $\underline{\mu}(\omega) = 1/(\bar{d}/2)$
18: Repeat all the steps for the other ω

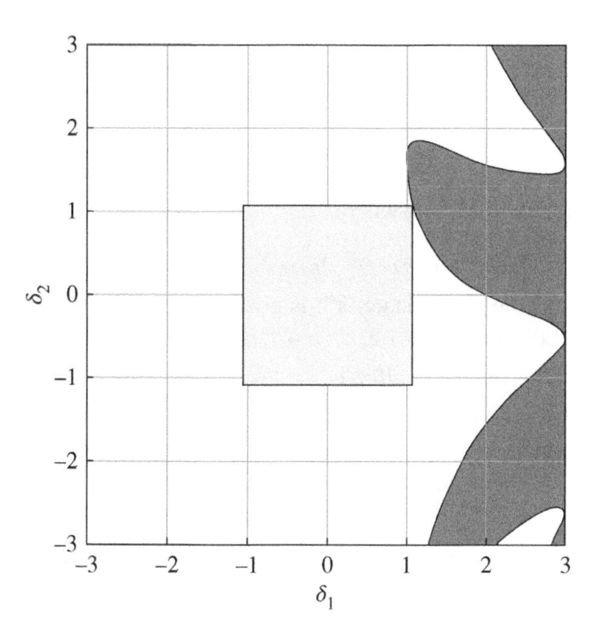

Figure 5.18 The square box contacts the unstable region shaded in the figure, where $\delta_1 + \sin(\delta_1\delta_2) \geq 2$.

5.2.2 Synthetic Circuits

The synthetic circuit designed in Section 5.1.1 reaches the following steady-state:

$$[\mathrm{E}]^{ss} = 0.1998, \quad [\mathrm{P}]^{ss} = 0.4999, \quad [\mathrm{ES}]^{ss} = 0.0002, \quad [X_1]^{ss} = 0.0558,$$
$$[X_2]^{ss} = 50.0415, \quad [\mathrm{S}]^{ss} = 50.0723, \quad [\Delta\mathrm{P}]^{ss} = 1.0001, \quad [X_{\mathrm{sensor}}]^{ss} = 0.4999$$

Introduce a small perturbation to the steady-state of E as follows:

$$[\mathrm{E}(t)] = [\mathrm{E}]^{ss} + \delta\mathrm{E}(t)$$

and similar perturbations in all other states. Derivative of the perturbed E(t) with respect to time is as follows:

$$\frac{d[\mathrm{E}(t)]}{dt} = \cancelto{0}{\frac{d[\mathrm{E}]^{ss}}{dt}} + \frac{d\delta\mathrm{E}(t)}{dt}$$

Substitute the perturbed states into the right-hand side of the following equation:

$$\frac{d[\mathrm{E}]}{dt} = -k_{\mathrm{on}}\,[\mathrm{E}][\mathrm{S}] + k_{\mathrm{cat}}\,[\mathrm{ES}]$$

and obtain

$$\frac{d\delta\mathrm{E}(t)}{dt} = -k_{\mathrm{on}}\left\{[\mathrm{E}]^{ss} + \delta\mathrm{E}(t)\right\}\left\{[\mathrm{S}]^{ss} + \delta\mathrm{S}(t)\right\} + k_{\mathrm{cat}}\left\{[\mathrm{ES}]^{ss} + \delta\mathrm{ES}(t)\right\}$$
$$= \cancelto{0}{-k_{\mathrm{on}}[\mathrm{E}]^{ss}[\mathrm{S}]^{ss} + k_{\mathrm{cat}}[\mathrm{ES}]^{ss}} - k_{\mathrm{on}}[\mathrm{S}]^{ss}\delta\mathrm{E}(t) - k_{\mathrm{on}}[\mathrm{E}]^{ss}\delta\mathrm{S}(t)$$
$$\cancelto{\approx 0}{-k_{\mathrm{on}}\delta\mathrm{E}(t)\delta\mathrm{S}(t)} + k_{\mathrm{cat}}\delta\mathrm{ES}(t)$$

where the higher order terms in the perturbations are negligible.

We write the procedures compactly. The kinetics is given by

$$\frac{dx_i}{dt} = f_i(\mathbf{x})$$

for $i = 1,2,\ldots,n$, where

$$\mathbf{x} = \begin{bmatrix} x_1 & x_2 & \ldots & x_{n-1} & x_n \end{bmatrix}^T$$

and the steady-state, $\mathbf{x}^{ss}$, is given by the solution of the n algebraic equations, $f_i(\mathbf{x}^{ss}) = 0$ for $i = 1,2,\ldots,n-1,n$. The perturbation dynamics is given by

$$\frac{d\delta\mathbf{x}(t)}{dt} = \left.\frac{\mathbf{df}(\mathbf{x})}{\mathbf{dx}}\right|_{\mathbf{x}=\mathbf{x}^{ss}} \delta\mathbf{x}(t) = A(\Delta)\delta\mathbf{x}(t)$$

where

$$\mathbf{a}_i = \left.\frac{df_i(\mathbf{x})}{\mathbf{dx}}\right|_{\mathbf{x}=\mathbf{x}^{ss}}$$

and $\mathbf{a}_i$ is the i-the column of A. It is the linearization process at the steady-state or the equilibrium point, and we may automatize it using symbolic calculations in MATLAB or Python.

5.2.2.1 MATLAB

Program5.5 obtains A using the *jacobian()* function in MATLAB. Because the largest real part of the eigenvalues is negative, the system is stable.

5.2.2.2 Python

Program 5.6 is the python script to obtain A using the *jacobian()* function in the method of Matrix object of the sympy package. *subs()* replaces the variables with the values. In line 66, the *numpy.array()* function converts the Matrix object of sympy to the numpy array, where we declare the data type by setting *dtype* as *np.float64*. If the data type is missing, we cannot calculate the eigenvalues in the next line.

```
1  clear
2
3  syms kon kcat kdg Kdeg kI eta gamma_G kP ks2 ks1 Pd X3 real;
4  syms E P ES X1 X2 S DP Xs real;
5
6
7  dE_dt = -kon*E*S + kcat*ES;
8  dP_dt = kcat*ES - kdg*P - Kdeg*X3*P;
9  dES_dt = kon*E*S - kcat*ES;
10 dX1_dt = kI*DP - eta*X1;
11 dX2_dt = -gamma_G*X2 + gamma_G*kP*DP;
12 dS_dt = ks2*X1 + ks2*X2 - ks2*S;
13 dDP_dt = ks1*Pd - ks1*DP*Xs - ks1*DP;
14 dXs_dt = -ks1*DP*Xs + ks1*P;
15
16 fx = [dE_dt; dP_dt; dES_dt; dX1_dt; dX2_dt; dS_dt; dDP_dt; dXs_dt];
17 state = [E; P; ES; X1; X2; S; DP; Xs];
18
19 dfdx = jacobian(fx,state);
20
21 %% Steady-state
22 Ess = 0.1998;
23 Pss = 0.4999;
24 ESss = 0.0002;
25 X1ss = 0.0558;
26 X2ss = 50.0415;
27 Sss = 50.0723;
28 DPss = 1.0001;
29 Xsss = 0.4999;
30
31 dfdx_at_ss = subs(dfdx,{E, P, ES, X1, X2, S, DP, Xs},{Ess, Pss,
       ESss, X1ss, X2ss, Sss, DPss, Xsss});
32
33 %% nomial stability with the nominal parameters
34 kP = 50;
35 kI = 5e-6;
36 gamma_G = 8e-4;
37 ks2 = 4e-4;
```

```
38 Kdeg = 1e-3;
39 X3 = 1;
40 KF = 3;
41 eta = 1e-4;
42 Pd = 1;
43 kon = 5e-5;
44 kcat = 1.6*2;
45 kdg = 8e-8;
46 ks1 = 3;
47
48
49 dfdx_nominal = subs(dfdx_at_ss, ...
50     {sym('kP'), sym('kI'),sym('gamma_G'),sym('ks2'),sym('Kdeg'),
       ...
51     sym('X3'),sym('KF'),sym('eta'),sym('Pd'),sym('kon'),sym('kcat')
           ,sym('kdg'),sym('ks1')}, ...
52     {kP,kI,gamma_G,ks2,Kdeg,X3,KF,eta,Pd,kon,kcat,kdg,ks1});
53
54 dfdx_nominal_val = eval(dfdx_nominal);
```

Program 5.5 (MATLAB) Nominal linear stability check using the jacobian

```
1 from sympy import symbols, Matrix
2
3 kon, kcat, kdg, Kdeg, kI, eta, gamma_G, kP, ks2, ks1, Pd, X3 =
      symbols('kon kcat kdg Kdeg kI eta gamma_G kP ks2 ks1 Pd X3')
4 E, P, ES, X1, X2, S, DP, Xs = symbols('E P ES X1 X2 S DP Xs')
5
6
7 dE_dt = -kon*E*S + kcat*ES;
8 dP_dt = kcat*ES - kdg*P - Kdeg*X3*P;
9 dES_dt = kon*E*S - kcat*ES;
10 dX1_dt = kI*DP - eta*X1;
11 dX2_dt = -gamma_G*X2 + gamma_G*kP*DP;
12 dS_dt = ks2*X1 + ks2*X2 - ks2*S;
13 dDP_dt = ks1*Pd - ks1*DP*Xs - ks1*DP;
14 dXs_dt = -ks1*DP*Xs + ks1*P;
15
16 fx = Matrix([[dE_dt], [dP_dt], [dES_dt], [dX1_dt], [dX2_dt], [dS_dt
      ], [dDP_dt], [dXs_dt]])
17 state = Matrix([[E], [P], [ES], [X1], [X2], [S], [DP], [Xs]])
18
19 dfdx = fx.jacobian(state)
20
21 # Steady-state
22 Ess = 0.1998
23 Pss = 0.4999
24 ESss = 0.0002
25 X1ss = 0.0558
26 X2ss = 50.0415
27 Sss = 50.0723
28 DPss = 1.0001
29 Xsss = 0.4999
```

```
30
31 dfdx_at_ss = dfdx.subs([[E,Ess],[P,Pss],[ES,ESss],[X1,X1ss],[X2,
       X2ss],[S,Sss],[DP,DPss],[Xs,Xsss]])
32
33 # nomial stability with the nominal parameters
34 kP = 50;
35 kI = 5e-6;
36 gamma_G = 8e-4;
37 ks2 = 4e-4;
38 Kdeg = 1e-3;
39 X3 = 1;
40 KF = 3;
41 eta = 1e-4;
42 Pd = 1;
43 kon = 5e-5;
44 kcat = 1.6*2;
45 kdg = 8e-8;
46 ks1 = 3;
47
48
49 dfdx_nominal = dfdx_at_ss.subs([
50       [symbols('kP'),kP],
51       [symbols('kI'),kI],
52       [symbols('gamma_G'),gamma_G],
53       [symbols('ks2'),ks2],
54       [symbols('Kdeg'),Kdeg],
55       [symbols('X3'),X3],
56       [symbols('KF'),KF],
57       [symbols('eta'),eta],
58       [symbols('Pd'),Pd],
59       [symbols('kon'),kon],
60       [symbols('kcat'),kcat],
61       [symbols('kdg'),kdg],
62       [symbols('ks1'),ks1]
63 ])
64
65 import numpy as np
66 dfdx_nominal_val = np.array(dfdx_nominal,dtype=np.float64)
67 [eig_val,eig_vec]=np.linalg.eig(dfdx_nominal_val)
```

Program 5.6 (Python) Nominal linear stability check using the jacobian

The μ-lower bound program implementing Algorithm 5.1 continues in Program 5.7 or 5.8 from Program 5.5 or 5.6. The two for-loops in the program that iterate over ω and the random samples, Δ, can be executed in a parallel computation architecture to speed up the computation. See page 119 for examples of parallel processing on multi-core CPU. The lower bound obtained by the geometric approach algorithm is shown in Figure 5.19. The empty circle at the top left corner indicates the lower bound value at $\omega = 0$. As the maximum lower bound is between 20,000 and 30,000, the magnitudes of destabilizing uncertainties

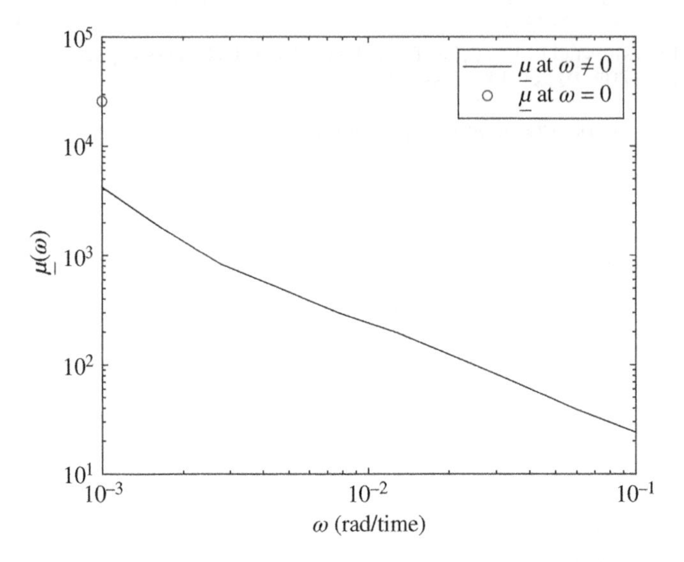

Figure 5.19 μ lower bound obtained by the geometric approach.

are at least smaller than 0.5×10^{-4}. It shows an extreme sensitivity and fragility of the synthetic circuits.

```
1  clear
2
3  syms kon kcat kdg Kdeg kI eta gamma_G kP ks2 ks1 Pd X3 real;
4  syms E P ES X1 X2 S DP Xs real;
5  syms d1 d2 d3 d4 d5 d6 d7 d8 d9 d10 d11 d12 d13 d14 d15 d16 real;
6
7
8  dE_dt = -(kon+d1)*E*S + (kcat+d2)*ES;
9  dP_dt = (kcat+d2)*ES - (kdg+d3)*P - (Kdeg+d4)*(X3+d5)*P;
10 dES_dt = (kon+d1)*E*S - (kcat+d2)*ES;
11 dX1_dt = (kI+d6)*DP - (eta+d7)*X1;
12 dX2_dt = -(gamma_G+d8)*X2 + (gamma_G+d9)*(kP+d10)*DP;
13 dS_dt = (ks2+d11)*X1 + (ks2+d12)*X2 - (ks2+d13)*S;
14 dDP_dt = (ks1+d14)*Pd - (ks1+d15)*DP*Xs - (ks1+d16)*DP;
15 dXs_dt = -(ks1+d15)*DP*Xs + (ks1+d16)*P;
16
17 fx = [dE_dt; dP_dt; dES_dt; dX1_dt; dX2_dt; dS_dt; dDP_dt; dXs_dt];
18 state = [E; P; ES; X1; X2; S; DP; Xs];
19
20 dfdx = jacobian(fx, state);
21
22 %% Steady-state
23 Ess = 0.1998;
24 Pss = 0.4999;
```

```
25 ESss = 0.0002;
26 X1ss = 0.0558;
27 X2ss = 50.0415;
28 Sss = 50.0723;
29 DPss = 1.0001;
30 Xsss = 0.4999;
31
32 dfdx_at_ss = subs(dfdx,{E, P, ES, X1, X2, S, DP, Xs},{Ess, Pss,
       ESss, X1ss, X2ss, Sss, DPss, Xsss});
33
34 %% nomial stability with the nominal parameters
35 kP = 50;
36 kI = 5e-6;
37 gamma_G = 8e-4;
38 ks2 = 4e-4;
39 Kdeg = 1e-3;
40 X3 = 1;
41 KF = 3;
42 eta = 1e-4;
43 Pd = 1;
44 kon = 5e-5;
45 kcat = 1.6*2;
46 kdg = 8e-8;
47 ks1 = 3;
48
49
50 dfdx_nominal = subs(dfdx_at_ss, ...
51     {sym('kP'), sym('kI'),sym('gamma_G'),sym('ks2'),sym('Kdeg'),
         ...
52     sym('X3'),sym('KF'),sym('eta'),sym('Pd'),sym('kon'),sym('kcat')
         ,sym('kdg'),sym('ks1')}, ...
53     {kP,kI,gamma_G,ks2,Kdeg,X3,KF,eta,Pd,kon,kcat,kdg,ks1});
54
55 dfdx_nominal_val = eval(dfdx_nominal);
56
57
58 %% mu-analysis
59 Ns = 5000;
60 eps = 1e-6;
61
62 num_state = 8;
63 num_delta = 16;
64 A0 =  eval(subs(dfdx_nominal,{d1 d2 d3 d4 d5 d6 d7 d8 d9 d10 d11
       d12 d13 d14 d15 d16},{zeros(1,16)}));
65
66 num_omega = 10;
67 omega_all = [0 logspace(-3,-1,num_omega)];
68 num_omega = num_omega + 1;
69
70 mu_lb = zeros(1,num_omega);
71
72 %% lower bound using geometric approach
73 for wdx=1:num_omega
```

```
 74     omega = omega_all(wdx);
 75     Mjw = inv(1j*omega*eye(num_state)-A0);
 76 
 77     d_lb = 1e-6;
 78     d_ub = 10;
 79     d_ulb = d_ub - d_lb;
 80 
 81     if omega==0
 82         size_check = 2;
 83     else
 84         size_check = 4;
 85     end
 86 
 87     while d_ulb > eps
 88 
 89         d = (d_lb+d_ub)/2;
 90 
 91         sign_all = [];
 92 
 93         for idx=1:Ns
 94             delta_vec = rand(1,num_delta)*d-d/2;
 95             rand_face = randi(num_delta,1);
 96             delta_vec(rand_face) = d/2;
 97 
 98             Delta = eval(subs(dfdx_nominal,{d1 d2 d3 d4 d5 d6 d7 d8
                   d9 d10 d11 d12 d13 d14 d15 d16}, ...
 99                 {delta_vec})) -A0;
100 
101             I_MD = det(eye(num_state)-Mjw*Delta);
102             fR = sign(real(I_MD));
103             fI = sign(imag(I_MD));
104 
105             sign_all = unique([sign_all; fR fI],'row');
106 
107         end
108 
109         if size(sign_all,1) == size_check
110             d_ub = d;
111         else
112             d_lb = d;
113         end
114 
115         d_ulb = d_ub - d_lb;
116 
117     end
118 
119     mu_lb(wdx) = 2/d_ub;
120 end
```

Program 5.7 (MATLAB) μ-lower bound calculation

```
1  import numpy as np
2  from sympy import symbols, Matrix
3
4  kon, kcat, kdg, Kdeg, kI, eta, gamma_G, kP, ks2, ks1, Pd, X3 =
       symbols('kon kcat kdg Kdeg kI eta gamma_G kP ks2 ks1 Pd X3')
5  E, P, ES, X1, X2, S, DP, Xs = symbols('E P ES X1 X2 S DP Xs')
6
7  d1, d2, d3, d4, d5, d6, d7, d8 = symbols('d1 d2 d3 d4 d5 d6 d7 d8')
8  d9, d10, d11, d12, d13, d14, d15, d16 = symbols('d9 d10 d11 d12 d13
       d14 d15 d16')
9
10 dE_dt = -(kon+d1)*E*S + (kcat+d2)*ES;
11 dP_dt = (kcat+d2)*ES - (kdg+d3)*P - (Kdeg+d4)*(X3+d5)*P;
12 dES_dt = (kon+d1)*E*S - (kcat+d2)*ES;
13 dX1_dt = (kI+d6)*DP - (eta+d7)*X1;
14 dX2_dt = -(gamma_G+d8)*X2 + (gamma_G+d9)*(kP+d10)*DP;
15 dS_dt = (ks2+d11)*X1 + (ks2+d12)*X2 - (ks2+d13)*S;
16 dDP_dt = (ks1+d14)*Pd - (ks1+d15)*DP*Xs - (ks1+d16)*DP;
17 dXs_dt = -(ks1+d15)*DP*Xs + (ks1+d16)*P;
18
19 fx = Matrix([[dE_dt], [dP_dt], [dES_dt], [dX1_dt], [dX2_dt], [dS_dt
       ], [dDP_dt], [dXs_dt]])
20 state = Matrix([[E], [P], [ES], [X1], [X2], [S], [DP], [Xs]])
21
22 dfdx = fx.jacobian(state)
23
24 # Steady-state
25 Ess = 0.1998
26 Pss = 0.4999
27 ESss = 0.0002
28 X1ss = 0.0558
29 X2ss = 50.0415
30 Sss = 50.0723
31 DPss = 1.0001
32 Xsss = 0.4999
33
34 dfdx_at_ss = dfdx.subs([[E,Ess],[P,Pss],[ES,ESss],[X1,X1ss],[X2,
       X2ss],[S,Sss],[DP,DPss],[Xs,Xsss]])
35
36 # nomial stability with the nominal parameters
37 kP = 50;
38 kI = 5e-6;
39 gamma_G = 8e-4;
40 ks2 = 4e-4;
41 Kdeg = 1e-3;
42 X3 = 1;
43 KF = 3;
44 eta = 1e-4;
45 Pd = 1;
46 kon = 5e-5;
47 kcat = 1.6*2;
48 kdg = 8e-8;
49 ks1 = 3;
```

```
50
51
52 dfdx_nominal = dfdx_at_ss.subs([
53         [symbols('kP'),kP],
54         [symbols('kI'),kI],
55         [symbols('gamma_G'),gamma_G],
56         [symbols('ks2'),ks2],
57         [symbols('Kdeg'),Kdeg],
58         [symbols('X3'),X3],
59         [symbols('KF'),KF],
60         [symbols('eta'),eta],
61         [symbols('Pd'),Pd],
62         [symbols('kon'),kon],
63         [symbols('kcat'),kcat],
64         [symbols('kdg'),kdg],
65         [symbols('ks1'),ks1]
66       ])
67
68 # mu-analysis
69 Ns = 5000
70 eps = 1e-6
71
72 num_state = 8
73 num_delta = 16
74 A0 = dfdx_nominal_val = dfdx_nominal.subs([
75       [symbols('d1'),0],
76       [symbols('d2'),0],
77       [symbols('d3'),0],
78       [symbols('d4'),0],
79       [symbols('d5'),0],
80       [symbols('d6'),0],
81       [symbols('d7'),0],
82       [symbols('d8'),0],
83       [symbols('d9'),0],
84       [symbols('d10'),0],
85       [symbols('d11'),0],
86       [symbols('d12'),0],
87       [symbols('d13'),0],
88       [symbols('d14'),0],
89       [symbols('d15'),0],
90       [symbols('d16'),0]
91     ])
92 A0 = np.array(A0,dtype=np.float64)
93
94 num_omega = 10
95 omega_all = np.hstack((0,np.logspace(-3,-1,num_omega-1)))
96
97 mu_lb = np.zeros(num_omega)
98
99 # lower bound using geometric approach
100 for wdx, omega in enumerate(omega_all):
101     Mjw=np.linalg.inv(1j*omega*np.eye(num_state)-A0)
102
```

```
103        d_lb = 1e-6
104        d_ub = 10
105        d_ulb = d_ub - d_lb
106
107        if omega==0:
108            size_check = 2
109        else:
110            size_check = 4
111
112        while d_ulb > eps:
113
114            d = (d_lb+d_ub)/2
115
116            for idx in range(Ns):
117                delta_vec = np.random.rand(num_delta)*d-d/2
118                rand_face = np.random.randint(0,num_delta,1)[0]
119                delta_vec[rand_face] = d/2
120
121                dfdx_nominal_val = dfdx_nominal.subs([
122                    [symbols('d1'),delta_vec[0]],
123                    [symbols('d2'),delta_vec[1]],
124                    [symbols('d3'),delta_vec[2]],
125                    [symbols('d4'),delta_vec[3]],
126                    [symbols('d5'),delta_vec[4]],
127                    [symbols('d6'),delta_vec[5]],
128                    [symbols('d7'),delta_vec[6]],
129                    [symbols('d8'),delta_vec[7]],
130                    [symbols('d9'),delta_vec[8]],
131                    [symbols('d10'),delta_vec[9]],
132                    [symbols('d11'),delta_vec[10]],
133                    [symbols('d12'),delta_vec[11]],
134                    [symbols('d13'),delta_vec[12]],
135                    [symbols('d14'),delta_vec[13]],
136                    [symbols('d15'),delta_vec[14]],
137                    [symbols('d16'),delta_vec[15]]
138                    ])
139
140                Delta = np.array(dfdx_nominal_val,dtype=np.float64) -
                       A0
141
142                I_MD = np.linalg.det(np.eye(num_state)-Mjw*Delta)
143                fR = np.sign(np.real(I_MD))
144                fI = np.sign(np.imag(I_MD))
145
146                if idx==1:
147                    sign_all = np.array([fR, fI])
148                else:
149                    sign_all = np.vstack((sign_all,[fR, fI]))
150                    sign_all = np.unique(sign_all, axis=0)
151
152            if sign_all.shape[0] == size_check:
153                d_ub = d
154            else:
```

```
155             d_lb = d
156
157         d_ulb = d_ub - d_lb
158
159         print(omega)
160         print(sign_all)
161         print(d_lb,d_ub)
162
163     mu_lb[wdx] = 2/d_ub
```

Program 5.8 (Python) μ-lower bound calculation

5.2.2.3 μ-Upper Bound: Geometric Approach

We can construct a μ-upper bound algorithm based on the geometric approach. The upper bound using the maximum singular value is too conservative for the enzyme–substrate network. The maximum singular value of $M(j\omega)$ at $\omega = 0$ is in the order of 10^{20}, which corresponds to the uncertainty magnitude to guarantee the stability is less than 10^{-20}. Given that the maximum μ-lower bound found is around 20,000, which corresponds to the uncertainty magnitude destabilizing the system around 0.00005, the upper bound provided by the maximum singular value is too big to estimate the true μ within the acceptable tolerance.

Figure 5.20 shows the concept of how to calculate the upper bound, i.e. the stability guaranteed bound, using the geometric approach. Inside the stable guaranteed box, we can find only three or fewer sign combinations for ω not equal to

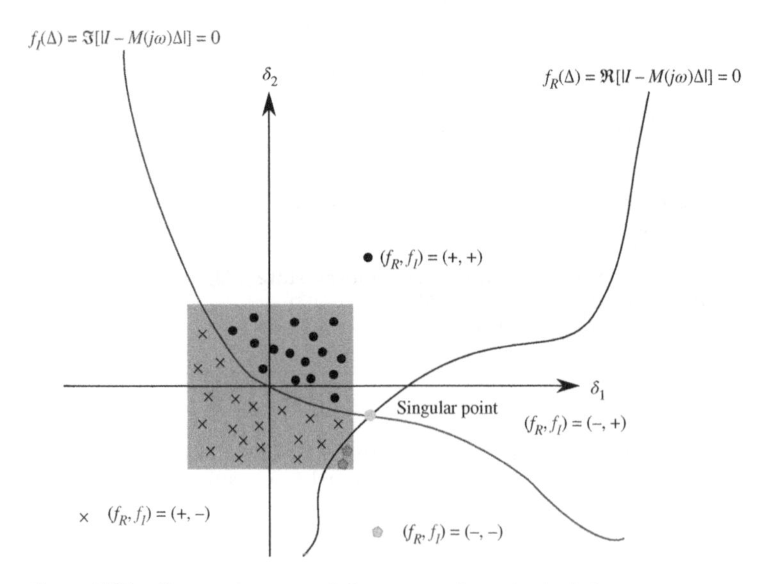

Figure 5.20 Geometric approach for μ upper bound calculation.

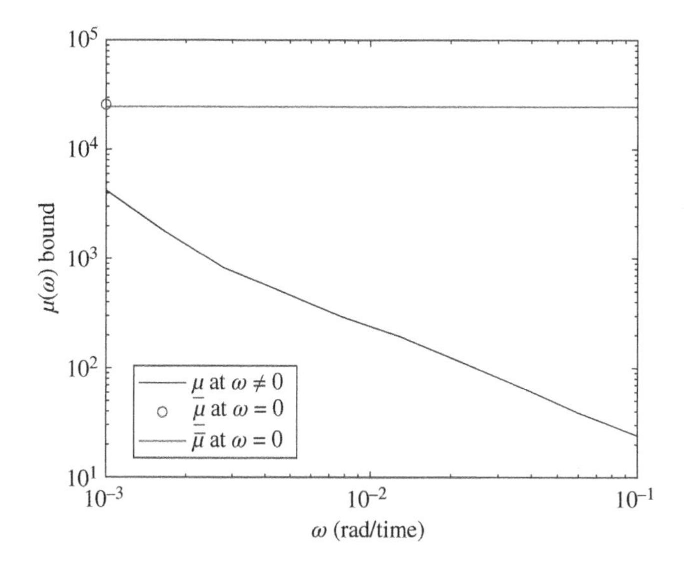

Figure 5.21 μ upper bound with the lower bound obtained by the geometric approach.

zero or one sign combination for ω equal to zero. In addition, the inverse of the half of the side length provides an upper bound. The upper bound calculated by this method is not deterministic as there is a chance that we miss finding all sign combinations. The upper bound, hence, only provides a probabilistic upper bound. It has a non-zero probability to be failed. Figure 5.21 shows the upper bound calculated using the geometric approach at $\omega = 0$, where the maximum lower bound occurs. The upper bound is close to the maximum lower bound, and we conclude that the true μ is between 20,000 and 30,000.

Exercises

Exercise 5.1 (MATLAB/Python) Implement the simulation program to produce Figure 5.11.

Exercise 5.2 Derive the differential equations corresponding to the following molecular interactions (Oishi and Klavins, 2011):

$$u^+ \xrightarrow{\alpha} u^+ + y^+\text{: catalysis}$$

$$u^- \xrightarrow{\alpha} u^- + y^-\text{: catalysis}$$

$$y^+ + y^- \xrightarrow{\eta} \varnothing\text{: annihilation}$$

where u^+ and u^- or y^+ and y^- represent the positive and the negative quantities corresponding to the signal $u = u^+ - u^-$ or $y = y^+ + y^-$, respectively.

Exercise 5.3 Using the differential equations derived in Exercise 5.2, derive the differential equations for u and y defined by

$$u = u^+ - u^-$$

$$y = y^+ - y^-$$

where the subtractions for u and y are not implemented. u and y are simply interpretations of the signals generated by u^+, u^-, y^+, and y^-.

Exercise 5.4 (MATLAB) Modify Program 5.3 and produce the results in Figure 5.13.

Exercise 5.5 (Python) Convert the MATLAB Program 5.3 to Python scripts and produce the results in Figure 5.13.

Exercise 5.6 (MATLAB/Python) Write the Program to produce Figure 5.15.

Exercise 5.7 (MATLAB/Python) Implement Algorithm 5.1 to calculate the μ-lower bound of (5.11) for $N_s = 1000$, $\varepsilon = 10^{-6}$, $\overset{d}{\longrightarrow} = 0.001$, $\bar{d} = 10$, and $\omega = 0$.

Exercise 5.8 (MATLAB/Python) Construct an μ-upper bound algorithm based on Figure 5.20 and implement the algorithm in MATLAB or Python to calculate the upper bound shown in Figure 5.21.

Bibliography

Gary J. Balas, John C. Doyle, Keith Glover, Andy Packard, and Roy Smith. *μ-Analysis and Synthesis Toolbox*. MUSYN Inc. and The MathWorks, Natick, MA, 1993.

Alexander P. S. Darlington, Jongrae Kim, and Declan G. Bates. Robustness analysis of a synthetic translational resource allocation controller. *IEEE Control Systems Letters*, 3(2):266–271, 2019. https://doi.org/10.1109/LCSYS.2018.2867368.

John Doyle. Analysis of feedback systems with structured uncertainties. In *IEE Proceedings D-Control Theory and Applications*, volume 129, pages 242–250. IET, 1982.

Andrea Fabrizi, Clément Roos, and Jean-Marc Biannic. A detailed comparative analysis of μ lower bound algorithms. In *2014 European Control Conference, ECC 2014*, pages 220–226, 06 2014. ISBN 978-3-9524269-1-3. https://doi.org/10.1109/ECC.2014.6862465.

Mathias Foo, Jongrae Kim, Jongmin Kim, and Declan G. Bates. Proportional–integral degradation control allows accurate tracking of biomolecular concentrations with fewer chemical reactions. *IEEE Life Sciences Letters*, 2(4):55–58, 2016. https://doi.org/10.1109/LLS.2016.2644652.

Gene F. Franklin, J. David Powell, and Abbas Emami-Naeini. *Feedback Control of Dynamic Systems*. Pearson, London, 2015.

Jongrae Kim, Declan G. Bates, and Ian Postlethwaite. A geometrical formulation of the μ-lower bound problem. *IET Control Theory and Applications*, 3(4):465–472, 2009.

NVIDIA Developer. CUDA GPUs — NVIDIA developer. https://developer.nvidia.com/cuda-gpus, 2021. Accessed: 2021-11-07.

Kevin Oishi and Eric Klavins. Biomolecular implementation of linear I/O systems. *IET Systems Biology*, 5(4):252–260, 2011.

W. H. Press, S. A. Teukolsky, W. T. Vetterling, and B. P. Flannery. *Numerical Recipes 3rd Edition: The Art of Scientific Computing*. Cambridge University Press, 2007. ISBN 9780521880688.

Clément Roos. Systems modeling, analysis and control (SMAC) toolbox: an insight into the robustness analysis library. In *2013 IEEE Conference on Computer Aided Control System Design (CACSD)*, pages 176–181, 2013. https://doi.org/10.1109/CACSD.2013.6663479.

Springer Nature Limited. Synthetic biology - latest research and news — nature. https://www.nature.com/subjects/synthetic-biology, 2021. Accessed: 2021-10-08.

P. M. Young, M. P. Newlin, and J. C. Doyle. Mu analysis with real parametric uncertainty. In *[1991] Proceedings of the 30th IEEE Conference on Decision and Control*, pages 1251–1256 vol. 2, 1991. https://doi.org/10.1109/CDC.1991.261579.

Yun-Bo Zhao, Jongrae Kim, and Declan G. Bates. LFT-free μ-analysis of LTI/LPTV systems. In *2011 IEEE International Symposium on Computer-Aided Control System Design (CACSD)*, pages 638–643, 2011. https://doi.org/10.1109/CACSD.2011.6044563.

6

Further Readings

Several important dynamic system modelling and recent algorithms worth including are beyond the scope of this book. Among them, we introduce Boolean network, network structure analysis, spatial-temporal modelling, reinforcement learning, and the deep learning neural network.

6.1 Boolean Network

In modelling biological networks, instead of tracking the concentrations of molecules in the networks, we would be only interested in the qualitative levels of the concentrations, i.e. high or low. Each molecule in the networks has the binary state, 1 (high) or 0 (low). Boolean networks model the dynamics of networks with binary states (Kauffman, 1969). For example Zhao et al. (2013),

$$\begin{aligned} x_1(k+1) &= x_2(k) \wedge x_3(k) \\ x_2(k+1) &= x_1(k) \vee x_3(k) \\ x_3(k+1) &= \neg x_3(k) \end{aligned}$$

where the binary state of x_i at $k+1$ step for $i = 1, 2, 3$ is the function of x_i at k step, $\wedge$ is the logical-and, which returns 1 only for both $x_2(k)$ and $x_3(k)$ equal to 1 and 0 otherwise, $\vee$ is the logical-or, which returns 1 for at least one of $x_1(k)$ or $x_3(k)$ equal to 1 and 0 otherwise, and $\neg$ is the negation, which returns the opposite of $x_3(k)$. In studying Boolean systems, how many stationary points exist and how large the state space converges to each has some important biological implications. Solving the following equation is to obtain stationary points:

$$\begin{aligned} x_1(k) &= x_2(k) \wedge x_3(k) \\ x_2(k) &= x_1(k) \vee x_3(k) \\ x_3(k) &= \neg x_3(k) \end{aligned}$$

Dynamic System Modelling and Analysis with MATLAB and Python: For Control Engineers,
First Edition. Jongrae Kim.

Companion Website: www.wiley.com/go/kim/dynamicmodeling

Finding all stationary points is not easy to guarantee without performing the exhaustive search, which requires a large computational cost even for the moderate size of networks, i.e. 2^n states for n molecular species in the network. One of the interesting approaches to solving the Boolean network analysis problems is the semi-tensor product approach in Cheng and Qi (2010). The semi-tensor approach represents the state transition using a series of matrix manipulations, which requires symbolic operations. An algorithm to perform the symbolic matrix manipulations for large-size networks is yet to be developed (Daizhan Cheng, 2021). It would significantly improve Boolean network analysis capability.

6.2 Network Structure Analysis

Large-scale networks are not only challenging to model but also to analyze. To remove the necessity to identify the dynamics of large-scale networks, we concentrate purely on the structural aspect of the networks. Consider the network with seven nodes and eight edges shown in Figure 6.1, where the network size is chosen to be small for the demonstration purpose. By examining the graph, we can see that there are two modules. One is a module by nodes 1, 2, 3 and the other is a module by nodes 4, 5, 6. Node 7 belongs to either module or neither module. We call it a grey node (Krishnadas et al., 2013).

By solving the following maximization problem, the module and the grey nodes are identified (Newman, 2006):

$$\mathbf{s}^* = \underset{\mathbf{s}\in\{-1,0,1\}^n}{\arg\max}\ Q = \frac{1}{4m}\mathbf{s}^T\left(A - \frac{\mathbf{k}\mathbf{k}^T}{2m}\right)\mathbf{s}$$

where $\mathbf{s}$ is the 7×1 vector, whose element is either $-1, 0$ or, 1, m is the total number of edges equal to 8 for the example network, $\mathbf{k}$ is the 7×1 vector, whose element is the number of edges connected to each node, i.e.

$$\mathbf{k} = \begin{bmatrix}2 & 2 & 3 & 3 & 2 & 2 & 2\end{bmatrix}^T$$

and A is the adjacency matrix. Q is the modularity defined by the difference between the actual connection, A, and the expected connection, $\mathbf{k}\mathbf{k}^T/(2m)$.

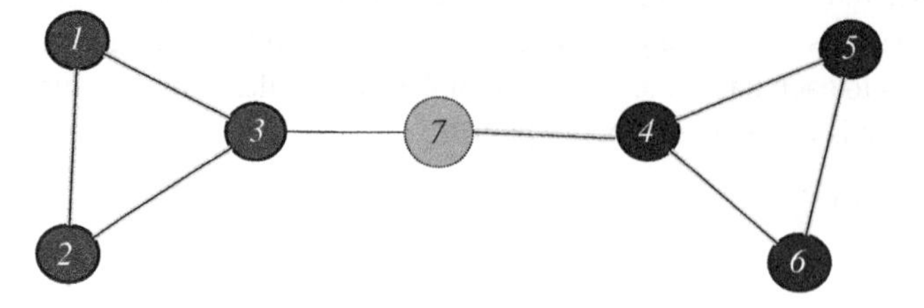

Figure 6.1 Two modules and a grey node in the network.

If we find more connections than the expected ones, i.e. Q is positive, then there exist modules. Among many possible module structures, we search the modules providing the maximum Q. For the network in Figure 6.1, the optimal solution is given by

$$\mathbf{s}^* = \begin{bmatrix} 1 & 1 & 1 & -1 & -1 & -1 & 0 \end{bmatrix}^T$$

which indicates two modules and one grey node.

Introducing perturbations in the network connections and analyzing how Q varies with respect to the perturbations provide the robustness of modularity. Kim and Cho (2015) present algorithms to calculate the upper and lower bounds of the worst perturbation to destroy the modularity.

6.3 Spatial-Temporal Dynamics

Partial differential equation models spatial-temporal dynamics. The following partial differential equation describes the mean firing rate of the neuron populations in the human brain (Detorakis and Rougier, 2014):

$$\frac{\partial u(\mathbf{x}, t)}{\partial t} = -\tau u(\mathbf{x}, t) + \alpha \int_{\mathbf{y} \in \Omega(\mathbf{x})} w(|\mathbf{x} - \mathbf{y}|) f[u(\mathbf{x}, t)] dy + i(\mathbf{x}, t)$$

where $u(\mathbf{x}, t)$ is the mean firing rate or the membrane potential activity of the neuron population in the volume, Ω, at the position, $\mathbf{x}$, in the brain, t is the time, τ is the temporal decay, α is the scale factor, $w(\cdot)$ is the difference between the short-range excitation and the long-range inhibition, which are functions of the distance from $\mathbf{x}$ to $\mathbf{y}$ in $\Omega(\mathbf{x})$, $f(u)$ is the firing function, and $i(\mathbf{x}, t)$ is the stimulus input.

functional Magnetic Resonance Image (fMRI) measures the localized oxygen level changes in the brain, which correlates with neuronal activities in the brain (Logothetis, 2008). fMRI measurements are noise corrupted and delayed signals of $u(\mathbf{x}, t)$. We establish the system identification problems as follows: given a particular stimulus, $i(\mathbf{x}, t)$, and the fMRI measurements infer localized neuronal activities and identify the unknowns in the dynamic model.

The molecular interactions studied in Chapters 4 and 5 are based on the spatial uniformity assumption. However, the enzyme and substrate molecules in the following reaction may spread unevenly over the reaction space:

$$\mathrm{E} + \mathrm{S} \underset{k_{\mathrm{off}}}{\overset{k_{\mathrm{on}}}{\rightleftharpoons}} \mathrm{ES}$$

and the reduction rates of [E] and [S] are not simply given as k_{on} [E][S] but are given in the form of partial differential equations, i.e. the reaction–diffusion equation (Smith and Grima, 2019). Kim et al. (2018) present a modelling method that includes spatial effect on molecular interactions using an additional parameter, δ, in the kinetic rate such as $k_{on}(1 + \delta)[E][S]$, where δ absolves the effect of spatial inhomogeneity of the molecular concentrations, [E] and [S].

6.4 Deep Learning Neural Network

Deep learning neural network is probably the most popular algorithm in the current artificial intelligence research (Goodfellow et al., 2016). The algorithm maps diverse forms of data to desired outputs and provides a certain level of generalization capabilities. Thanks to the fast glowing computation speed, its capabilities and application areas are growing. Deep learning neural network would model the unknown structures of uncertainty dynamics using its general function approximation capability (Eldan and Shamir, 2016). The fragile robustness of the algorithm to the input data perturbation has been revealed (Papernot et al., 2016, Su et al., 2019), and the ways to robustify the algorithms have been studied (Gu and Rigazio, 2014, Madry et al., 2017).

6.5 Reinforcement Learning

Reinforcement learning (Sutton and Barto, 2018) is a new control algorithm that demonstrates the potential to solve many difficult control problems that are non-linear, complex, and uncertain. Reinforcement learning has two modes called *exploration* and *exploitation*. The exploration mode learns the environment. The value function and the action value function store and update the expected rewards for the given states and actions. The exploitation mode chooses the optimal action to maximize the reward based on the current action-value function. The two important aspects of constructing a reinforcement learning algorithm are as follows (i) how to balance between the time spent on exploration or exploitation mode and (ii) how to design the rewards. The state action space could be high dimensional. Deep learning neural network has been introduced recently to express the value function and the action-value function (Mnih et al., 2013, Lillicrap et al., 2015). The robustness of reinforcement learning embedded systems is yet to be fully investigated.

Bibliography

Daizhan Cheng. STP toolbox for Matlab/Octave. http://lsc.amss.ac.cn/dcheng/stp/STP.zip, 2021. Accessed: 2021-11-15.

Daizhan Cheng and Hongsheng Qi. A linear representation of dynamics of Boolean networks. *IEEE Transactions on Automatic Control*, 55(10):2251–2258, 2010.

Georgios Is Detorakis and Nicolas P. Rougier. Structure of receptive fields in a computational model of area 3B of primary sensory cortex. *Frontiers in Computational Neuroscience*, 8:76, 2014.

Ronen Eldan and Ohad Shamir. The power of depth for feedforward neural networks. In *29th Annual Conference on Learning Theory, volume 49 of Proceedings of Machine Learning Research* (Vitaly Feldman, Alexander Rakhlin, and Ohad Shamir, Eds.), pages 907–940, Columbia University, New York, USA, 23–26 Jun 2016. PMLR. https://proceedings.mlr.press/v49/eldan16.html.

Ian Goodfellow, Yoshua Bengio, and Aaron Courville. *Deep Learning*. MIT Press, 2016. http://www.deeplearningbook.org.

Shixiang Gu and Luca Rigazio. Towards deep neural network architectures robust to adversarial examples. *arXiv preprint arXiv:1412.5068*, 2014.

Stuart A. Kauffman. Metabolic stability and epigenesis in randomly constructed genetic nets. *Journal of Theoretical Biology*, 22(3):437–467, 1969.

Jongrae Kim and Kwang-Hyun Cho. Robustness analysis of network modularity. *IEEE Transactions on Control of Network Systems*, 3(4):348–357, 2015.

Jongrae Kim, Mathias Foo, and Declan G. Bates. Computationally efficient modelling of stochastic spatio-temporal dynamics in biomolecular networks. *Scientific Reports*, 8(1):1–7, 2018.

Rajeev Krishnadas, Jongrae Kim, John McLean, David Batty, Jennifer McLean, Keith Millar, Chris Packard, and Jonathan Cavanagh. The envirome and the connectome: exploring the structural noise in the human brain associated with socioeconomic deprivation. *Frontiers in Human Neuroscience*, 7:722, 2013. ISSN 1662-5161. https://doi.org/10.3389/fnhum.2013.00722. https://www.frontiersin.org/article/10.3389/fnhum.2013.00722.

Timothy P. Lillicrap, Jonathan J. Hunt, Alexander Pritzel, Nicolas Heess, Tom Erez, Yuval Tassa, David Silver, and Daan Wierstra. Continuous control with deep reinforcement learning. *arXiv preprint arXiv:1509.02971*, 2015.

Nikos K. Logothetis. What we can do and what we cannot do with FMRI. *Nature*, 453(7197):869–878, 2008.

Aleksander Madry, Aleksandar Makelov, Ludwig Schmidt, Dimitris Tsipras, and Adrian Vladu. Towards deep learning models resistant to adversarial attacks. *arXiv preprint arXiv:1706.06083*, 2017.

Volodymyr Mnih, Koray Kavukcuoglu, David Silver, Alex Graves, Ioannis Antonoglou, Daan Wierstra, and Martin Riedmiller. Playing atari with deep reinforcement learning. *arXiv preprint arXiv:1312.5602*, 2013.

M. E. J. Newman. Modularity and community structure in networks. *Proceedings of the National Academy of Sciences of the United States of America*, 103(23):8577–8582, 2006. ISSN 0027-8424. https://doi.org/10.1073/pnas.0601602103. https://www.pnas.org/content/103/23/8577.

Nicolas Papernot, Patrick McDaniel, Somesh Jha, Matt Fredrikson, Z. Berkay Celik, and Ananthram Swami. The limitations of deep learning in adversarial settings. In *2016 IEEE European Symposium on Security and Privacy (EuroS P)*, pages 372–387, 2016. https://doi.org/10.1109/EuroSP.2016.36.

Stephen Smith and Ramon Grima. Spatial stochastic intracellular kinetics: a review of modelling approaches. *Bulletin of Mathematical Biology*, 81(8):2960–3009, 2019.

Jiawei Su, Danilo Vasconcellos Vargas, and Kouichi Sakurai. One pixel attack for fooling deep neural networks. *IEEE Transactions on Evolutionary Computation*, 23(5):828–841, 2019.

Richard S. Sutton and Andrew G. Barto. *Reinforcement Learning: An Introduction*. MIT Press, 2018.

Yin Zhao, Jongrae Kim, and Maurizio Filippone. Aggregation algorithm towards large-scale Boolean network analysis. *IEEE Transactions on Automatic Control*, 58(8):1976–1985, 2013. https://doi.org/10.1109/TAC.2013.2251819.

Appendix A

Solutions for Selected Exercises

A.1 Chapter 1

Exercise 1.4

The three reactions affecting ligand concentration are interpreted as

$$\mathrm{R} + \mathrm{L} \xrightarrow{k_{\mathrm{on}}} \mathrm{C} \Rightarrow \frac{d[\mathrm{L}]}{dt} \propto -[\mathrm{R}][\mathrm{L}]$$

$$\mathrm{C} \xrightarrow{k_{\mathrm{off}}} \mathrm{R} + \mathrm{L} \Rightarrow \frac{d[\mathrm{L}]}{dt} \propto [\mathrm{C}]$$

$$f(\mathrm{t}) \xrightarrow{1} \mathrm{L} \Rightarrow \frac{d[\mathrm{L}]}{dt} \propto [f(t)]$$

Using the reaction rate constants, construct the following differential equation:

$$\frac{d[\mathrm{L}]}{dt} = -k_{\mathrm{on}}[\mathrm{R}][\mathrm{L}] + k_{\mathrm{off}}[\mathrm{C}] + [f(t)]$$

Similarly, $d[\mathrm{C}]/dt$ can be obtained.

Exercise 1.5

The function passed to *odeint* is 'RLC_kinetics'. Its first and second arguments are 'time' and 'state'. The default argument order assumed in *odeint* is 'state' and 'time'. The optional argument, *tfirst*, indicates whether the time is the first argument of the function to pass to the integrator. By setting the optional argument equal to *true*, we can pass the same function to *solve_ivp* or *odeint*.

Dynamic System Modelling and Analysis with MATLAB and Python: For Control Engineers,
First Edition. Jongrae Kim.

Companion Website: www.wiley.com/go/kim/dynamicmodeling

A.2 Chapter 2

Exercise 2.5

We have the direction cosine matrix between the body frame and the reference frame provided by the Kinematic equation of the body frame: C_{BR}. Using the sensor configuration in the body frame, we obtain the direction cosine matrix between the sensor frame and the body frame: C_{SB}. For the configuration of the sensor given in Figure 2.28

$$C_{SB} = \begin{bmatrix} 1 & 0 & 0 \\ 0 & -1 & 0 \\ 0 & 0 & -1 \end{bmatrix}$$

For the given configuration, the sensor on the body, C_{SB}, is a constant matrix. Therefore, the direction cosine matrix between the sensor frame and the reference frame, C_{SR}, is obtained as

$$C_{SR} = C_{SB}C_{BR}$$

Hence, the following equation converts $\mathbf{r}^1$ in the reference frame to the one in the sensor frame:

$$\mathbf{r}_S^1 = C_{SR}\mathbf{r}_R^1 = C_{SB}C_{BR}\mathbf{r}_R^1$$

A.3 Chapter 3

Exercise 3.1

The attractive force in the positive **y**-direction is obtained by

$$-\frac{\partial U_a}{\partial y} = -\frac{\partial}{\partial y}\left(\frac{1}{2}k_a\rho_a^2\right) = -\frac{\partial}{\partial \rho_a}\left(\frac{1}{2}k_a\rho_a^2\right)\frac{\partial \rho_a}{\partial y} = -k_a(y - y_{\text{dst}})$$

and the i-th repulsive in the positive **y**-direction for $\rho_r^i \leq \rho_o^i$ is

$$-\frac{\partial U_r^i}{\partial y} = -\frac{\partial}{\partial y}\left[\frac{1}{2}k_r\left(\frac{1}{\rho_r^i} - \frac{1}{\rho_o^i}\right)\right] = \frac{1}{2}k_r\left(\frac{1}{\rho_r^i}\right)^2\frac{\partial \rho_r^i}{\partial y} = \frac{k_r(y - y_{\text{ost}}^i)}{(\rho_r^i)^3} \tag{A.1}$$

Exercise 3.6

As shown in Figure 3.14, the body frame and the reference frame correspond to the following direction cosine matrix:

$$C_{BR} = \begin{bmatrix} \cos\phi & \sin\phi \\ -\sin\phi & \cos\phi \end{bmatrix}$$

The following multiplication converts the control input in the reference frame to the body frame:

$$\mathbf{u}^B = \begin{bmatrix} u_x^B \\ u_y^B \end{bmatrix} = C_{BR}\mathbf{u}^R = \begin{bmatrix} \cos\phi & \sin\phi \\ -\sin\phi & \cos\phi \end{bmatrix} \begin{bmatrix} u_x \\ u_y \end{bmatrix} = \begin{bmatrix} u_x\cos\phi + u_y\sin\phi \\ -u_x\sin\phi + u_y\cos\phi \end{bmatrix}$$

A.4 Chapter 4

Exercise 4.1

$$Y(s) = \int_{t=0}^{t=\infty} x(t-\tau)e^{-st}\, dt$$

Let $v = t - \tau$ and $dv = dt$

$$Y(s) = \int_{v=-\tau}^{v=\infty} x(v)e^{-s(v+\tau)}\, dv = e^{-s\tau}\int_{v=0}^{v=\infty} x(v)e^{-sv}\, dv = e^{-s\tau}X(s)$$

where $x(v) = 0$ for $v \in [-\tau, 0)$.

Exercise 4.2

Use the integration by parts, $\int u\dot{v}\, dt = uv - \int \dot{u}v\, dt$,

$$sY(s) - y(0) = \int_{t=0}^{t=\infty} \dot{y}(t)e^{-st}\, dt$$

where $u = e^{-st}$ and $v = \dot{y}(t)$, then the integral becomes

$$\int_{t=0}^{t=\infty} \dot{y}(t)e^{-st}\, dt = e^{-st}y(t)\Big|_{t=0}^{t=\infty} + s\int_{t=0}^{t=\infty} e^{-st}y(t)dt = -y(0) + sY(s)$$

Exercise 4.7

Pseudo-code for the robustness analysis is given by

1: Set $t_0 = 600$ minutes, $t_f = 1200$ minutes, $\Delta t = 0.1$ minutes and $p_\delta = 2\%$
2: **while** p_δ is not the smallest **do**
3: Solve the minimization problem in (4.30) and obtain δ^*
4: Check the time history of [ACA] for δ^* if oscillation exists
5: **if** there is oscillation **then**
6: Increase p_δ; use the bisection method
7: **else if** no oscillation **then**
8: Decrease p_δ; use the bisection method
9: **end if**
10: **end while**

A.5 Chapter 5

Exercise 5.2

Deduce the differential equations as follows:

$$\frac{du^+}{dt} = 0$$
$$\frac{du^-}{dt} = 0$$
$$\frac{dy^+}{dt} = \alpha u^+ - \eta y^+ y^-$$
$$\frac{dy^-}{dt} = \alpha u^- - \eta y^+ y^-$$

Exercise 5.3

As the definitions for u and y are given by

$$u = u^+ - u^-$$
$$y = y^+ - y^-$$

the differentiations are equal to

$$\frac{du}{dt} = \frac{du^+}{dt} - \frac{du^-}{dt} = 0$$
$$\frac{dy}{dt} = \frac{dy^+}{dt} - \frac{dy^-}{dt} = \left(\alpha u^+ - \eta y^+ y^-\right) - \left(\alpha u^- - \eta y^+ y^-\right)$$

Therefore,

$$\frac{du}{dt} = 0$$
$$\frac{dy}{dt} = \alpha u$$

i.e. y is the integration of u with the integral gain equal to α.

Index

Dynamic System Modelling and Analysis with MATLAB and Python: For Control Engineers,
First Edition. Jongrae Kim.

Companion Website: www.wiley.com/go/kim/dynamicmodeling

Titles in the IEEE Press Series on Control Systems Theory and Applications

Series editor: Maria Domenica Di Benedetto

The series publishes monographs, edited volumes, and textbooks which are geared for control scientists and engineers, as well as those working in various areas of applied mathematics such as optimization, game theory, and operations.

Autonomous Road Vehicle Path Planning and Tracking Control
Levent Guvenc, Bilin Aksun-Guvenc, Sheng Zhu, Sukru Yaren Gelbal
ISBN: 9781119747949 December 2021

Embedded Control for Mobile Robotic Applications
Leena Vachhani, Pranjal Vyas, Arunkumar G. K.
ISBN: 9781119812388 August 2022

Merging Optimization and Control in Power Systems: Physical and Cyber Restrictions in Distributed Frequency Control and Beyond
Feng Liu, Zhaojian Wang, Changhong Zhao, Peng Yang
ISBN: 9781119827924 August 2022

Dynamic System Modeling and Analysis with MATLAB and Python: For Control Engineers
Jongrae Kim
ISBN: 9781119801627 October 2022

Model-Based Reinforcement Learning: From Data to Actions with a Python-based Toolbox
Milad Farsi, Jun Liu
ISBN: 9781119808572 December 2022

In Production

Control over Communication Networks: Modeling, Analysis, and Design of Networked Control Systems and Multi Agent Systems over Imperfect Communication Channels
Jianying Zheng, Liang Xu, Qinglei Hu, Lihua Xie
ISBN: 9781119885795

Disturbance Observer for Advanced Motion Control with MATLAB/Simulink
Akira Shimada
ISBN: 9781394178100

Advanced Control of Power Converters: Techniques and MATLAB/Simulink Implementation
Hasan Komurcugil, Sertac Bayhan, Ramon Guzman, Mariusz Malinowski, Haitham Abu-Rub
ISBN: 9781119854401

The Impact of Automatic Control Research on Industrial Innovation: Enabling a Sustainable Future
Silvia Mastellone, Alex van Delft
ISBN: 9781119983613

Distributed Space Systems: Dynamics, Navigation, and Control of Multi-Satellite Missions
Simone D'Amico, Matthew B. Willis
ISBN: 9781119808954

Dynamic System Modelling and Analysis with MATLAB and Python: For Control Engineers
Jongrae Kim
ISBN: 9781119801627

Printed and bound by CPI Group (UK) Ltd, Croydon, CR0 4YY

16/04/2025

14658601-0002